Deep Oceans

Deep
Oceans

Edited by
Peter J. Herring and
Malcolm R. Clarke

Praeger Publishers
New York · Washington

BOOKS THAT MATTER
Published in the United States of America in 1971
by Praeger Publishers, Inc., 111 Fourth Avenue,
New York, N.Y. 10003

Library of Congress Catalog Card Number: 69-10356

Printed in Great Britain

Contents

Foreword

The sea is a world apart, with its own characteristics, life in both its depths and superficial layers, states of fury and perfect calm, rapacity when it engulfs ships, treasures and lives, and liberality when it supplies energy, fish and petroleum.

Thirsty, it drinks the flow of all the world's rivers; munificent, it distills them and gives forth the rain to the benefit of the whole surface of the earth. By its enormous mass and its inertia, it regulates the climates in its vicinity; by its size it separates enemies, but at the same time brings together those who wish to meet and trade. Over the centuries the sea has been considered of little account in itself, yet he who ruled it, ruled the world; few journeys of consequence could help traversing it. The sea let pass those who wished, but fixed its toll in shipwrecks. For a million years it inspired terror, for though it was not understood it could not be ignored.

At the end of the last century it was decided to find out more about the sea. Various great expeditions, such as those of Prince Albert of Monaco, the *Valdivia*, the *Challenger* and others, ploughed much of its surface, and in some places even probed its depths by means of newly conceived instruments. The reports caught the public imagination, though they were little more than an appetiser for the great revelations which were to come.

Always impartial, the sea submitted to this scrutiny and only claimed the lives of a few among those who probed its mysteries. Initially of interest only to a few academics and to the fishermen who then believed they understood it from long experience, this slow unveiling of the sea has since continually accelerated. Progressively, as revelations about the sea became greater and more frequent, and hypotheses became more attractive and more promising, all sciences took part: physics, chemistry, biology, geology and even cosmogony; men looked to the sea for the beginnings of all human problems: the origin of life, the birth of the earth; knowledge of the sea helped to explain space, and the bottom of the greatest ocean was once seen as the birthplace of the moon.

As the problems became more exacting, the tools for investigation had, of necessity, to become ever more complex. Since 1930 a start has been made in replacing artificial probes by investigators who themselves go down into the sea for some hundreds of metres. William Beebe in his cable-suspended bathysphere was a successful pioneer. But even then the explorer was not free; a line still linked him to the surface, that level of the sea to be mistrusted most. The next step— a decisive one – was taken by Professor Auguste Piccard when he invented and constructed his Bathyscaphe; an entirely autonomous apparatus which remains the only means available to an observer to descend to any required depth. Several countries have used it – Belgium, Switzerland, France, Italy, the USA and Japan. The fascinating results and the harvest of information gleaned from under water inspired other investigators to devise new methods for underwater research; new submersibles, each having its own particular qualities, its advantages and certainly its limitations. Even Switzerland herself did not lag behind. Besides the bathyscaphes of Professor Piccard, from Switzerland came two so-called Mesoscaphes, submersibles for use in medium depths. The chief feature of this type of submerged vehicle lies in its static stability, which enables it to hover freely between two water masses. It was this principle of free buoyancy which led to the use of the Mesoscaphe in 1969 for the submerged drift expedition in the Gulf Stream, during which it covered a track of some 1,500 miles whilst submerged. This close contact with the sea was an unforgettable lesson in oceanography for the six privileged participants in the drift expedition.

Technology today offers such a diversity of tools to study the sea that it appears nothing can prevent its complete unveiling. Yet there is one limit: to put into practice such plans is costly, and there are many other claims on available funds. But the sea now offers, to those who know how to harvest them, such riches in minerals, in potential food supplies, and in nigh-inexhaustible energy, that a whole new group of people are attracted to it, namely the businessmen.

They have enterprisingly invested in the sea not for love of its beauty or its grandeur, but quite simply because they wish to make a profit.

Thus, have we gone a full circle now? Will the material gains provide the capital for the investigations of the scientists, the capital to tap food resources demanded by an ever increasing and starving humanity, to find sources of energy which industry needs in order to grow and, in turn, to support more research? Will these new riches suffice to protect humanity's capital of health, work, benefits and faith which permit us to pursue our aim towards a peak ever higher and more attractive but always more distant and more difficult to attain?

One must hope so. A work such as we now present will contribute in large measure by offering the reader results from a thousand investigations and unceasing endeavour.

JACQUES PICCARD
(Translated from the original French text)

Foreword

La mer est un monde à part, avec ses phénomènes propres, sa vie intérieure profonde et sa vie superficielle, ses colères furieuses et ses moments de grand calme, son égoïsme quand elle engloutit navires, trésors ou vies, et sa générosité quand elle donne énergie, poissons ou pétrole. Avide, elle boit toutes les rivières du monde; désintéressée, elle les distille et rend la pluie, bienfaisante pour toute la surface de la terre. Par sa masse énorme et par son inertie, elle régularise le climat partout où elle est présente; par sa grandeur, elle sépare les frères ennemis, mais en même temps elle rapproche et réunit ceux qui veulent se rencontrer ou échanger leurs biens. Pendant des siècles, elle n'a rien été en soi, mais celui qui la possédait le monde, car on ne pouvait aller presque nulle part sans passer par elle. Elle accordait le passage à qui voulait, mais elle choisissait elle-même son tribut et fixait le péage en réglant les naufrages. Pendant un million d'années, elle inspira la terreur, parce qu'on ne la connaissait pas et parce qu'on ne pouvait l'ignorer.

A la fin du siècle dernier, on décida d'apprendre à la connaître. Quelques grandes expéditions, celles du Prince Albert de Monaco, de la *Valdivia*, du *Challenger*, et d'autres, égratignèrent partout sa surface et la pénétrèrent même profondément en quelques points, grâce à toutes sortes de sondes nouvellement conçues. Ce qu'on en rapporta fit grand bruit, bien que ce ne fût encore guère qu'un avant-goût des révélations qui ne devaient plus tarder.

Bonne joueuse, la mer se laissa étudier et ne fit que peu de victimes parmi ceux qui l'analysaient. Cette lente découverte de la mer alla en accélérant sans cesse; au début, seuls quelques savants s'y intéressaient; les pêcheurs, quant à eux, croyaient déjà la connaître de longue date. Progressivement, comme les révélations se faisaient plus grandes et plus fréquentes, les hypothèses plus séduisantes et plus prometteuses, toutes les sciences se mirent de la partie: physique, chimie, biologie, géologie, et même cosmogonie; on alla alors chercher dans la mer l'origine de tous les problèmes humains; naissance de la vie, naissance de la terre; par la mer, on expliqua l'espace, on y vit même, dans le fond de son plus grand océan, le berceau de la Lune.

Comme les problèmes devenaient plus ardus, les moyens d'investigation devaient devenir plus perfectionnés; dès 1930, on commença à remplacer les sondes artificielles par des explorateurs qui descendaient eux-mêmes à quelques centaines de mètres. William Beebe et sa bathysphère suspendue à un câble fut un précurseur heureux. Mais, même alors, l'explorateur n'était pas libre; un lien encore le retenait à la surface et la surface était et est toujours le seul endroit de la mer dont il faille essentiellement se méfier. Le pas suivant, décisif, fut franchi par le Professeur Auguste Piccard, lorsqu'il inventa et construisit le Bathyscaphe, appareil totalement autonome, qui reste encore le seul moyen offert à l'observateur pour descendre à n'importe quelle profondeur. Plusieurs pays l'utilisèrent: la Belgique, la Suisse, la France, l'Italie, l'Amérique et le Japon. Les résultats fascinants et la moisson d'informations glanées sous l'eau invitèrent d'autres savants encore à créer de nouveaux moyens de recherche, de nouveaux engins sous-marins ayant chacun ses qualités propres, ses avantages et, bien sûr aussi, ses limites. La Suisse même ne resta pas en arrière. Elle créa de son côté, outre les Bathyscaphes du Professeur Piccard, deux Mésoscaphes, ou sous-marins pour moyenne profondeur. La caractéristique principale de ce type d'appareils est stabilité propre, qui lui permet de flotter librement entre deux eaux; c'est sur ce principe que fut conçue, et réalisée en 1969, l'expédition-dérive dans le Gulf Stream, au cours de laquelle le Mésoscaphe parcourut en plongée quelque 1500 milles. Cette prise de contact avec la mer a été une inoubliable leçon d'océanographie pour les six hommes qui ont eu le privilège d'y participer.

La technique offre maintenant un tel éventail de moyens pour l'étude de la mer qu'aucune limite ne semble plus s'opposer à sa découverte totale. Et pourtant il y en a une: la réalisation pratique coûte cher et les crédits sont sans cesse disputés par les autres administrations. Mais la mer offre maintenant à qui sait les prendre de telles richesses en minéraux, en potentiel alimentaire, en énergie quasi inépuisable, qu'une nouvelle catégorie d'intéressés est attirée par elle: les hommes d'affaire; beaucoup ont entrepris d'investir dans la mer, non qu'ils se soucient de sa beauté ou de sa grandeur, mais parce qu'ils veulent tout simplement faire fructifier leurs deniers.

La boucle sera-t-elle bouclée ainsi ou la spirale amorcée? Ce nouvel afflux de moyens matériels permettra-t-il d'effectuer les recherches souhaitées par les savants, de puiser les ressources alimentaires demandées par une humanité toujours plus nombreuse et affamée, de trouver les sources d'énergie que l'industrie ne demande qu'à consommer au fur et à mesure qu'on les lui offre? Et ces nouvelles richesses suffiront-elles à leur tour à garder à l'humanité un capital suffisant de santé, de travail, de richesses et de foi qui lui permette de poursuivre sa marche vers un sommet toujours plus haut, toujours plus beau, mais toujours plus lointain et plus problématique aussi? Il faut le souhaiter. Une oeuvre comme celle que nous présentons aujourd'hui au lecteur y contribue largement par l'offrande qu'elle fait au public des résultats de mille recherches et de labeurs sans cesse renouvelés.

JACQUES PICCARD

Preface

The oceans comprise the most extensive environment on the surface of the earth, and their very presence has profound effects upon the characteristics of the aerial and terrestrial realms. Biological and technological limitations have hitherto restricted mankind to a peripheral interest in the oceans, but in recent years many of the limitations have been overcome, and sciences whose primary interest was once terrestrial or aerial have been extended to include the oceanic realm.

It would be as impossible to cover comprehensively all aspects of oceanic science in a single book as it would all aspects of terrestrial science. Nor can the two be separated for the principles do not differ in the two realms, merely the applications. Specialists from different disciplines have therefore been invited to give an account of how their branch of science has approached the deep oceans, and something of the discoveries made so far.

The approaches of the several authors naturally differ in emphasis and outlook, for each chapter is in essence an independent account. None the less they are to a great extent interdependent, because advances in one subject greatly facilitate those in another. Nowhere is this more apparent than in the reliance on technological advances for the acquisition of the information each approach seeks to obtain. Oceanography, the study of the oceans, owes its present success and advancement to the enthusiasm and ability of the many people of different nationalities, beliefs and interests who are united in bringing their own particular approach to bear upon the problems of the oceans.

In this book emphasis has been laid particularly upon the deep oceans, because the relatively shallow waters fringing the continents are already reasonably well known and well documented. It is the deep oceans that cover most of the earth.

EDITORS

1 The Past, the Present and the Promise

Peter J. Herring

Man is not an aquatic animal. His early fish-like ancestors were once wholly marine, but took the improvident step of crawling ashore. In so doing they relinquished 71 per cent of the earth's surface for the unknown 29 per cent, and only now are we, their beneficiaries, able to take our first faltering steps back into the realm whose potentialities it has taken us so long to grasp. The ocean which was for so long regarded merely as an inconvenience between land masses, is now appreciated as the repository of not only the effluent of affluence but also almost untapped riches, economic, social and strategic.

The seas surrounding man's limited domain have always been a world ripe for investigation, the more acute pressures being those of hunger for food and other natural resources, the more chronic those of curiosity. The information whose collection these pressures have inspired is the basis for the composite science of oceanography. A composite science, for under its accommodating wing are grouped all those disciplines involved in the study of the oceans, their contents, composition and boundaries. Specialists in the multitude of scientific subjects involved are united, too, in having to overcome the manifold problems of investigating an environment so very different from that to which they are accustomed.

The world's oceans abut on to the land as a relatively shallow fringing shelf, known as the *continental shelf,* which slopes almost imperceptibly from the shore to a depth of about 200 m. The average width of the shelf is 48 km (30 miles) for the world as a whole, though it may be absent off mountainous coasts and islands and very much wider off glaciated coasts and the mouths of large rivers. This adjacent area is most accessible to terrestrial man, and is relatively well known and well documented. However, such areas comprise only 7·6 per cent of the total area of the oceans and seas, and only 3·1 per cent of that of the major oceans. It is with the less accessible and relatively unknown deep oceans that extend beyond the shelf and cover most of the globe that this book is primarily concerned.

Some idea of the extent of the deep oceans and their

TABLE I
Some facts about the world oceans

AREA
World's surface area: 70·8% sea   29·2% land
Northern hemisphere: 60·7% sea   39·3% land
Southern hemisphere: 80·9% sea   19·1% land
Total area covered by sea: 361×10^6 km² ($139·5 \times 10^6$ square miles)

VOLUME
Total volume of the sea: 1370×10^6 km³ (328×10^6 cubic miles)
This is 1 85% of the total water on the earth's surface, in its rocks, and in its atmosphere; and is 98·5% of the free water and ice on its surface
  2 40000 times the annual total river inflow
  3 0·13% of the volume of the earth ($1·083 \times 10^{12}$ km³)
For comparison, the volume of land above sea level is 130×10^6 km³ ($31·5 \times 10^6$ cubic miles)

MASS
Total mass of the sea: $1·5 \times 10^{18}$ tons
This is 0·25% of the mass of the earth ($5·98 \times 10^{27}$ g or $5·9 \times 10^{21}$ tons)

DEPTH
Maximum recorded depth: 11,515 m (37,780 ft)
This was recorded in the Mindanao Trench (near the Philippines) in 1962 by HMS *Cook*
Average overall depth: 3800 m (12,460 ft)
Average depth of Atlantic ocean: 3926 m (12,900 ft)
Average depth of Pacific ocean: 4282 m (14,100 ft)
Average depth of Indian ocean: 3963 m (13,000 ft)
If spread over the whole surface of the earth the average depth would be 2440 m (7000 ft)
75·9% of the ocean area has a depth of between 3000 and 6000 m (9840 – 19,680 ft)
For comparison, the mean height of land above sea level is 840 m (2760 ft), and mean radius of the earth 6371 km (3950 miles)

SALINITY
Highest in open ocean: northern subtropical Atlantic surface, 37·5‰
Highest locally: a pocket at the bottom of the Red Sea, 270‰

TEMPERATURE
Highest at the surface: Persian Gulf, over 32°C
Highest locally: pocket at the bottom of the Red Sea, 55·6°C

It is almost impossible to calculate some of the above figures to a high degree of accuracy and they must generally be regarded as informed estimates rather than precise statements.

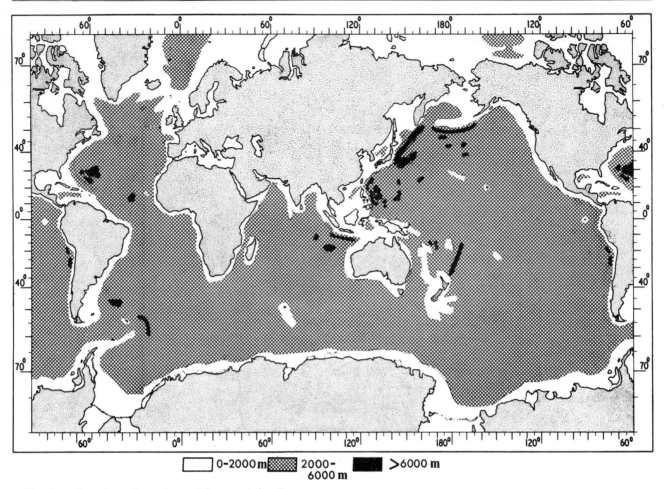

0–2000 m ▨ 2000–6000 m ■ >6000 m

1. Depth regions throughout the world oceans (after Bruns 1958)

separate regions may be gained from the bare statistics (Table 1, figure 1). The maximum depth is about ten km (seven miles) but some 75 per cent of the total area has a depth of three to six km (1·9–3·7 miles). The mean height of land above sea level is about 800 m (0·5 mile) whereas the mean depth of the seas (including continental shelf regions) is 3·8 km (2·4 miles). The total weight of sea water has been estimated at $1·5 \times 10^{18}$ tons, or five hundred million tons for every man, woman and child on the earth. The immensity of this watery blanket hides a terrain every bit as awe-inspiring as the world above. From the continental shelf the sea-floor inclines more steeply down the continental slope to a gently sloping area known as the continental rise (figure 2). At the bottom of the continental rise lie the abyssal plains, large areas of almost flat bottom. The terrain of the continental slope and rise may be dissected by submarine canyons or punctuated by isolated sea-mounts (elevations of the sea-floor of 1,000 m or more) in addition to local ridges and gullies. In places the abyssal plains are cleft

by trenches or V-shaped channels with steep sides plunging to a flat floor ranging in width from a few hundred metres to several kilometres. Such trenches may be several hundred kilometres long and it is in these trenches, or certain sub-sections of them known as 'deeps', that the deepest soundings have been made. The curious distribution of these trenches is described in Chapter 10 where the far from peaceful nature of the sea-floor and its vital importance in our understanding of the formative processes of the earth's crust is emphasized.

The exploration and investigation of the oceans present a number of severe problems to those who would attempt them, and it is perhaps salutary at the start of a book such as this to outline briefly some of the difficulties of gaining the information that is set out in subsequent chapters, and some of the ways in which they are overcome. The first problem is simply that of scope, for it can be readily appreciated that the difficulty of investigating adequately an area 2·5 times that of the total land mass is daunting indeed. This

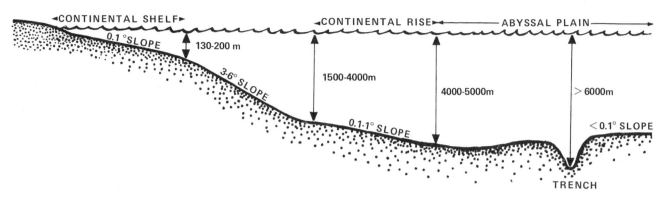

2. Diagrammatic representation of the realms into which the oceans are divided

problem is compounded by the three-dimensional nature of the oceans, so that points on the surface of the open ocean have an average depth of water of 4·2 km (2·6 miles) beneath them, and very many features change with the depth in the water as well as with the geographical area. The surprising feature is not so much the enormity of the problem as that so much has been learned in so short a time.

The first major deep-water oceanographic expedition was that of HMS *Challenger* (figure 3), and in less than 100 years since then our knowledge of the oceans in all their aspects has advanced at a prodigious rate. However it is only recently that we have begun to understand some of their dynamic aspects and have been able to correlate the mass of information available in order to predict some of their internal changes. The *Challenger* sailed in December 1872 and returned in 1876 after a journey of 68,890 miles; this is still the longest continual exploration by a scientific vessel. Hitherto she had been a naval steam corvette of 2,306 tons displacement, but a complete refit, including the removal of all but two sixty-four pounders of her eighteen guns, equipped her with laboratory space and deck equipment for all the intended work. The expedition was under the leadership of Wyville Thomson and had the aim of investigating the 'Conditions of the Deep Sea throughout all the Great Oceanic Basins'. Despite the far-reaching importance of her project she only sailed after Parliament had been assured that the cost of the expedition would be no more than that of keeping the ship in commission. In addition to the crew of almost two hundred and forty (of whom sixty-one deserted during the voyage) there were the six scientists whose collections and work were the foundation of the whole of modern oceanography as well as of the fifty volumes of *Challenger Reports* that appeared by 1895. Their equipment included

5,600 fathoms of two to three inch dredging and trawling rope and 176,000 fathoms of wire, sounding line and current line, in addition to more esoteric items such as birdlime and hawk-, fox-, rabbit-, rat- and mouse-traps. The scope of the work included the investigation of many islands and land areas, and during the course of one such visit the ship's dog was pecked to death by penguins. Despite such vicissitudes the ship's complement returned home with 715 new genera and 4,417 new species of organism, in addition to the mass of accumulated information not only about the physics, chemistry and geology of the sea and land, but also about the anthropology of many of the places they visited. Although the type of oceanographic voyage has changed considerably since then (Chapter 11) and the equipment available has become highly sophisticated, many of the problems faced by the *Challenger* expeditionaries are still as important as ever.

It is no mean feat to get the oceanographer and his equipment to the area in which he wishes to work. The distances involved in travelling to a point several hundred or more kilometres from the nearest land, let alone from the home port, are obvious enough. The oceanographer requires a ship to get him there, a laboratory in which to work when finally on the spot, time and suitable conditions to carry out his experiments, and the specialised equipment for his particular field of interest. To work in a tolerable degree of comfort in mid-ocean for any length of time requires a ship of reasonable size. Too small a ship means insufficient stability for many measurements, and lack of heavy equipment. Large ships are expensive, not only to build but to run. With cruising speeds of 10–15 knots and daily costs of at least several hundred pounds, it is uneconomic for the ship to spend too much of its time in transit, and cruises of several months are therefore necessary to provide sufficient relative

3. HMS *Challenger,* previously a naval corvette, embarked on the first major oceanographic expedition in 1872 (from the original in the possession of the Challenger Society)

4. The National Institute of Oceanography's research vessel RRS *Discovery* (commissioned 1962, gross tonnage 2665), carries 40 crew and 21 scientists

working time in distant areas. To get the maximum amount of useful work out of an oceanographic cruise the scientific staff may change during the period, those whose work is completed returning home, while new workers fly out to join the ship. Budgetary pressures are obviously of importance when planning a cruise, the more so since the type of work undertaken rarely produces immediate results, such as the delimitation of a new fishing ground or a potential oilfield, that might be regarded as immediate justification for the expenditure of what are in most cases public funds. Lengthy cruises put further pressures on the design of research ships, and the 'average' oceanic research ship is therefore a compromise of generally about 1,000–3,000 tons displacement and carrying a total complement of forty to eighty with one scientist to two or three crew members (Table 17, figures 4, 5 and 6); i.e. similar in size to the *Challenger* but much less crowded! No longer do ships set off for a single voyage of several years with the same scientific staff throughout, and with periods of up to several weeks in port, as did the *Challenger*. Present research ships are rarely away for as much as a year, the scientific staff often change during a cruise, and the ship comes into port for no more than a few days to refuel and reprovision before setting off for several more weeks at sea.

Once at sea the ship is dependent upon the clemency of the weather for completion of its programme, though most can carry out a large proportion of their work in conditions up to wind force six or seven on the Beaufort scale, albeit with increased hazards both to safety and success during operation of equipment. Once at the point, or 'station', at which the ship is to work, the scientists will wish to measure a wide variety

of items at various depths: salinity, temperature, light penetration, current velocities, chlorophyll, oxygen and nutrient concentrations, the composition of the sediments, the quantity and composition of the animal and plant life, and a host of other factors. Most of these measurements depend simply upon the use of a measuring or sample-collecting instrument either hung on a wire over the side, dropped to float freely in the sea, or tethered to the bottom. Apart from the actual measurement or sample it is also necessary to know the depth from which it came, and most instruments therefore incorporate a depth measuring device. Despite such devices it is still essential to have most instruments which are suspended from the ship on as near-vertical a wire as possible, and to utilise ship time most advantageously two or more wires with different instruments at different depths may be suspended from the ship at once. With perhaps the surface current in one direction, the wind in another, deeper layers moving in different directions and the limited manoeuvrability of a large ship, it is not easy to avoid tangles. Despite the provision, in most research ships, of subsidiary bow propellers or active rudders to increase their near-stationary manoeuvrability, it still requires very considerable skill on the part of the officer on watch to manoeuvre the ship to prevent such tangles and the consequent loss of expensive and sometimes irreplaceable equipment. For this reason most new research ships have subsidiary ship controls on special extensions of the bridge, from where the officer can see everything that is happening both fore and aft, and from where he has direct control of the ship's engines.

It is axiomatic that as soon as an instrument is

6. *top* The German Hydrographic Institute's research ship *Meteor* (commissioned 1964, gross tonnage 2615), has a crew of 55 and accommodation for up to 24 scientists

5. *Atlantis II,* the largest of the Woods Hole Oceanographic Institution's research ships (commissioned 1963, gross tonnage), carries 28 crew and 25 scientists

placed in an environment, the environment is altered. Such alterations include the contamination of collected sea-water samples by the material of the water-bottles (Chapter 2), the interference of water flow by a current meter and the changes in distribution of animals brought about by net avoidance. Though such problems are overcome by the physicist and chemist, either by design of equipment or allowing time for instruments to equilibrate, it is not so easy for the biologist. He must chase very alert animals, that are often highly mobile and adept at avoiding active predators, with nets whose high drag properties determine that progress through the water must of necessity be slow (Chapter 2).

By the use of a wide variety of mesh sizes the biologist can be confident of having caught a representative selection of most of the species in the catching area, though the smallest plants will pass through the finest mesh available (Chapter 5). However, animals that apparently occur only sporadically in the nets may in fact be present in the water in considerable numbers and merely be adept at avoiding the sampler. Thus, among the larger animals it can be fairly claimed that nets catch only the slow, the stupid or the greedy, and many animals (especially the larger squids), are best known from the stomachs of whales or sharks. On the other hand, the use of bait may attract from surrounding areas animals which can then be caught or photographed (figure 244), whereas the chemist cannot determine the salinity of water from a distance by attracting it into his water-bottle! Despite limitations set by the presently available techniques, biologists are now building up a more and more detailed picture of the distribution of the teeming animal and plant life of the oceans. This is a particularly laudable achievement since, apart from all the other difficulties of three-dimensional sampling, the animals themselves often undergo very extensive daily vertical movements that further confuse the already complex pattern of zoogeographical, seasonal and vertical distribution. However, what is known of the distribution of animals and plants (Chapters 5 and 7) is but the first step towards an understanding of why their distribution is so variable and why certain definite distributional boundaries occur in the oceans, separating quite discrete groups of organisms found in adjacent water masses apparently differing in only relatively trivial physico-chemical properties. To understand such anomalies much more needs to be known about the physiological requirements and tolerances of the relevant species (Chapter 8). This raises the question (that must always be in mind when evaluating apparent cause and effect relations) whether the oceanographer is measuring the factors important to living organisms; for example perhaps vitamin concentrations are more important to animals than salinity and temperature. As another example, it has recently been suggested that certain nutrient-rich water is unsuitable for phytoplankton growth because of the absence of compounds that normally combine with the nutrient salts to form complexes which are acceptable to the plant.

The penetration of light into the sea is negligible in comparison with its depth; almost all ocean water is, to all intents and purposes, dark below about six hundred metres. The oceanographer uses sound as his underwater illumination, for sound waves penetrate the ocean rapidly and with little attenuation. Echo-sounding provides a rapid means of mapping the ocean floor, and the fine detail it and its modifications, such as sonar, provide (Chapter 2), show the complexity of the bottom to a degree that would have been quite impossible with the old lead-line method (though *Challenger* achieved a sounding of 8,190 m with this method). Not only does it show the bottom, but any other sound-scatterers. Shoals of fish and the curious 'deep scattering layer' (figures 48 and 49), DSL, are readily identified. The DSL varies in appearance and probably in composition in different areas (Chapters 7 and 8) but is believed, from the evidence of both net hauls and direct observation from submersibles, often to consist of migratory layers of fish and siphonophores (Chapters 2 and 6). Sound, from explosive devices, is also used to investigate the structural composition or profile of the sea-floor sediments (Chapter 10). The use of 'pingers' (Chapter 2) to track floating instruments and to monitor the depth and other details of performance of a wide variety of instruments is now commonplace. The advantages of knowing precisely what is happening to an instrument at any time during its use are inestimable, not least for the fact that steps can often be taken to correct any change in performance or position of the instrument, and result in greater accuracy of both use and the information obtained.

Much of the laborious work of recording the physico-chemical or navigational information is now undertaken by shipboard computers (Chapter 2) into which many instruments record directly. Such continuous analysis of the cruise results makes the programme more flexible, for one day's work may now determine what is attempted on the next. It is regrettably impractical (as yet) for most biological determinations to be wholly automated because ultimately the organisms must still be sorted and identified by specialists. Botanists utilise chlorophyll measurements as a short cut to the growth potential of plants in a certain area, and the continuous plankton recorder achieves continuous sampling, although the animals still need to be identified. It may take biological oceanographers a long time to draw conclusions from a cruise's net catches, since identification presents many problems. Though the net hauls may have been made particularly to catch fish (or any other group), nothing caught after so much effort can be discarded and the other animals will be preserved until other specialists have time to examine them – which may be many years. Any such subsequent identification can only enhance the value of the original observations for which the haul was made.

Automatic recording of measurements is the most obvious means of increasing the amount of work that can be achieved in limited ship time. Many such instruments are now available, either recording the data on tape or transmitting it directly back to the ship, including a probe for continuously measuring depth, salinity and temperature (Chapter 2) and a machine for automatically analysing concentrations of nutrient salts in samples of sea water. Such advances are a tribute to the designers and engineers who produce the equipment, for hard though it may be to build it for shore-based laboratories it is still harder for deep-sea work where pressure, temperature and corrosion are among a few of the additional problems to be faced (Chapter 2). The problems of long cruises have been partially overcome by automated instrumentation. Instead of a large expensive ship spending weeks at a time in one place, instruments can be left on buoys by a smaller ship which then returns home or goes elsewhere. Such instruments may be suspended from or attached to an anchored surface buoy (which is subject to the hazards of waves, shipping and piracy), or to lines buoyed up from the sea bed by subsurface floats. After any desired period the ship can return to the spot and collect the surface buoy or release the subsurface buoyed line from its anchor (by means of a sound trigger to an explosive bolt) (Chapter 2). Surface buoys can transmit their information directly to the shore or to a satellite. Buoys are now increasingly used for continuous monitoring of the ocean, often in networks amassing vast amounts of data. Off-shore towers can be used for near-coastal observations, automatic instruments being attached to the legs and superstructure of such constructions. A very considerable amount of data can be obtained by highly sensitive probes on, or photography from, aircraft and both manned and unmanned satellites. Such measurements have been much used for tracking the meanderings of the Gulf Stream, and the development of techniques for space research has resulted in very significant scope for the acquisition of oceanographic information on a global scale.

Recent development of submersible research vehicles (Chapter 2) has greatly increased the scope of man's marine explorations. No longer is the information obtained remotely, it can now be acquired by the man on the spot. The scientific crew remain completely independent of the mother-ship throughout the period of the dive, and can collect biological, geological and water samples while *in situ*. Naval submarines have relatively limited diving capabilities, today's nuclear submarines probably not going deeper than about five hundred metres, but research submersibles may be designed for almost unlimited depth ranges. At present they are extremely expensive and dependent upon the continuous support of a mother-ship to which they return at the end of each dive. The greatest rewards from their use come in the fields of geological sampling and the observation of deep-sea marine animals going about their normal business, but many more applications, including commercial examination of gas and oil pipelines and deep-water salvage, are becoming commonplace.

It is not always appreciated how difficult it is to navigate with accuracy in the rather featureless seascape of mid-ocean waters. Calculations based on sun and star sights are sufficiently accurate for most ships, but it is often necessary for an oceanographic vessel to return to a precise spot, and accuracy of within a mile or so is not good enough. Use is made of detailed echo-sounder maps of the bottom where the sea-floor topography has sufficient distinctive features, or artificial acoustic beacons are laid on the bottom. Accurate though such land-based systems as Decca and Loran are, they do not reach many oceanic areas. Shipboard computers and the release of communications satellites for general purposes is likely to bring satellite navigation into routine use. The accuracy to which the position of lunar vehicles and planetary probes can be determined will perhaps one day become available for ships at sea.

Early oceanographic observations were largely exploratory and as such useful in posing specific scientific questions, but not always in answering them. However, the growing awareness of the multi-disciplinary nature of marine problems, and the development of new sensing instrumentation, data processing and analytical techniques, have enabled more recent oceanographic work to concentrate upon specific aspects of the marine environment, and upon the development of many of the potential resources hitherto locked in the sea. Much, both scientific and sensational, has been written about resources in and under the oceans, but the dividing line between the potential resources and the economically available resources is dependent largely upon the state of the art of marine technology, which in turn requires the oceanographic information for its terms of reference.

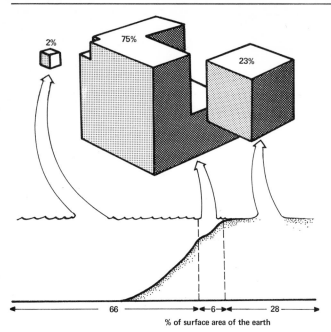

2%

75%

23%

66 ◄───────► ◄─6─► ◄────── 28 ──────►
% of surface area of the earth

7. The percentages of the total aquatic resources, obtained in 1967 from freshwater, the continental shelf and the deep ocean, show the high proportion obtained from shelf waters in relation to their small area

The potential resources may be divided into mineral, chemical and organic. The mineral resources include the naturally precipitated minerals such as phosphorite and manganese nodules and the detrital or deposit minerals. Phosphorite, a complex phosphate deposit, is particularly abundant in shallow areas where upwelling of phosphate-rich deep water occurs, but is unlikely to be economically mined from the sea-bed for several decades (though licences have been issued for deposits in over 1,000 m of water), for there are large land deposits of easier access. Manganese nodules (Chapter 4) appear to be a more likely source of economic potential in the future, despite their predominantly deep-sea distribution. Detrital minerals are those from weathered or eroded rocks, derived mainly from land, and are therefore generally restricted to shallow coastal zones. Sand and gravel are the most important at present, and these only from shallow deposits. The cost of mining other materials such as gems, tin, gold, platinum, and the ores of titanium, iron and zirconium from deep water are completely prohibitive at present, though monazite, a rare earth ore, may be more practicable because of its high unit value. Chemical resources of the deep-sea are manifold, and it has been calculated that every cubic kilometre of sea water contains 40 million tons of dissolved solids with a value of more than $1,000 million if extracted. Unfortunately, with the exceptions of salt, bromine and magnesium,

none can be economically extracted (Chapter 4). The fresh-water content itself is of enormous potential value though only economically extractable in certain arid areas where other sources of supply are inadequate. Organic deposits of gas, oil, coal and sulphur in the deep-sea floor will undoubtedly provide a major source of these materials in the near future, since the technological problems of their exploitation have already been largely overcome, and licences have been issued for areas lying under 500 m of water. Sulphur deposits are produced by bacteria in salt domes, and the simplicity of extraction and great demand for the material from shallow-water deposits makes it an important potential deep-sea resource.

By far the most important resources of the sea for mankind are its living resources (figure 7), contained in the various ecological realms and zones used to classify the sea as a habitat, (figure 8, Table 2). These resources have supplied at least a portion of man's food since time immemorial and there is every prospect of their relative importance becoming greatly increased by careful exploitation. The basic requirement from the sea is one of animal protein. Although the production of plant life in the oceans is so great (Chapter 5), not only are the concentrations too low for economic harvesting but mankind's nutritional requirement is for animal protein, for only animal proteins supply all the amino acids essential for human nutrition. Present world fisheries probably take sixty million tons a year from the sea, and are one of the few industries whose performance has consistently exceeded predictions. In 1949 a maximum yield of twenty-two million tons was suggested, a level exceeded within a year of publication of the prediction, and the level estimated by FAO in 1953 to be achieved by 1960 was in fact exceeded in 1956. Recent estimates suggest a potential harvest of about 150–200 million tons a year could be reached. Such a figure would theoretically supply the minimal animal protein requirements of five thousand million people, the estimated world population in 1990. Such a harvest could be achieved not only by the exploitation of as yet undiscovered stocks of fish (as occurred in the Peruvian anchovy fishery developed in the last decade) but also by full exploitation of presently known stocks. Most commercial fisheries are based on the shallow continental shelf areas, and the relatively small proportion of the fisheries harvest at present taken from deep ocean areas (figure 7) can probably be dramatically increased by the exploitation of new species and new fishing grounds (Chapter 9). A particularly encouraging feature is that the rate of increase of the fisheries harvest has consistently exceeded that of the world's population. The importance to mankind of the living resources of the sea lies not

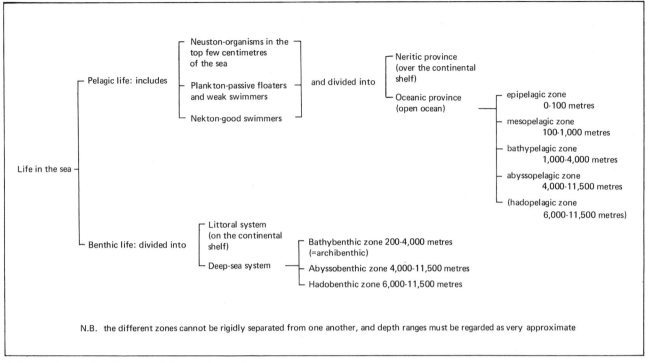

8. A descriptive classification of the marine environment

TABLE 2

Characteristics of the major ecological zones in the ocean

Ecological zone	Upper limit (m)	Lower limit (m)	Temp. range °C (in temperate zones)	Area (% of total ocean bed)	Light
Littoral	Shoreline	100– 300	25–5°C	8	Well-lit
Bathyal	100– 300	1000–4000	15–5°C	16	Almost dark
Abyssal	1000–4000	6000–7000	< 4°C	75	Dark
Hadal	6000–7000	11500	3·6–1·2°C	1	Dark

only in the supply of animal protein, for more and more biochemical constituents of marine organisms are proving to be of great potential value to medical science. Powerful antibiotics have been obtained from such diverse organisms as dinoflagellates and sponges, and recent investigations suggest that steroids found in certain echinoderms may be of value in the chemotherapy of malignant tumours. Other substances isolated from marine animals have antiviral, nerve-blocking or heart-stimulating properties in laboratory experiments, and it is certain that the future harvest from the sea will include many important drugs.

The uncontrolled exploitation of marine resources can easily have disastrous effects. This has already been demonstrated in the cases of many marine mammals, particularly the baleen whales (Chapter 9), and there is a very good chance that similar stock reduction or even possibly extinction will apply to unprotected fish populations. Conservation of such resources is vital, and already a limit has been set to the annual catch of the Peruvian anchovy to prevent over-exploitation. Unfortunately the effects of tampering with even the shore and shallow water shelf environment are almost completely unknown, as was amply demonstrated by the *Torrey Canyon* and Santa Barbara oil spillages, and the effects on the deep ocean can only be guessed at. This is at least partly because there is still insufficient information to establish what the 'normal' environment is really like, particularly in the deep oceans. It may already be too late in many instances, for artificial pollution is already of almost global extent (Chapter 4). Until the normal composition, changes, cycles and fluctuations are known, there is little prospect of recognising, let alone predicting, the effect on these of man's intervention. Terrestrial agriculture is based on a sound knowledge of the environment, and, until this can be achieved in the marine sphere, marine harvesting must remain a hazardous experiment. The short- and long-term effects of industrial projects in the ocean must be a matter of conjecture in the present state of our knowledge. The effect on the biological populations on either side of the projected sea-level canal across the isthmus of Panama is quite unpredictable. The long-term effects of steps suggested to improve or alter the weather (Chapter 3) are unlikely to be predictable when even short-term weather predictions are restricted by the lack of sufficient ocean surface-layer observations. It is to be hoped that national governments will remember Theodore Roosevelt's dictum that 'The Nation behaves well if it treats the natural resources as assets which it must turn over to the next generation increased and not impaired in value'. The public alarm over the dumping in the oceans of surplus nerve gas (and other munitions of war), and over the leakage of crude petroleum from the Santa Barbara offshore oilfields, shows a growing awareness of the degree and variety of pollution. The urgency of some pollution problems is high-lighted by Thor Heyerdahl's reports of how his papyrus raft Ra II was forced to sail through a vast sea of asphalt and crude oil in mid-Atlantic.

The establishment of a United Nations committee for the study of the peaceful exploitation of the resources of 'the sea-bed and ocean floor, and subsoil thereof underlying the high seas beyond the limits of present national jurisdiction' recognised the need for international co-operation for control of oceanic development. Among the first stumbling-blocks for effective international control is the absence of any accepted legal jurisdiction by anyone over any of the deep ocean. By the terms of the 1958 Geneva Conventions a nation has sovereign rights over its continental shelf to a water depth of 200 m, or beyond to where technology can reach, subject only to the right of passage. The so-called 'exploitability clause' gives coastal nations the opportunity to claim jurisdiction over more and more of their neighbouring sea-bed, further restricting the area freely available to other nations. There is no single law of the open sea, merely a motley assemblage of codes, international agreements, unilateral declarations, precedents and traditional practice, some already rendered obsolete by the new

capabilities of marine technology. The gradual encroachment of nations into the deep ocean without legal impediment will give rise to increasing dispute. Many regard the geographical quirk of ownership of a national coastline as no justification for the exclusive exploitation of the resources beyond it, to the economic detriment of nations less favourably sited. However, while the question of legal ownership remains in abeyance, commerce is unlikely to invest large amounts of capital in projects whose long-term ownership cannot be guaranteed and whose profits may ultimately be reaped by someone else. Such considerations are as applicable to living resources as they are to mineral resources, and enrichment of an area of open sea aimed at increasing the fisheries harvest is not practicable when another nation may catch the fish. Opinions are crystallising that a three mile limit of full sovereignty and a further nine miles of limited sovereignty (chiefly exclusive fisheries rights), is at present the most equitable arrangement. There are, however, at least 8 United Nations signatories claiming territorial waters out to a distance of two hundred miles. To quote the Director of the Woods Hole Oceanographic Institution and his associates, 'The old concept of the freedom of the seas was fine so long as the ocean was considered nearly worthless, except for cheap transportation or national defence. Management of the ocean for the exploitation of the natural resources that it contains, or that could be produced, demands a clarification of ownership' (Fye and others, 1968).

Although emphasis has been laid upon some of the problems facing oceanographers, in both research and development, the ingenuity and enthusiasm required to overcome them is present in abundance, as can be seen from the subsequent chapters of this book. Technological advances in other fields and military pressures have provided the present-day oceanographer with unprecedented opportunities to increase our understanding of the oceans and their interrelations with the rest of the world environment. As the terrestrial world becomes more crowded and its resources more impoverished so will man turn to the unexplored oceans to help redress the balance. It is the ultimate aim of all involved in the science of the ocean that their work will in some way contribute to man's future ability to utilise, without destroying, the world at whose edges he is at present merely wading. It is worth remembering that at the start of what the President of the United States of America called 'an historic and unprecedented adventure—an International Decade of Ocean Exploration for the 1970s', man has walked upon the surface of the moon, but not upon the bed of the deep ocean.

2 Hardware and Software

Engineering for the Ocean *F. Pierce*

The marine scientist is dependent upon the expertise of the engineer and his design staff, not only for the comfort and convenience of the research ship, but also for the reliability and performance of a bewildering variety of specialised instruments upon whose successful operation the very value of the research ship itself often ultimately depends. It is the engineer whose task it is to marry the often apparently conflicting requirements of delicacy of measurement with robustness of construction, flexibility of application and consistency of performance, all within the confines of a far from limitless budget.

Many of the difficulties presented by the design of marine instruments are general, and a brief account of corrosion and materials as applicable to the marine engineer precedes the amounts of the instruments themselves. The important features of winches and wires so vital to oceanographic work are then described.

It would be impossible to describe, or even to catalogue, in so limited a space the immense variety of oceanographic hardware now generally available, in addition to the multitude of ingenious devices thought up by individuals for their own specific requirements. Instead, an attempt has been made to give an account of the kind of problems presented to an engineer by the design and construction of a large deck installation (the helical warping capstan), an underwater instrument attached to the ship's hull (sonar), an instrument suspended from the ship on a wire (a camera), and an independent system (pop-up system).

With the insight into problems presented by these examples, the reader will be more readily able to appreciate the ingenuity and subtleties of design of the selection of equipment in use at sea illustrated in the second section of this chapter. The ever-increasing scope for the use of electronics in oceanographic work, and hence of computers, and the specialised requirements and achievements of biological sampling gear and the deep submergence research vehicles, are dealt with separately in their turn.

CORROSION AND MATERIALS

One of the greatest problems facing the marine engineer is that of corrosion (figure 9). Indeed it has recently been stated that 'sea-water is one of the most insidious corrosive environments for metal structures immersed in it or exposed to the atmosphere in its vicinity'. This is no exaggeration, and every effort must be made to ensure that compatible materials are used in the construction of all equipment that has to function reliably, whether on deck, floating at the sea surface or immersed at any depth.

Corrosion may be regarded as the destruction of a material by chemical or electro-chemical reaction with its environment, while the term erosion is used for similar destruction arising from abrasive physical action. A third related problem is that of fouling of materials by the attachment of marine organisms to their surfaces. Not all corrosion is to be regarded as detrimental for in many cases the initial corrosion consists of chemical oxidation and a film of unreactive oxide is deposited over the surface of metal rendering it immune to further corrosive attack. Use is made of this in the artificial production of such films on aluminium, zirconium, titanium and beryllium, a process known as anodising. This oxide coating is also of particular importance in the prevention of further corrosion in nickel-chrome alloys and titanium.

The major types of corrosion/erosion may be divided into four. *Impingement attack* is caused by the collision of gas bubbles with a metal surface, eroding away any protective film and preventing it reforming. Similar, more abrasive, damage may occur through the action of suspended and/or sediment particles in moving sea-water, especially close to the bottom. *Cavitation* damage occurs only in rapidly flowing liquid environments brought about by rotation, vibration or any alteration of flow pattern increasing the relative speed of the liquid over the surface. It occurs where regions of low pressure are produced, the liquid consequently 'boils', and vapour cavities are

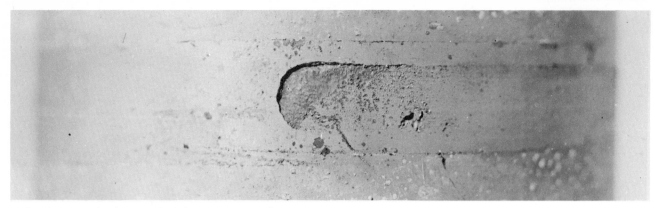

9. Crevice corrosion has attacked the outer casing of this current meter where the clamp encircled it. After 5 weeks at 30 m in the North Sea corrosion has eaten 1 mm deep into the metal.

formed whose subsequent collapse with the release of considerable energy causes the damage. *Stress corrosion* occurs when a metal subjected to a tensile stress, either internal or external, is corroded at the same time. It only occurs in specific corrosion environments and in specific materials such as aluminium, magnesium and titanium alloys, stainless steels and some brass. *Crevice corrosion* usually occurs at the junction of materials and is due to deficiency of oxygen. In conditions of low oxygen the protective oxide films break down and the underlying metal is subject to corrosion. Crevices provide such conditions since the oxygen in such stagnant sites is rapidly depleted and not easily replenished. It is of particular importance in multistrand steel wire rope, where such conditions are common, though a single strand of similar material would be unaffected. The use of more than one material in a design for marine use greatly increases the likelihood of corrosion. This is because two dissimilar metals separated by a conductor (such as sea-water) will act as a simple battery if they are connected to each other, and one will suffer rapid electro-chemical corrosion. The relative rapidity of corrosion under these conditions depends on which metals are used, and by recourse to a table of the galvanic series of metals and alloys in sea-water (Table 3) the designer can predict the rapidity of corrosion in given circumstances. The anodic or base metal is corroded in the system, and the more separated in the series the two metals are, the more rapid is the corrosion. Thus coupling of a base metal such as magnesium or zinc with a noble metal like titanium or gold would be disastrous. Use of combinations giving rapid corrosion of the anode can, however, be made in the construction of pop-up system release mechanisms.

The observed effects of corrosion may be uniform over the surface of the material or restricted to certain areas, as occurs in pitting and stress corrosion. Prevention of corrosion is vital in the design and construction of marine equipment and is generally achieved either by modifying the metals' properties, applying a protective coating or by 'cathodic protection'. Modification of the metals' properties is most simply achieved by alloying, although the method of manufacture may also have considerable effect.

Thus the addition of chromium, nickel or molyb-

TABLE 3
Galvanic series of metals and alloys in sea-water

CATHODIC PROTECTED END
NOBLE

Platinum, Gold
Titanium
Silver
Monel (67% Ni, 30% Cu, 1·2% Mn, 1·2% Fe + C & Si)
*Passive stainless steel (18% Cr 8% Ni)
Inconel (80% Ni, 13% Cr, 6·5% Fe)
Nickel
Copper
Brass (70% Cu, 30% Zn)
Tin
Lead
*Active stainless steel (18% Cr, 8% Ni)
Cast iron
Mild Steel
Aluminium
Zinc
Magnesium

ANODIC CORRODED END
BASE

*Passive stainless steel is covered with an oxide film and resists corrosion. Active stainless steel is of the same composition but the oxide film has been destroyed and the alloy behaves in much less noble manner.

denum to steel (giving stainless steel) allows the engineer to utilize the excellent mechanical properties of steel while providing greatly increased corrosion resistance (by the formation of an oxide film). Protective coatings may be metallic, generally provided by cladding or galvanising, or non-metallic. Cladding is the sandwiching of one metal between two sheets of another, as in duralumin which combines the mechanical strength of the central alloy with the corrosion resistance of the outer aluminium layers. In galvanising, electrolytic deposition of very thin layers of protective metals occurs on the surface of the susceptible material, as in the deposition of zinc on iron. The deposition of very thin layers allows the use of very expensive noble metals in applications of this sort, but if noble metals are to be used, the deposited layer must be completely gap-free, as very rapid corrosion will occur in any exposed underlying base metal due to the separation in the galvanic series of the two metals, and this is accentuated by the concentration of corrosion on the relatively small exposed anodic area relative to the large cathodic area of undamaged covering. Non-metallic coatings are of many types. Chemical protection occurs by the formation of films of reaction products such as the unreactive oxide layers already mentioned. To be effective it is essential that the oxide film has a very low electrical conductivity and thus acts also as an insulator for the underlying metal. Other artificially produced insoluble reaction products are films of lead sulphate on lead and magnesium fluoride on magnesium. Mechanical protection is provided by coverings of cements, glasses, glossy enamels and resins, all of which must of course be uncracked. Painting provides not only mechanical protection but by suitable choice of paint also forms corrosion inhibitors and is probably the most widely used type of corrosion protection. Cathodic protection is utilised in conditions where the juxtaposition of materials renders galvanic corrosion inevitable, but by providing a sacrificial anode in the form of a block of base metal such as zinc or magnesium the corrosion is preferentially concentrated on the block which can be replaced when necessary.

Biological fouling does not usually affect the mechanical properties of materials except for those subject to attack by boring organisms. It may, however, provide much more favourable micro-environments on the surface of the material for the initiation of corrosion, especially crevice corrosion, at such sites as the base of barnacles (figure 11). It may also destroy any protective coating already present. Prevention can be achieved by the use of copper-containing alloys that slowly release copper, an element that is toxic to most fouling organisms. Alternatively the provision of

10. Structural weakness in this deep-sea float caused it to implode under the pressure of water

poisons in the paint to kill settled animals, or vibration, gentle heating or rapid waterflow to prevent the initial settlement, will usually control the problem. It should be noted, however, that the presence of some fouling may help to prevent corrosion by the reduction of water velocity at the surface of a material covered by a thin film of bacterial slime or algae.

The exact location of equipment greatly affects the corrosion that is likely to occur. Apart from the obvious effects of abrasion in silty water, anything buried in mud or sediment may suffer from corrosive micro-environments in the mud. The presence of a protective oxide film on many metals requires oxidising conditions, and in certain oxygen-deficient waters the film breaks down and is not self-healing as in oxygenated water. Deep waters are generally lower in oxygen than well-aerated surface waters and pitting of aluminium alloys may be more prevalent under deep-water conditions. On the other hand, the lower temperature in deep water decreases the rate of corrosion. Differential aeration resulting in the rapid alternation of oxide film breakdown and formation also accelerates corrosion. A further important environmental effect is that of water speed because the higher the water speed, the greater the chances of protective film removal and turbulence damage.

The choice of materials to be used in the construction of marine equipment is not easy, for there are many conflicting factors to be taken into account. On large units, such as ship hulls, economics require the use of a low-cost material, but this is not always such an overriding factor in smaller pieces of equipment where less material is to be used. Apart from the budget within which he must work, the engineer must consider, among other factors, not only the corrosion propensities of his material but its weight, strength in tension

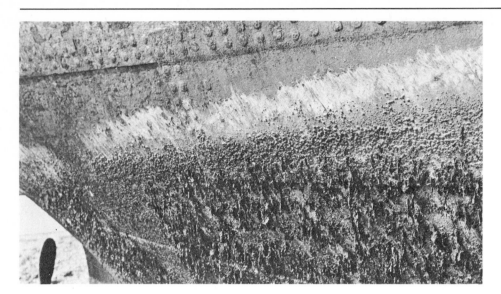

11. Encrustation of various fouling organisms on the hull of a dredger

and compression, its thermal expansion and its compressibility. The latter two factors are of particular importance in deep-sea equipment whose efficiency must be unaltered by rapid changes in temperature and pressure. The drop in temperature as a piece of sealed equipment is lowered to the depths may cause the deposition of dew over the delicate electronic components it contains, and thus some provision must often be made for keeping the equipment dry.

Of metallic materials, alloys can now be almost tailor-made to suit the design requirements. Thus magnesium and aluminium alloys provide both good corrosion resistance and high strength; indeed some aluminium alloys have been shown to suffer little or no loss of tensile strength after up to five years' immersion. Copper-nickel alloys such as monel provide similar capabilities and look promising materials for the construction of wire ropes. The problems of crevice corrosion in stainless steel wire ropes have already been mentioned, and this is at present overcome by the provision of cupro-nickel, zinc or aluminium coatings. The velocity of the wires through the water can rapidly cause the loss of the latter two coatings, however, and the threat of galvanic corrosion is sometimes decreased by the provision on the wire of sacrificial zinc anodes, efficient even at intervals of up to 1,000 m. In other equipment sealing joints with zinc-containing grease helps prevent crevice corrosion. Lead is used to cover undersea cables, being corrosion-resistant due to the protection of a sulphate or carbonate film and has the advantage of being easily extruded, an important feature in the manufacturing considerations of wire and cables. Cathodic protection has been utilized on the DSRV *Aluminaut* with great success, in combination with protective coatings of four coats of polyurethane. Such vehicles require very high strength

– low density materials, and titanium is another such material much used, its oxide film being even more resistant than that of aluminium. Unfortunately, many materials of such type are susceptible to stress corrosion, and therefore have their limitations. This is equally the case in materials needed for the construction of pressure-testing vessels that must endure high tensile stresses. The simultaneous use of otherwise incompatible metals or alloys can be achieved without fear of corrosion providing they are suitably insulated from each other, and this can be done by using plastic, resin or synthetic rubbers. Most non-metallic materials do not approach the mechanical strength of metals and their alloys, but some are unaffected by electro-chemical corrosion, and most are easily moulded or formed, lightweight and cheap. As such, they are an attractive proposition to the designer in conditions requiring only limited mechanical strength. Nylon, terylene and polypropylene are used for ropes, and are greatly superior to cotton and manila in submerged application, for they are unaffected by the biological damage that rapidly occurs in the natural materials. Similar bacterial damage, or rotting, occurs in nets of natural materials, and the greater strength and wear of the artificial fibres has revolutionized net construction. Nylon, teflon, polystyrene, polythene, vinyl tubing and polyvinylchloride have been shown to be subject to some damage from boring organisms and are therefore not suitable for long-term submerged equipment. They are invaluable for such purposes as water-sampling bottles; the determination of trace amounts of metallic elements in sea-water necessitates the use of samplers which will not provide their own supply of such elements.

Glass reinforced plastics (fibreglass) are used for many purposes such as streamlining towed bodies and

12. The main electric trawling winch on the afterdeck of RRS *Discovery*. The centre drum contains 10000 m of tapered wire, and the two end drums each accommodate 1000 m of 2.25 inch circumference wire for use when twin towing warps are required, as in the operation of large commercial trawls. Traversing rollers in front of each drum lay the wire evenly, and the large paired sheaves guide and measure the wire as it is payed in and out

sonar domes and are easily moulded and repaired. They are corrosion-resistant but little is yet known about their strength or fatigue characteristics at high pressures. They can suffer mechanical deterioration from absorption of moisture by the glass fibres at cut ends or joints, and from abrasion, and terylene overlays are sometimes used. PTFE is an admirable material for bearings, while synthetic rubbers are used for 'O' ring seals and diaphragms.

Glass, or ceramics, are very promising materials for they are not corroded or fouled and are particularly attractive for deep-sea work in that though they have little tensile strength they become stronger under compression and have a very favourable strength/weight relationship. They thus have great potential for deep-sea floats and buoys (figure 10), and submersibles, and much effort is going into their further development for such equipment, because in such applications they would greatly increase the buoyancy and hence the pay-load.

Future developments of carbon fibre materials may well prove invaluable for the oceanographic engineer in the construction of submersibles where the strength/weight relationship is at such a premium. The designer and engineer have an exceedingly difficult task in the marine environment, for failure of any of their products may easily result, at best, in the effective loss of months of previous work, and at worst, hazard the life and limb of the oceanographer and his crew, particularly since the economics of research ship time often necessitate the use of equipment, especially wires and cables, at close to their design limits.

WINCHES

Development of winches used in commercial fishing facilitates progress in the design of oceanographic winches, but far greater lengths of wire must be accommodated on the latter and this presents additional problems.

Let us consider some of the problems if a large orthodox oceanographic winch is to be designed. First, one drum must hold 15,000 m of wire with a circumference of about $1\frac{1}{2}$ inches if it is to be usable for plankton tows at depths of 6,000 m. Then, if a very large trawl is also to be used, two further drums must accommodate lengths of trawling wire of $2\frac{1}{2}$–$3\frac{1}{4}$ inches circumference and the several drums must be independently controlled. For this the winch needs to be large and very heavy. To reduce its effect on deck space and ship's stability it may be placed below decks but then needs remote control. It must have the power to pull in the weight of wire with the drag of a net or instrument at speeds of 0·4–4 m/sec. The low speeds are necessary for controlled raising of some instruments while the high speeds can save much valuable ship time during deep hauls. In winding such long wires on to a winch under load, great pressures are exerted laterally on the end plates of the winch barrel as the outer turns of the wire force between the inner turns. If the barrel and end plates are not cast as one unit or are insufficiently thick, the end plates may be forced off the barrel (figure 12).

The source of power is somewhat dictated by the ship's power. When vessels were steam driven, steam winches were easy to install and they had infinitely variable speed with constant torque over the whole speed range as well as rapid acceleration. Unless special boilers are fitted, steam winches can no longer be used and other methods have to be considered. Electric power is easily obtained on diesel-electric ships, and research ships usually have ample electric power. Hydraulic power, where oil from a pump is used to drive the winch barrel round, has the advantage of good control of speed, speedy acceleration and automatic lubrication. It is, however, rather less efficient than direct electrical power.

Oceanographic work is often necessarily carried out near tolerance limits and, with prototype gear, past experience does not always cover the eventualities which arise. Safety and reliability of the winch are therefore very important. Rotating machinery must be enclosed as much as possible, the winch operator must be in a position to observe the run of the wire the whole time, the system must be entirely reliable during long periods without maintenance and, where possible, safety devices should be incorporated to allow the winch to pay out if excessive loads occur. Corrosion can be reduced by having automatic cleaning and oiling systems for the wire and enclosed machinery; this also allows a greater choice of materials for the machinery because they need not be selected according to the galvanic series (see CORROSION). Pinching of the wire, resulting in reduction of the breaking load and corrosion, is very much reduced by automatic spooling so that the wire's progression across the drum is controlled during hauling.

During paying out and while the wire is out, it is necessary to control or stop the rotation with a brake. This is often a band round the edge of the drum operated by a screw and handwheel, a system which does not allow slip when the bottom is fouled and produces considerable heat and excessive wear. If an air cylinder is introduced it allows the brake tension to be read from a pressure gauge as well as adjusting for shrinkage of the drum as it cools after paying out. Hydraulic disc or band brakes pressured by a small pump are also used. Another system is to design the driving gears to allow regenerative braking when paying out, by which the winch becomes the driving force which feeds energy back into the winch engine and provides resistance on the barrel. This may effectively be combined with disc brakes.

There are several advantages in separating the traction unit from the wire storage spool; the weight of the wire can be placed at a lower level in the ship, so reducing problems of stability, deck space and corrosion; the storage drum need not be as heavy because there will be little lateral thrust on the end plates; the wire is stored under low tension; storage drums may be changed so that different gauges of wire can be used and wire may be changed easily. Independent traction units usually consist of two grooved drums around which the wire passes several times. As mentioned later in this chapter, a helical warping capstan or a 'caterpuller' system are particularly suitable for faired cable.

Design of smaller winches handling 4–6 mm wire involve many of the problems discussed above, but weight and bulk are not among them.

The future will certainly see further use of constant tension devices. In addition, electrical circuits and associated hydraulic linkages can be used to give a servo-control loop which permits the insertion of any desired control function such as depth, temperature or distance from the bottom. Thus the future winch may assume powers of data collection unattainable at present by manually controlled winches.

WIRES

Lowering instruments or nets on a long wire from a vessel is a hazardous undertaking unless the strain on the wire is properly considered. In research operations where diverse gear is developed and used, performance of wires and handling gear must frequently be re-assessed. During some operations, such as dredging, weak links may be put at the bottom of the wire so that the gear may be jettisoned before the wire breaks. Clearly it is essential that the wire should break at a lower strain than the sheaves, the gallows or the winch, if serious accidents are to be prevented.

If the wire has been used for a long time and wear or corrosion have affected it, the wire may break before the specified nominal breaking load has been reached. In addition, breaks may easily occur if the elastic limit has been reached during previous use. The wire's yield point occurs at a load equivalent to about 60 per cent of the breaking load; at that point the wire is stretched but does not resume its former condition of length or strength when the load is removed. A similar limit may be reached if the wire is kinked during such operations as dredging or grabbing. Wire damage can also result from pinching of the wire on the winch drum or cutting by grooves on worn blocks and rollers. Some of these dangers can only be reduced by continual maintenance and checking, and normal wire engineering practice prescribes a safe working load of one fifth of the specified breaking load. Adoption of such a limit would seriously hamper much oceanographic work and reliance must be put on good load-measuring devices and an awareness of the wire limitations by the operator.

The load on the wire is monitored by a 'dynamometer' which operates by the wire pressing against springs or a load cell. It registers the total strain on the wire caused by the weight of the instrument in water, the weight of the wire in water, the drag of the wire and instrument, and the effect of the jerks on the wire by the ship's up and down motion, the 'live loading'. The weight of the instrument is usually of minor importance. A long wire can be broken by its own weight if it is

13. Hydrographic winch operation with a plankton net just coming inboard. The spring accumulator damps out the 'live loading' and keeps a relatively even tension on the wire. The dials on the winch tell the operator how much wire there is still outboard

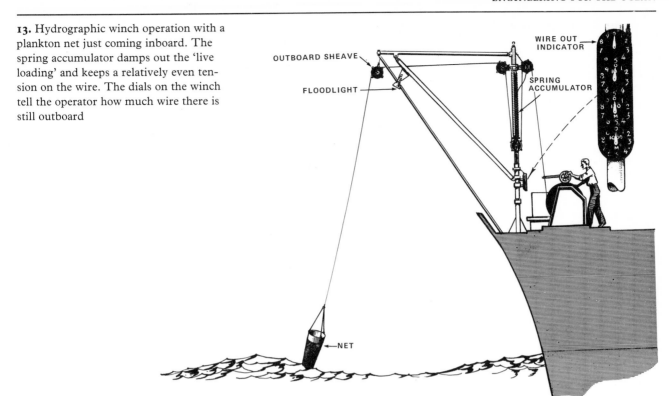

thick and hangs vertically. To avoid such a disaster wires have been developed which have an elevated elastic limit; others are gradually tapered along their length so that their weight and drag are reduced. Drag of the wire has been reduced by using streamlining or 'fairing' consisting of solid rubber or hair-like trailing polyurethane filaments attached to the wire. Solid fairing introduces handling difficulties (see below) and hair-fairing is more easily used on short electric cables than on trawling wires. Live loading is extremely dangerous in worsening weather conditions, when heavy gear often needs to be retrieved, particularly when the gear has a large drag as in the case of a net. This load is often reduced by 'accumulators' so that sudden pulls on the wire are absorbed by large springs and sudden slack in the wire is taken up by release of the springs. Servo-control of winches by which the speed of hauling is regulated automatically by the load on the wire is now coming into use. A further strain which is not registered upon a dynamometer may be caused by excessive bending of a wire round a sheave. Sheaves should always be more than 25 to 42 times the diameter of the wire depending upon its construction (figure 14). Wire rope construction is described by the number of strands and the number of wires per strand. The more strands and wires there are the more flexibility and fatigue resistance but the lower the resistance to abrasion and corrosion. A popular construction is 6 × 19, which has good all-round properties (figure

14). The strands are wound round a core of either fibre or metal. The greater strength of the latter is usually favoured for oceanographic work.

The most usual material for oceanographic wires is plow steel with tensile strengths ranging between 100 and 130 tons per sq. inch. Stainless steel and monel have poorer fatigue resistance and are much more expensive.

HELICAL WARPING CAPSTAN
Development of deck machinery for oceanographic work poses several problems which may be illustrated by the helical warping capstan developed to handle a thermistor chain.

A thermistor chain is, in essence, an electric cable with temperature sensitive units along its length, stretched out by a heavy streamlined weight at its lower end and towed at relatively high speeds. To reduce drag of the cable it has rubber stream-lined 'fairing' clipped to it. Handling of this faired cable introduces two main problems for the engineer. First, the system must not damage the rubber fairing, and secondly the electric wire should not be stored under too high a tension. The use of a conventional winch would result in damage to the rubber fairing and clips, especially when storing long lengths of faired cable in multiple layers. The damage is caused by the top layers pressing through those beneath them under the high compression loading occasioned by storing under

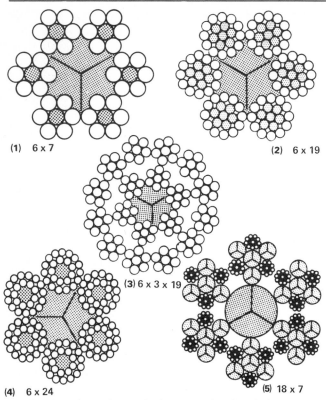

(1) 6 x 7   (2) 6 x 19   (3) 6 x 3 x 19   (4) 6 x 24   (5) 18 x 7

14. Cross sections of several wire ropes showing their construction: 1. 6 × 7, good abrasion qualities but poor flexibility and fatigue life; 2. 6 × 19, popular in oceanography since it has good average abrasion flexibility and fatigue life; 3. 6 × 3 × 19, spring lay for mooring and towing, flexible; 4. 6 × 24, for mooring lines; absorbs shock, strong and flexible; 5. 18 × 7, non-rotating rope for lifting loads; difficult to handle (from Meals 1969)

fairing, and the helical warping capstan was specially devised to spread the turns so that there is room for the fairing, and to prevent any lateral movement of the turns along the drum.

The principle of this capstan has been used in the textile industry for some time and was first adapted for marine use by the US Naval Research Laboratory. Overriding turns are prevented by having two identical skeleton drums each consisting of an end disc with sixteen axially mounted bars. The bars of the two drums intermesh and rotate in much the same space that would be occupied by a conventional drum (figure 15). Each skeleton drum rotates on bearings mounted eccentrically on a fixed shaft so that the drum axes are slightly offset and canted with respect to one another. Thus, a cable round the drums is carried by the bar members on one drum for one half a revolution and by bars of the other drum for the next half revolution, and the offset relationship of the bars allows the cable to be transferred laterally along the drum without slipping. Each element of the cable follows an identical helical path across the capstan bars. By having four turns on this capstan the tension on the take-up drum is only one twentieth of the outboard load on the cable (figure 15).

Next, a means of turning the capstan and the storage drum must be selected. Research ships usually have ample electric power, and the gear described here is most efficient if a single prime mover for the capstan and storage drum is electric or electro-hydraulic. Turning of the storage drum must be regulated so that the tension in the wire remains about the same, even though the effective diameter of the drum increases as outer turns of the wire are laid upon inner turns. To do this, its speed of rotation must decrease as each layer is added, and this is accomplished by use of a slipping clutch. The most suitable is a fluid magnetic clutch which consists of two discs rotating in a fluid containing spherical magnetic particles within a variable magnetic field provided by an electro-magnet. By varying the current through the electro-magnet the number of particle chains between the clutch plates is varied, and this influences the torque applied to the storage drum. The mechanical load within the clutch is kept as low as possible by automatically balancing the clutch with a variable speed gearbox via a differential gear.

For flexibility, the storage drum must be readily interchangeable so that the capstan can handle either faired cable or wires. For use with electric cables, the drum is fitted with low noise level sliprings so that signals can be passed from the cable's inner end via the sliprings and into the laboratory.

Materials for this machinery must be selected with both strength and corrosion in mind. Flying spray and

a tension equivalent to the weight of the cable, the streamlined weight and the drag of the gear through the water. To avoid this, the pulling machinery must be separated from the storage drum. This may be done by using 'two drum traction units', 'caterpullers' (a type of tracked vehicle), or a capstan. Of these, the first has not been successfully used with 'faired' cable, and the second tends to damage the clips attaching fairing to the cable. The capstan seems the most promising for the purpose but in its simple form it has several disadvantages. The cable with the weight on the end runs inboard, takes three or four turns round the drum of the capstan and then runs under little tension to the storage drum. The large load outboard is held by a small pull inboard because of the friction of the turns round the capstan. Unfortunately, as the capstan rotates, the wire rides along the drum and overrides itself unless it is constantly attended by an operator. Such overriding would seriously damage the

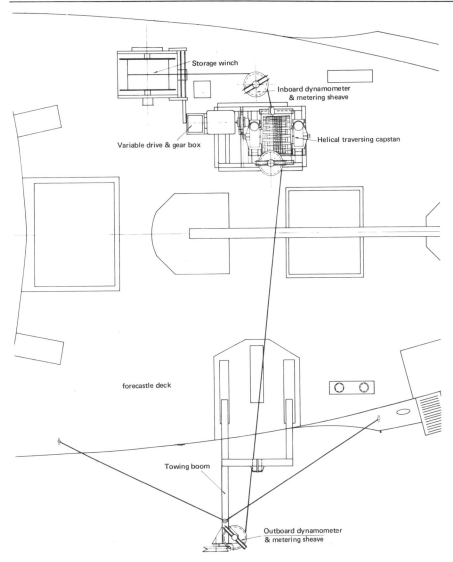

16. A diagrammatic view from above of the deck layout of the helical warping capstan and its associated units

Storage winch

Inboard dynamometer & metering sheave

Variable drive & gear box

Helical traversing capstan

forecastle deck

Towing boom

Outboard dynamometer & metering sheave

15. The thermistor chain is brought inboard on the helical warping capstan. In the foreground is the capstan with its interlocking bars carrying four turns of faired cable and a thermistor unit on top; in the left background is the storage drum

water off the cables necessitates materials which are both compatible with one another and resistant to salt water, and all gears must be totally enclosed.

Weight of the gear is important because it may effect the stability of the ship and hence the siting of the machinery. Strengthening of the deck must also be possible at the site selected.

Finally, the machinery must be as compact as possible to justify its place in the limited space of a ship's deck (figure 16).

THE DEEP-SEA CAMERA

Much valuable information about the ocean floor has been obtained by remote sampling or sensing methods, ranging from the earliest collections of bottom samples by means of a lump of tallow in the bottom of a sounding lead to the highly sophisticated sub-bottom profiling techniques now in use. However, many interesting features can only be recognised by visual means, and though it has recently become possible to observe the bottom directly from the security of a deep submersible vehicle, photographs still provide the most economic, convenient and informative data on the ocean bed.

The deep-sea camera is typical of all instruments lowered from the ship on a wire in that it must be a totally self-contained unit with its own built-in trigger or response to the environment, it must be unaffected by the large and rapid changes of temperature and pressure regularly experienced during its journeys to and from the depths of the ocean, and above all, it must be simple, easy to handle and flexible in application. Its weight must be sufficient to allow it to sink rapidly but not so great as to make handling hazardous, or to require the use of excessively thick wire to take the strain of hauling it to the surface. It must be robust enough in construction to withstand the inevitable knocks during lowering and recovery procedures, which may well take place in awkward sea conditions, as well as accidental bumping on the sea bed. A typical camera comprises a light source, a photographic unit (or two for stereoscopic work) and a triggering mechanism. These units are mounted on a framework of stainless steel or aluminium alloy tubing providing lightness with strength, freedom from corrosion, minimum drag and simple alteration of the units' arrangement. The watertight housings of the light and camera units are cylindrical, for although a sphere is the strongest and simplest of pressure-resistant shapes, a cylinder is more readily packed with equipment and its flat ends are suitable for windows and access ports. The housings are usually of aluminium alloys or steel, for these materials are sufficiently strong to withstand the great pressures to which they will be subjected without requiring massively thick walls, thus greatly increasing the weight. They are also relatively free from corrosion problems and cheap. Electrical connectors between the units are required for synchrony of operation, and these must be made with care for they must also be free from pressure-induced leakage and could become points of mechanical weakness in the housing assemblies. It is of course essential to test all the units to their operating pressures before even attempting to use the whole assembly in the sea. The ends of the cylinders are usually threaded as screw caps, the final seal being provided by an 'O'-ring of rubber. Since it is essential to force the end cap down on to the seal the screw threads must be sufficiently slack to allow this slight movement to occur.

The light source is an electronic flash unit protected by a hemispherical glass dome (glass has great strength in compression but not in tension). An electronic flash has the advantage of very short duration thereby stopping the relative movement of camera and object inevitable with a small camera dangling on the end of a mile or more of wire. The short period flash and absence of ambient light eliminate the need for a shutter, except in shallow sunlit water; many flashes can be obtained at precise intervals from a small battery supply and the colour of the light is similar to daylight. A silvered reflector spreads the light evenly over the area to be illuminated.

Since light in water suffers both absorption and scattering, these features will affect the brightness and clarity of the object. Absorption can be compensated for by increasing the flash intensity or by reducing the object distance, but scattering can only be reduced by illumination of the object from an angle. This is achieved by careful alignment of the geometry of the system on its framework, and this is essential from other points of view. The properties of water, particularly its refractive index, are such that a camera system which takes perfect pictures in air will not do so in water. The camera needs to be focussed for a distance in air of three-quarters the actual distance in water; it will have a smaller angular field of view, and an object produces a larger image than it would in air. In addition there is some dispersion of white light, since the refractive index varies with the colour, and chromatic aberration will occur in the outer part of the camera's field of view. There is also distortion of the geometrical pattern of light rays reflected from an object in water when forming an image in air. By suitable alignment, the engineer can ensure that perfect pictures are routinely obtained, chromatic aberration being overcome by the use of correcting lenses.

The deep-sea camera developed at the National Institute of Oceanography (U.K.) has the camera itself specially designed to fit into its cylindrical housing, and, as in most other such instruments, uses 35 mm film for ease of processing. The camera has no shutter, the exposure being predetermined by the flash intensity, and the film is wound on automatically after each exposure by a battery operated motor. Originally five metres of film, or one hundred exposures, could be loaded for a single run, but recent designs have greatly increased these figures. Two interchangeable lenses have angular fields of respectively 37° and 47°. The window of the camera housing is a truncated cone of perspex, fitting into an accurately machined recess in the end cap and sealed by an 'O'-ring. This arrangement provides mechanical support for the plastic cone which would otherwise become appreciably deformed by the pressure. Flat glass windows are less prone to pressure distortion but their strength is not easily predictable.

17. On the end of the wire is a probe that will continuously record temperature, salinity and depth as it is lowered through the water. Just above is a current meter that will be used to measure the currents at various depths during the descent; the use of two such instruments on a single wire greatly increases the amount of data that can be obtained in a given time

18. The 70 cm diameter net has a mesh fine enough to catch the smaller zooplankton as it is hauled vertically through the water. Round its canvas neck can be seen the loop of rope by which the net mouth is throttled when the net is closed before being brought back to the surface

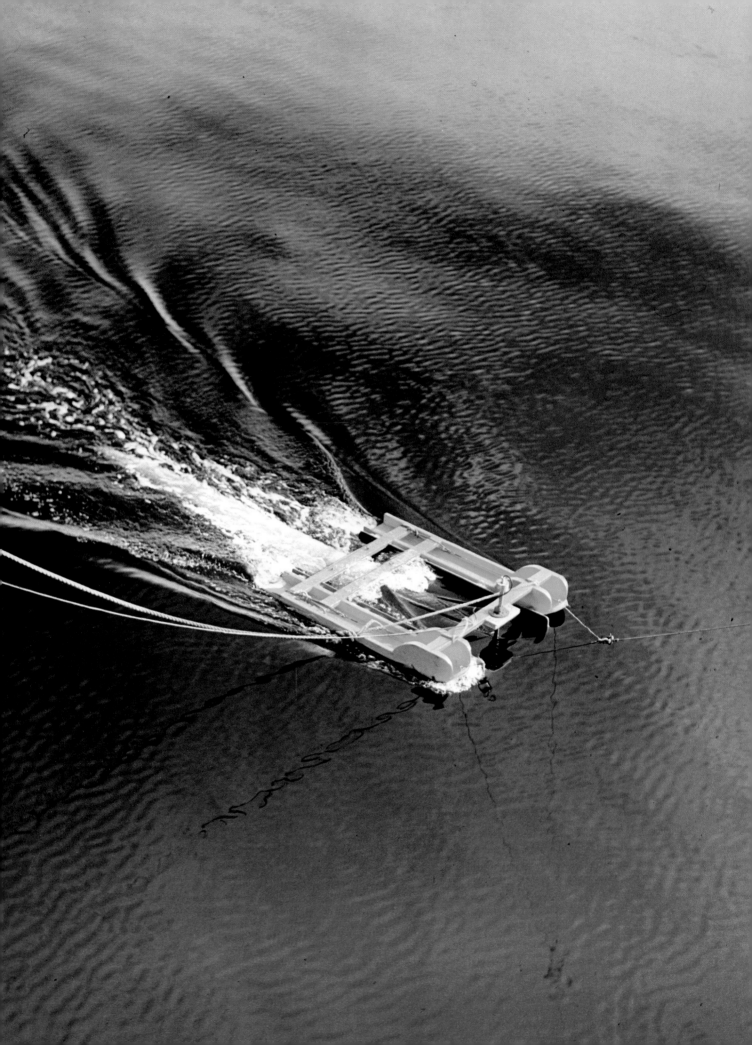

While the deep-sea camera is most often used for bottom photographs, the provision of suitable trigger mechanisms enable it to be used for other purposes. For bottom photographs a weight below the camera closes a micro-switch when it hits the bottom, and this supplies a six-volt triggering pulse to the flash unit after which the film is wound on one frame. The camera can then be raised a few metres and lowered again to take another picture, and so on as the ship drifts. Alternatively the camera may be mounted just above the bottom on a sled or floats and take sequential pictures, providing a photographic profile of the bottom over which it is towed (figures 20, 27). The micro-switch is protected from the pressure by filling its thin perspex case with a relatively incompressible non-corrosive mineral oil. In theory the camera and flash units could also be oil-filled but the practical problems of an oil-immersed film are obvious, and a pressure-resistant housing is simpler. An alternative mode of operation allows the film to wind on without each frame being individually triggered, the delay between frames being preselected by gearing the wind-on motor. An added sophistication is the coupling of the camera to a pinger allowing, by reflection of the ping from the bottom, continuous monitoring of the camera's depth, and by changes in pulse rate, initiated by operation of the bottoming switch, informing the listener on board ship that the camera has operated correctly. The electronic components of the pinger are housed in one of the framework tubes. For photography of mid-water animals a variety of ingenious triggers have been devised from the mechanical pull of an animal tugging at bait (figures 21, 22, 23, 24, 25) to its interruption of infra-red or sonic beams or by the utilisation of the animal's luminescent flash (figure 26). The simultaneous inclusion in the picture of a clock, depth meter and/or data board greatly simplifies the subsequent problems of identifying exactly when and where each photograph was taken.

While there are many permutations of the above systems in operation at present, the future development of deep-sea photography seems likely to be based on more direct control of a camera suspended by armoured electrical cable, thereby overcoming many of the problems of an instrument at present forced to be largely pre-programmed, and enormously increasing its flexibility of operation. In particular, view-finding improvements utilising television or high-resolution sonar will then be much more readily achieved.

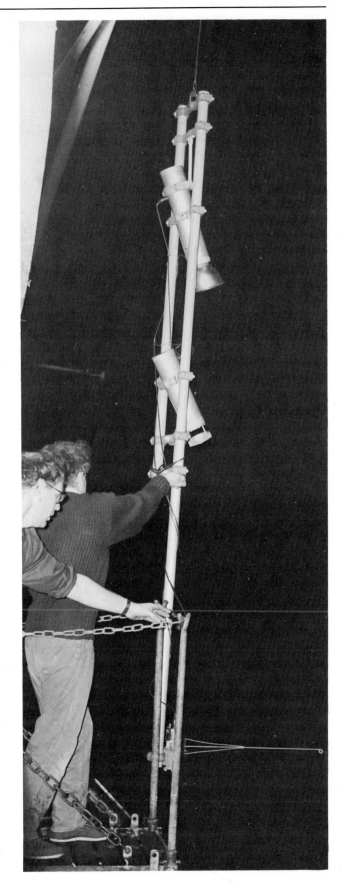

20. Camera arranged for photography of midwater animals is prepared for lowering at night. The bait is suspended from the end of the trigger arm projecting out to the right

19. *opposite* A net attached to a frame mounted on two ski-boards is used to catch the neuston, the fauna of the uppermost few centimetres. The very deep blue of ocean water is well illustrated in these very calm conditions

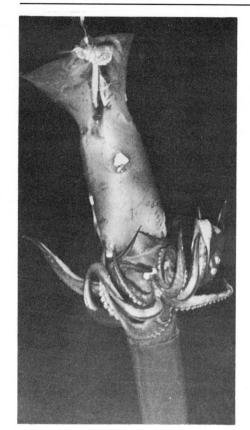 

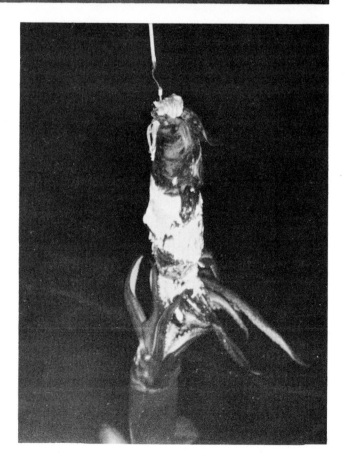

21, 22, and **23.** A squid 45 cm long that has become 'foul-hooked' by the tail is attacked and partly eaten by other individuals of the same species; these pictures are taken from a series of 30 recording the events

24 and **25.** A squid about 45 cm long that has taken the bait on the midwater camera struggles to free itself, its cloud of ink showing up feebly in the light of the flash

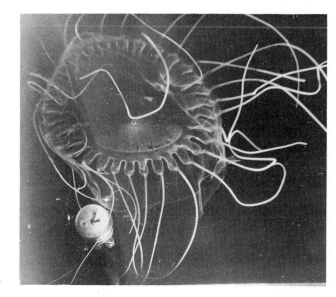

26. This rare 12 cm medusa *Solmissus incisa* took its own photograph at 1000 m; its own luminescence was picked up by a phototube which triggered the electronic flash. The 2 cm watch provides both a scale and the moment of exposure

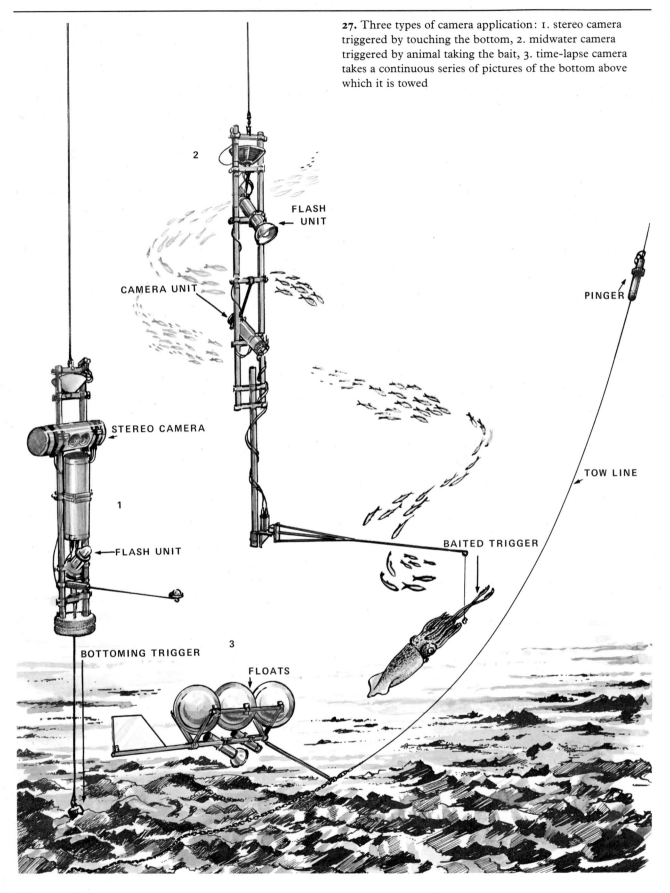

27. Three types of camera application: 1. stereo camera triggered by touching the bottom, 2. midwater camera triggered by animal taking the bait, 3. time-lapse camera takes a continuous series of pictures of the bottom above which it is towed

FLASH UNIT

PINGER

CAMERA UNIT

STEREO CAMERA

TOW LINE

FLASH UNIT

BAITED TRIGGER

BOTTOMING TRIGGER

FLOATS

POP-UP SYSTEMS

Acquisition of data by instruments dangled from a ship is an expensive procedure largely because of the cost of ship time. Main engines and most of the men on board are under-employed during the process, and the ship must make expensive movements to and from ports during a long survey at a particular spot. If the instruments are suspended from moored buoys, data collection will be cheaper and much long-term data may be acquired from an array of buoys in cases where an array of ships would be prohibitively expensive. Unfortunately, moored buoys have proved very liable to loss from breaks of the mooring cable as they toss in high seas. Modern pirates who are attracted by the buoy, curious about the instruments and acquisitive by nature have caused further waste. Such factors have led to the development of unattractive surface buoys and of pop-up systems by which the buoyancy is submerged and can be popped up by signals from the surface.

The basic requirements of a pop-up buoy are buoyancy sufficient to bring the weight of all the instruments to the surface but which will not implode at the depth to which it sinks, weight sufficient to hold the buoyancy and instruments firmly moored to the bottom, a connection between the weight and the float, and a release which will detach the weight at an acoustic command from the surface or after a specified period of time. The instruments are either housed within the float or are attached to a wire joining the float and weight (figure 28).

An example of such a system with a wire bearing four current meters will serve to illustrate the main problems. First, similar problems will arise to those for any instruments which are lowered into the sea. Materials must be selected to resist corrosion and which are very compatible with one another, because they may be in the sea for a year or more. Instrument casings and floats must withstand both the required pressure and the rapid temperature changes experienced when they are quickly lowered or dropped from a deck in the hot sun to depths where the water may be at only 2°C (figure 29).

Secondly, certain problems specific to a pop-up system must be overcome. There must be a careful balance between the total weight, the drag of the whole system and the buoyancy, so that the wire is held as near vertical as possible. This can only be achieved by weighing all the components in water before going to sea, finding the buoyancy of the floats accurately and calculating the drag of the float, wire and instruments in various conditions of current (figure 28). The instrument cases are made positively buoyant where possible and the area exposed to

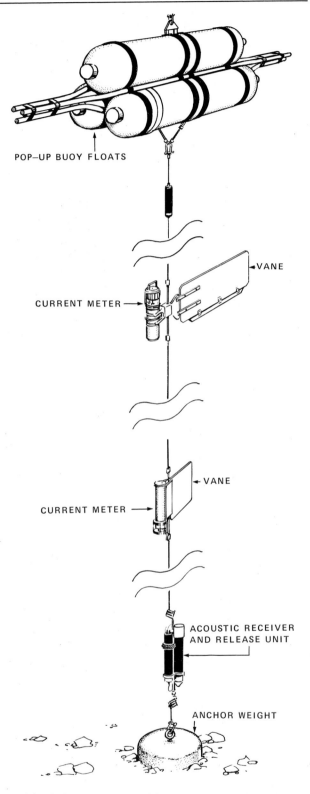

POP-UP BUOY FLOATS

VANE

CURRENT METER

VANE

CURRENT METER

ACOUSTIC RECEIVER AND RELEASE UNIT

ANCHOR WEIGHT

28. A typical pop-up array with current meters situated at intervals down the wire. At the bottom is the anchor weight which is jettisoned by the release unit on receipt of the appropriate acoustic command. The subsurface floats then bring the current meter array to the surface for recovery

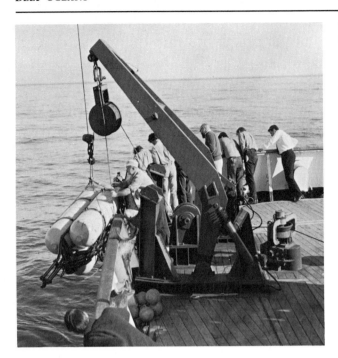

29. The subsurface floats for a pop-up system being lowered over the side

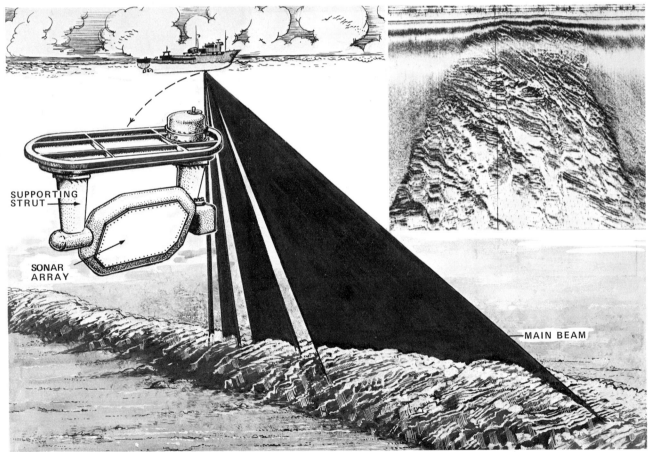

30. The plate bearing the narrow beam sonar array is mounted on the ship's bottom, and the array can be trained to any desired angle. The sound emerges in one major and several minor beams or lobes. A reflector on the sea-bed will give an echo from each beam, but that from the major beam will give the most detailed information. In the record (top right) the thin traces at the top are from the minor beams, the detailed one at the bottom from the major beam

horizontal currents is usually kept small to minimise drag (not in figure 28). The wire itself is a very important contributor to drag, so it must be as thin and strong as possible and its ultimate breaking load must be at least twice the maximum load anticipated. Steel is most suitable but galvanised mild steel cannot be relied on for more than three months. Stainless steel lasts longer but is prone to crevice corrosion. Steel wire, of course, is heavy, and to hold the weight as it is being lowered and the buoyancy when in position, it usually requires 4 or 6 mm wire. A combination of these can be used so that the upper wire is stronger and the lower wire is lighter and has less drag. Man-made fibre rope such as terylene can be used deeper than about 2,000 m, but shallower than that there is considerable risk that it will be severed by fish bites. While this is neutrally buoyant and eliminates weight problems, further complications arise because it stretches much more than wire.

Various types of release are available for pop-up systems. There is usually a hydrophone which picks up the sound signal from a ship overhead and throws a switch when a particular coded signal is received. The switch is used to activate an electric motor or to fire an explosive charge, either of which releases a bolt holding the weight. Explosive charges are less likely to stick because of fouling and have the advantage of being smaller. Short term release can be achieved by using a magnesium link or a link of metals which are far apart on the galvanic scale (Table 3), so the metal corrodes so rapidly that it dissolves altogether after a few hours or days. The time taken to dissolve is rather unpredictable, however, because it varies with temperature and pressure. Where it is necessary to release a weight at a particular depth as an instrument descends, a release using the hydrostatic pressure to operate a piston which shears a metal pin may be used. These can be made to operate very close to the depths selected.

NARROW BEAM SONAR

Sonar, which is usually called asdic from the initials of 'Allied Submarine Detection Investigation Committee', is a device which detects underwater objects by timing the interval between transmission of a sound impulse and the return of its echo from the object. In practice, the time is registered as a distance on recording chart paper. Sonar can be directed downwards to act as a means of recording depth, called echo sounding, but it is normally used at an oblique angle to the vertical. A sonar device producing a beam which is broad vertically but narrow in the horizontal plane, has proved extremely valuable in geological studies of the sea bed. With the ship under way the sonar beam scans to

port or starboard of the ship's course and builds up a detailed picture of the sea bed which is displayed as an acoustic picture on a facsimile recorder (figure 30).

To produce a narrow beam in the horizontal plane, the array of transducers must be many wave lengths long, and to make the picture clear the array must be stabilised so that the rolling of the ship does not cause the sonar beam to sweep up and down. So that the array can be trained and aeration avoided it must be held clear of the ship's hull. The sonar device may be mounted on a plate at the bottom of a floodable trunk running vertically down through the ship; by this means, the sonar can be maintained without the necessity for dry docking the ship. The transducer array is contained in a gunmetal casting; this casting together with the supporting struts are streamlined as much as possible to reduce both drag and water noise which quenches the echo.

The array is immersed in oil to maintain electrical insulation and has an acoustic window of fibre glass sealed into the gunmetal casting. Expansion and contraction of the oil is allowed for by a large rubber diaphragm mounted opposite the acoustic window and exposed to sea water for pressure balancing. Organisms living in sea water are known to attack rubber, and the ingress of sea water is arranged through holes small enough to exclude their eggs. Fouling is no problem on the fibre-glass window, and the transducer housing is gunmetal which contains a large proportion of copper and fouls only slightly. The journal bearings supporting the array are of laminated plastic material and are sea-water lubricated; the frictional losses with this type of bearing are greater than that for ball or roller bearings but the weight of the array is not sufficient to make this a limiting factor. Cables entering the array are long enough to allow for twisting as the array turns on its axis.

The struts are manganese bronze castings long enough to hold the transducer clear of the bubble layer which passes along the hull and this increases the forces acting on the struts. The greatest loads (up to four tons) are experienced when the ship is rolling through maximum amplitude and the asdic is turned to present its broad face sideways. The struts are hollow to contain the electric cables and the drive shaft for training the array. The array is gyro stabilised to ensure that the rolling motion of the ship is fully compensated. The gyro unit senses the ship's roll and operates a split field servo motor which drives through the gear train and maintains the array in a fixed position. With the ship rolling heavily a maximum torque of 400 lbs feet is needed to train the array; gears give a 900:1 servo motor to array ratio. The position of the array is indicated by magslip transmitters and

31. Divers check the streamlined outer hull of the asdic vehicle *G.L.O.R.I.A.* during trials. Inflated bags provide buoyancy for the heavily armoured cable while the vehicle is at the surface

receivers and the array can be trained through an angle of 240°, to provide scanning over 120° on the port or starboard side of the ship.

GEOLOGICAL LONG RANGE INCLINED ASDIC (G.L.O.R.I.A.)

As a result of the successful use of the narrow beam Sonar to depict the form of the sea floor at continental shelf depths, it was decided at the National Institute of Oceanography (U.K.) in 1965 to design a similar device for the deep ocean. The new sonar has a range of about ten miles and is designed to produce acoustic pictures of the sea floor in depths of up to 8,000 m. Unlike the shallow water version, the transducer array is housed in a glass fibre vehicle which can be towed by a ship at a depth of about 200 m. This allows the sound to be launched below the near surface thermocline with a reduction in the bending of the sound ray so that long ranges can be obtained. The resulting hydrostatic pressure at this depth allows a large amount of acoustic power to be radiated without cavitation. Other advantages gained include quietness for listening for the received echoes, and decoupling from the ship's yaw so that the amount of vehicle yaw can be kept to a minimum. This last requirement is particularly important since the sonar is trying to detect geological features out to ten miles with a horizontal beam width of only about 2°. The above requirements called for a very large array which resulted in a vehicle about 32 feet long and $5\frac{1}{2}$ feet in diameter which weighs nearly seven tons in air (figure 31). A vehicle of this size presents many shipboard handling problems which need not be considered here.

The vehicle is towed from an armoured electric cable with an ultimate breaking load of about thirty tons. The cable carries about 140 cores, two of which carry 65 kilowatts to power the transducer, while the remainder provide power for the other devices within the vehicle. The towing cable has a diameter of 1.8 inches and requires fairing over its full immersed length to reduce drag and vibration. At intervals along the cable, cylindrical modules are swaged to the armouring, and to these modules are attached twin towing bridles, which consist of short lengths of flexible steel ropes attached to long lengths of nylon ropes. These ropes are anchored to the ship's side and then pass over rollers to give an effective length of 300 feet of rope which will stretch about 20 feet under the maximum towing load. The result of this accumulator system takes the live loading (mainly due to ship's pitching) out of the towing cable. The vehicle weighs $3\frac{1}{2}$ tons in water but when flooded takes in a further $21\frac{1}{2}$ tons of water, and without an accumulator system the sudden acceleration of a mass of 25 tons may break the towing cable. The system is provided with dynamometers which measure the tensile loading on deck and at the point of attachment of the cable to the vehicle. To prevent excessive bending of the cable near the vehicle, the cable passes through a flexible polyurethane tube and into a ball joint.

The electrical wires enter a junction box from which a large array of harnesses emerge to provide power to the array and the various devices and auxiliary equipment within the vehicle.

The vehicle is streamlined to reduce drag and flow noise. The towing depth decouples the vehicle to some extent from the motion of the ship. The fineness ratio, central towing position, and large shrouded tail assembly all help to reduce the yaw. Yaw is further controlled by an active rudder system operating through servo loops from a gyro. This is backed up by a beam steering system which keeps the acoustic beam locked on the strip of ocean floor indicated. The body of the vehicle needs to be acoustically transparent and

the most appropriate material is glass fibre. The central tubular section is filament wound glass fibre while the nose and tail are chopped strand glass fibre. Aluminium alloy watertight bulkheads are provided at each end of the centre section. These bulkheads carry the self-aligning bearings in which the array rotates. The bulkheads also provide a mounting for the junction boxes and air bottles with associated control valves. The frame of the array is also of aluminium alloy and is strengthened by being designed in the form of a series of square cells into which the transducers fit. A servo system to control the beam angle in the vertical plane is not required because the amount of roll is slight, but a training motor is used to tilt the array to launch the sound at any desired angle. The angular position of the array is indicated by means of a magslip transmitter and receiver.

The transducers must operate in water, and this means that the vehicle needs to be flooded during operation. The faces of the transducers are protected from corrosion by a thin resin covering. The great weight of the flooded vehicle makes it necessary to force the water out before lifting it aboard the towing vessel. For this purpose, large air vessels are housed in the nose and tail, containing air at a pressure of 4,000 p.s.i.

When required, the air vessel blowing valves are operated from the control console aboard the towing vessel, and air from the bottles forces the water out through Kingston valves at the bottom of the vehicle.

Finally, let us remember that all oceanographic research ultimately rests on such minute details of design as those discussed above.

Electronics in Oceanography *R. Bowers*

As in everyday life electronics play an ever increasing role, so it is in oceanography. There are very few instruments now used in oceanographic research which do not use electronic circuitry. There are two reasons for this: first the need to measure parameters in the ocean much more precisely, and secondly, development of very reliable, low power, low cost and compact electronic devices such as transistors and integrated circuits.

Before making any measurements at sea one needs to know where one is. Position fixing is normally done with a sextant and an accurate clock. However, this method only gives position to within about three miles, and whilst this is generally satisfactory for sailing from one port to another, for much oceanographic research

work or for positioning a 'Polaris' submarine it is much too inaccurate. Thus, a number of radio aids have been developed. For use in the open ocean these normally work on the principle that a master transmitting station and two slave stations transmit pulses of radio signals at the same instant in time. Then, at any point between the stations, the difference in the arrival times of the pulses may be used to calculate position. Under ideal conditions this can give a position to within about 200 yards, but such systems suffer from the different propagation modes of the radio signal i.e. the radio waves can come to the ship either as a ground wave (through the air above the surface of the earth), or as a sky wave (a wave that goes upward from the transmitter and is reflected down to the ship from the

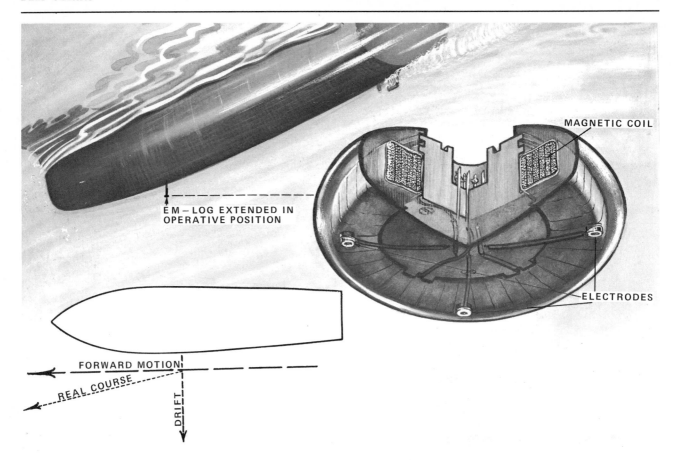

EM – LOG EXTENDED IN
OPERATIVE POSITION

MAGNETIC COIL

ELECTRODES

FORWARD MOTION

REAL COURSE

DRIFT

32. The electromagnetic log measures both the forward speed and the sideways drift of the ship, enabling the resulting course to be determined accurately. The measuring head is mounted beneath the ship's hull and consists of a coil and four elec- trodes. Sea water flowing through the magnetic field generated by the coil produces voltages proportional to the forward and sideways components of the water velocity, which are picked up separately by the two pairs of electrodes

ionosphere). These two waves can interfere with each other and lead to errors. The latest navigation system which is currently being adopted is satellite navigation. In this system a series of satellites are going around the earth on 690 miles high polar orbits, each orbit taking 90 minutes. The system works on the principle that, if two positions of a satellite in space are known, and the difference in range of an object (the ship) on earth from each of these two positions can be found, the object will be on an hyperbola whose foci are the two known positions. If the process can be repeated with another two known positions, the object will lie on another hyperbola. Where these hyperbolae intersect is the position of the object. Normally three hyper- bolae are obtained to get a position fix. To ascertain the position of the satellite, the ship's navigator receives from it by radio complete details of its own orbital parameters and precise time. To obtain the range difference, the navigator measures the doppler shift

(the apparent change of frequency of the radio source due to its movement relative to the observer) of the satellite's transmitting frequency, due primarily to the satellite's motion through space. Finally, to calculate his own ship's position, the navigator feeds all this information into a computer, together with his own estimated position, course and speed. The computer then works out the complicated equations and finally prints out the ship's position. There are three satellites in operation at present which give a fix every 36 minutes to a ship in the North Atlantic and every 138 minutes on the equator. These fixes are claimed to be accurate to 200 yards.

To find a ship's position between these spot fixes, a navigator relies on dead reckoning based on the ship's course and speed. The course is obtained from a gyro compass and the speed from the ship's log. A type of log which is becoming widely used is the electro-magnetic log. This offers very good accuracy over a wide speed

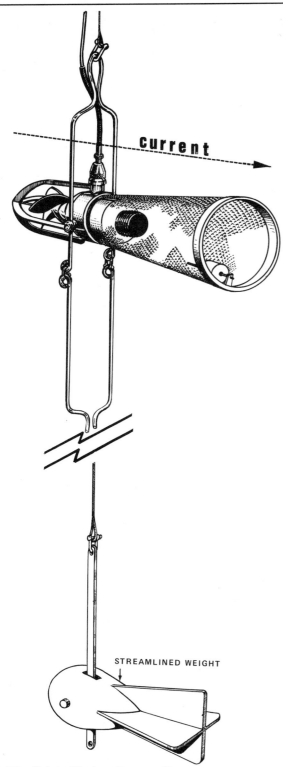

current

STREAMLINED WEIGHT

33. The Kelvin-Hughes direct reading current meter can measure both the speed, by the rate of turn of the propeller, and direction of the current, by a similar system to that in aircraft compasses. Although they can be used singly, they are usually used in pairs, one of which is at the surface, so that currents in deeper water are measured relative to that at the surface

range. It works on the principle that a conductor (sea water) passing through a magnetic field has a voltage generated across it proportional to the rate of flow through the field. This voltage is picked up on two electrodes in the sea water and integrated with respect to time to give the distance travelled through the water. A more complex version has two sets of electrodes to give both fore and aft speed and athwartships speed. The difference in position between the dead reckoned position and the satellite position may be used to calculate the surface currents through which the ship has passed (figure 32).

Surface currents are also measured with direct reading current meters suspended by conductor cored cables from the survey ship (figure 33). They have a propeller which rotates in the current and generates a voltage proportional to the rate of rotation and thus to the current, and a compass which gives out a voltage proportional to the bearing. Both these voltages are brought up the cable and measured in the ship. Normally two current meters are used so that the surface current may be measured relative to some deeper, more stable layer of water.

For long-term current measurements of a month or so, self-contained recording current meters are used. A widely used type has a water-driven rotor which produces a voltage proportional to the number of revolutions that the rotor has made (figure 28). This voltage is sampled at regular intervals (every five minutes for example), converted to a digital form and recorded on magnetic tape. Thus, each reading gives the number of revolutions of the rotor since the previous reading. Each meter also contains a compass and the direction of this is recorded after each corresponding current reading. In the meter described, the conversion into digital form is done with mechanical switches and electronics (although purely electronic conversion must surely be used soon); in other meters the sensors are coupled to transparent discs which have on them opaque positions arranged so that if a light shines from one side of the disc to a fixed array of light-sensitive pick-ups on the other, the output from the pick-ups is a digital signal proportional to the angular rotation of the disc. The pick-ups used are fibre-optic tubes, the outputs of which are recorded on light-sensitive film; in yet others the current and direction are not digitized but recorded as arcs of varying lengths on film.

These current meters are hung on buoy moorings with an anchor at the bottom and a surface or subsurface float at the top. Surface floats have the disadvantage that they are sometimes stolen and are strongly affected by wind and waves, but they have the advantage of being able to carry a radio transmitter

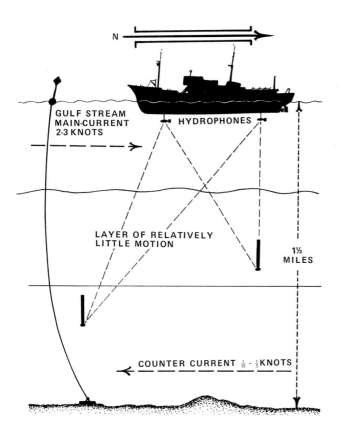

N

GULF STREAM
MAIN-CURRENT
2-3 KNOTS

HYDROPHONES

LAYER OF RELATIVELY
LITTLE MOTION

1½
MILES

COUNTER CURRENT ⅒ - ⅓ KNOTS

34. Neutrally buoyant floats designed to float at calculated depths contain a pinger by which they are tracked from the ship. The different velocities of the water masses at various depths, as in the case of the Gulf Stream and its counter-current, can thus be determined relative to an anchored buoy

which may telemeter data back to shore. In this case the current meters used would be hung on a conductor cored cable and be capable of being interrogated by the surface buoy on receipt of a command radio signal from shore; then if the buoy is lost the data up to the time of loss has already been recovered, while in a self-recording system the data is only recovered providing the mooring can be brought up from the ocean bed.

Sub-surface floats are occasionally hauled up accidentally by mid-water trawls, but on the whole they are very reliable since they are moored deep enough to avoid most wave motion. However, in order to recover them either a small auxiliary buoy on the surface is used, or a release gear is installed below the lowest current meter which releases the anchor when appropriate signals are sent to it. The latter system is coming into widespread use. In order that the release units should not trigger on spurious noises in the water they

are designed to operate only on receipt of a complex coded signal, whereupon they activate a release device such as an explosive bolt and the buoy and mooring rise to the surface.

Deep currents are also measured using neutrally buoyant floats (figure 34). These work on the principle that aluminium tubing can be made to be less compressible than sea-water. Thus a sealed tube which just sinks at the surface will sink until the pressure is such that the water density makes it neutrally buoyant. At this depth it will then drift along with the water. By the addition of weights it can be made to sink to any required depth. An acoustic transmitter is built into the tube, and as the float drifts along it sends out pings which may be tracked from a surface ship using hydrophones (underwater microphones), and in this way its trajectory, and thus that of the water carrying it, may be plotted. Floats of this kind have sometimes been followed for weeks.

Currents in the ocean may also be calculated from pressure gradients. In order to measure pressure gradients, the salinity and temperature of the water at various depths must be found. This has been done with strings of water bottles hung on a vertical wire which simultaneously take a water sample and record the temperature on mercury in glass thermometers at discrete depths (figure 120). The bottles are brought to the surface and the salinity of the water samples measured on a laboratory salinometer. This measures salinity in terms of conductivity, i.e. the electrical resistance of the water sample is measured, with a precision of 1 in 10^6. The temperature, salinity and the depth from which the sample came being known, the density of the water can be calculated. From these densities the pressure at a given depth may be worked out, and if this is done at a number of stations the currents due to the pressure gradients may be calculated.

Water bottles have the advantage that they bring up samples of water which may then be analysed for oxygen, nutrients etc., but they have the disadvantage that they only measure at discrete intervals. By using bottles as close together as was practicable, results have been obtained which indicate that there might be vertical stratification of temperature and salinity on a scale smaller than had previously been appreciated. This has been borne out by results obtained with continuous profiling temperature-salinity-depth probes (figures 17 and 35). These devices measure temperature, depth (as pressure), and salinity (as conductivity corrected for temperature and pressure effects), and uses each of these parameters to control the frequency of an oscillator. The three resultant frequencies are mixed together and telemetered up a conductor cored

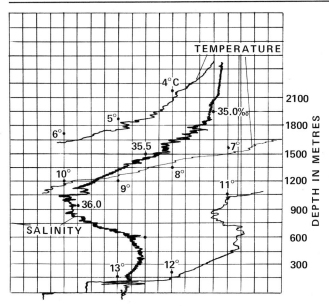

35. The fine structure of the temperature and salinity layering of the sea is well illustrated by this portion of the record of a continuously recording temperature, salinity and depth probe

36. A bathythermograph consists of temperature and pressure (or depth) sensors in a heavy streamlined housing. The instrument is allowed to fall through the water, and is then hauled in again, the sensors having recorded their information on a smoked glass

cable (which also supplies the power to the probe), filtered out into the three original signals, converted to voltages proportional to the frequencies and recorded on pen recorders (figure 35). In order to take samples of water when the record indicates that the probe is in a layer of particular interest, multiple water samplers have been developed which can be attached to the probe and which take a water sample on receipt of a command signal down the cable from the surface.

The variation of sea temperature with depth is also measured because of its effect on sound velocity in the sea, and is thus important in the detection of under-water objects. Salinity also affects the sound velocity but in most areas salinity variations are sufficiently small to have a negligible effect compared with temperature. Until recently most sea temperature-depth profiles were taken with a mechanical device (the bathythermograph) which produced a trace on a smoked slide (figure 36). These records could only be taken at modest speeds (up to three knots), but the situation has changed with the introduction of the expendable bathythermograph. In this instrument a thermistor (a bead of semiconductor material whose resistance changes markedly with temperature) is mounted in a streamlined housing which falls freely in the water unspooling a very fine 3-core wire behind it (as it falls). Simultaneously, the other end of the wire unspools from the ship as it steams away. The resistance of the thermistor is continuously plotted

against time on a recorder as the instrument falls through the water. On the assumption that the rate of fall of the streamlined housing is constant, a plot of temperature v. depth is obtained (figure 37). This system may be used at speeds up to 30 knots, and can thus be used on passage without slowing down. The velocity of sound may be measured directly using a sound velocity meter. Typically, this would have a piezo-electric transducer (a material which changes its dimensions when a voltage is applied across it) which transmits a pulse of acoustic energy towards a fixed reflector about ten cm away. When the reflected pulse is received at the transducer a circuit is triggered which makes the transducer transmit another pulse. In this way the pulse repetition frequency is proportional to the sound velocity in the water column between the transducer and reflector. Such a device, on being lowered on a conductor cored cable yields a vertical profile of sound velocity against depth.

In the ocean there is often a layer of warm water floating on the colder denser water below. The boundary between these layers is a density discontinuity and can oscillate with waves on it just as the surface-air discontinuity does. The easiest way of looking at the oscillations of the layer boundary–*thermocline*–is to study the temperature structure. One way in which this has been done is to do repeated bathythermograph dips at a spot or to have a thermistor probe on a servo-controlled winch which automatically raises and

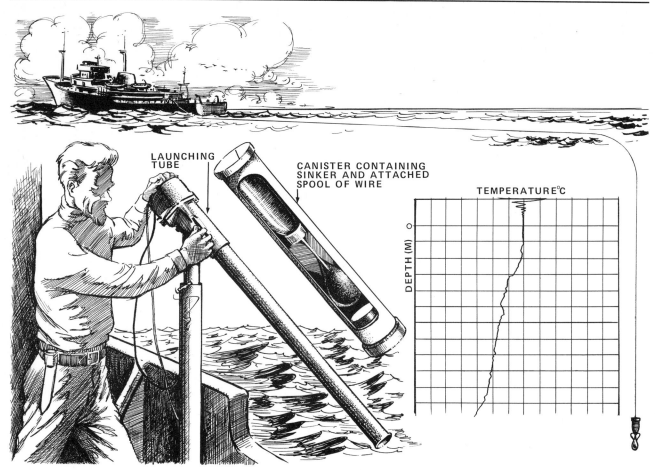

37. In the expendable bathythermograph a streamlined probe containing a thermistor, and attached to a coil of fine wire, is launched into the sea. From one end of the coil the probe falls vertically unspooling the wire, while the other end unspools as the ship steams along, allowing measurements to be made at full speed. The temperature is recorded up the wire to the ship, and the known rate of fall of the probe enables the depth to be determined simultaneously

lowers the probe in such a way as to keep the thermistor resistance constant and thus keeps the thermistor on an isotherm (line of constant temperature). The output from a pressure transducer attached to the probe then plots out the depth fluctuations of the isotherm. In order to get a spatial coverage of the internal waves, devices have been built which can take measurements from ships under way at speeds of up to ten knots. One such system has a series of thermistors connected to a multicore faired cable 600 feet long, with a one ton streamlined weight on the end. As this is towed through the water the resistance of each thermistor is used to produce a continuous profile of the temperature structure of the water (figure 38). Another ingenious system for studying internal waves has been developed using an undulating vehicle. This has only one temperature sensor but the towed vehicle has moving fins which cause it to ascend and dive alternately in a regular manner between pre-set depths. In this way the single thermistor monitors temperature over a cycling range of depths, and from the data obtained the isotherms may be plotted.

The pressure effects of travelling surface waves are reduced with depth to the extent that the pressure fluctuation is very small at a depth equal to the length of the wave. Thus, to measure waves on the deep ocean, surface buoys are usually used. These normally have an accelerometer in them which measures the vertical acceleration of the buoy as it passes over the wave. The output from the accelerometer is processed electronically to give an output proportional to the height of the wave, and this signal is telemetered by cable or radio to a ship where it is recorded. Some measure of the direction of the waves may be obtained by measuring the tilt of the buoy with respect to some form of compass. More detailed measurements of the direction, frequency and height of waves require complex buoys with three floats mounted in a framework, each float in universal mountings. Each float measures tilt in two directions and the output from these give the curvature of the wave which, taken in conjunction with the pitch, roll and heave of the mounting frame, gives the required information (figures 39, 40). Another type of buoy has a pressure transducer hanging beneath it at such a depth that the pressure changes due to the surface waves are negligible. The transducer thus measures its own depth below the mean (or average) surface level. Thus, when the surface buoy then goes up and down on the waves, the pressure transducer moves up and down relative to the mean surface level (just as if the transducer were lifted up and down by a helicopter) and measures a change of pressure proportional to the wave height. This is telemetered and recorded as described before. Waves may be measured from a ship by having pressure transducers mounted below the water line. Two are used – one on each side of the ship – and the average of the two outputs is taken (figure 41). This cancels out any effect due to the ship's rolling and also any shielding effect the ship may have on waves which are short in length compared with the ship. Waves which are long compared with the ship's length move the ship bodily up and down so that the pressure transducers stay the same distance below the surface and do not detect any pressure change. To detect these long waves an accelerometer is mounted adjacent to each pressure transducer and the average output converted (as in the buoy) to wave amplitude. This system does not work well when the ship is under way. To overcome this, a wave recorder has been developed using an inverted echo-sounder mounted on a spar sticking out in front of the ship to measure its height above the wave, with an accelerometer to give the vertical movement of the echo-sounder so that the ship's pitching movement may be allowed for.

The measurement of tides in the deep ocean is difficult and has only been achieved satisfactorily in recent years because the water height must be measured very accurately in depths of up to 6,000 m. This is now done with a very accurate pressure transducer

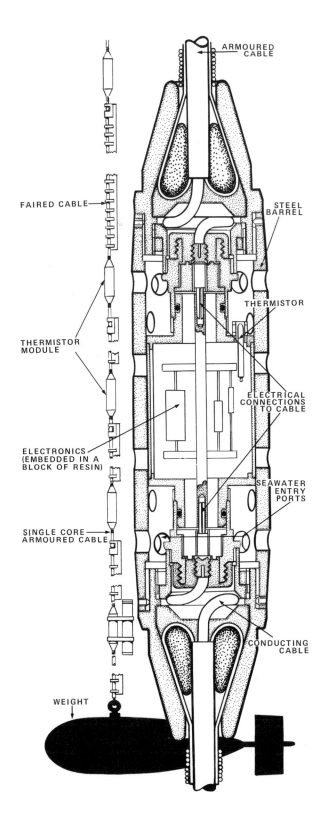

38. A thermistor chain (left) consists of a series of temperature sensors at intervals along a faired armoured cable. Each thermistor is housed in a free-flooding steel casing, which also takes the strain on the cable, and while towed through the water records the temperature changes direct to the ship via the conducting core of the cable

ARMOURED CABLE

FAIRED CABLE

STEEL BARREL

THERMISTOR

THERMISTOR MODULE

ELECTRICAL CONNECTIONS TO CABLE

ELECTRONICS (EMBEDDED IN A BLOCK OF RESIN)

SEAWATER ENTRY PORTS

SINGLE CORE ARMOURED CABLE

CONDUCTING CABLE

WEIGHT

39. Launching a cloverleaf buoy for measuring waves on the surface

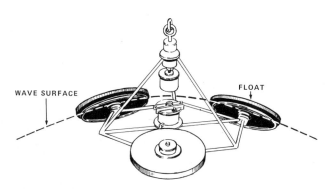

40. The different directional tilt of the three floats of a cloverleaf buoy measure the curvature of a surface wave

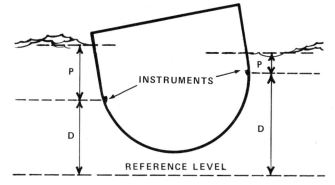

41. The ship-borne wave recorder measures the height of the water surface relative to a point on the ship's hull, and adds this to the height of the point relative to an imaginary reference level. The first quantity, P, is measured by a pressure transducer, the second, D, by the vertical component of an accelerometer. Accuracy is increased by averaging the results from two instruments on opposite sides of the ship

42. *opposite* The broad V of the lower depressor plate of the Isaacs-Kidd midwater trawl is clearly visible as the net is lifted over the stern of RRS. *Discovery*. A canister containing a depth recorder is being attached to the bottom of the plate

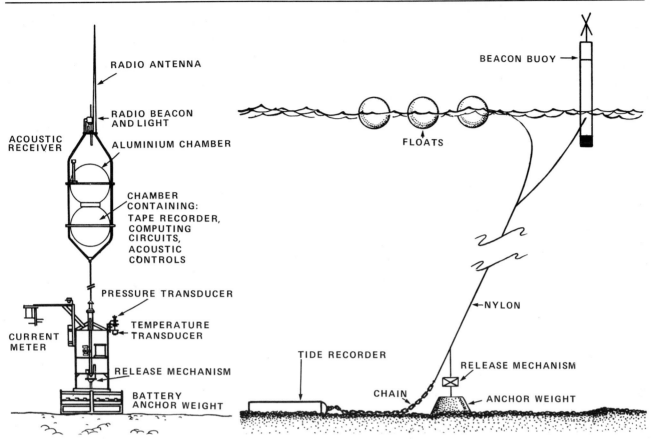

45. Deep-sea tide recorders: left, Scripps Institution of Oceanography recorder designed to pop up to the surface on command; right, a French assembly anchored to the bottom but with surface floats and beacon buoy. In both instruments the frequency of a taut vibrating wire is related to the pressure of water, and thus the tide height, above it (after Eyriès 1968, and Snodgrass 1968)

mounted in a housing on the sea-bed which can resolve changes of water height of 1 cm in 6,000 m depth. In order to measure this accurately, the temperature of the pressure transducer has to be measured so that corrections for the change in output of the pressure transducer with temperature can be made. A 'resolution' of one cm in the depth of water may only be obtained because the environment is very stable and any changes are very slow. However, due to inherent errors in the system, the long term 'accuracy' is of the order of one metre (figure 45).

The depths of the oceans are measured with sound pulses. An echo-sounder transmits a pulse at a frequency of 10 to 12 kHz and measures the time for the pulse to be reflected by the sea-bed back to the ship. These transmitted and reflected pulses are monitored on a recorder in which a stylus traverses the recording paper at a very accurate rate (typically one sweep per second). The output pulse is transmitted at the start of the sweep and the reflected pulse recorded when it is received. In this way the distance across the paper is a measure of depth and each successive reflected pulse records beside the previous pulses, thus drawing out a profile of the sea-bed (figure 46). Using such recorders, the depth may be measured to an accuracy of 2 m in 5,000 m depth.

When doing a survey on a specific area the echo-sounder may be used to navigate relative to an array of transponders which may be laid on the sea-bed or on buoy wires. These are devices which emit an acoustic pulse immediately they receive one. Thus, if the echo-sounder is transmitting, it receives back not only an echo from the sea-bed directly beneath the ship, but also pulses from the transponders. These pulses also line up on the recorder and, by measuring the time between transmission and reception from each transponder, the ship's position relative to the transponders may be worked out. Echo-sounders may also be used as a passive listening instrument with 'pingers' when it is required to put something very close to the bottom. A pinger is a device which emits an acoustic pulse at a rate equal to the scan rate of the echo-sounder (e.g.

43. *opposite above* The research submersible *Deep-Star* 4000
44. *below* The diver support vehicle *Deep-Diver*

46. Profile of a rugged sloping bottom appearing on the precision echo-sounder record. The dark band on the extreme left of the trace is a combination of surface noise and near-surface sound scatterers

one per second). This is attached to the device which is to be placed on or above the sea-bed (e.g. a corer or water bottle) and lowered on a wire. Since it pings at the same rate as the echo-sounder sweeps, it draws lines on the recorder both with the direct pulse and also with its pulse reflected from the bottom. In this way the pinger is lowered until the two traces are the required time separation apart i.e. the pinger is the required distance off the bottom. Using this system, gear may be positioned 2 m above the sea-bed in 5,000 m of water. All acoustic sounding and ranging equipment uses the technique of the travel time of the pulse to obtain range information. Sub-bottom profiling for studying the strata beneath the sea-bed is done in the same way as ordinary echo-sounding except that the frequency of the transmitted pulse is much lower (figure 47). Typically, the frequencies used are 20 Hz to 1 kHz. These low frequencies are used because lower frequencies travel further in the bottom sedi-

ments than do higher frequencies of the same energy. Thus a 100 Hz signal will penetrate twice as far as a 200 Hz signal for the same attenuation. Various sources are used to produce these low frequency pulses including explosives, very powerful electric sparks, and compressed air. The signals returned from deep sub-bottoms tend to be very weak and barely detectable in the background noise which is always present in the sea (due mainly to wave motion and ship noise). In order to detect these weak signals very complex equipment may be used. These use correlation techniques where, simply, the received signals are continuously compared with a replica of the transmitted pulse, and the size of the recorded signal depends on how well they match up. Thus a pulse of signal reflected from a sub-bottom, although it may be very weak, tends to match up well whilst the background noise, although larger, tends to be random in nature and does not match up well.

Narrow beam asdics (or side scan sonars) have

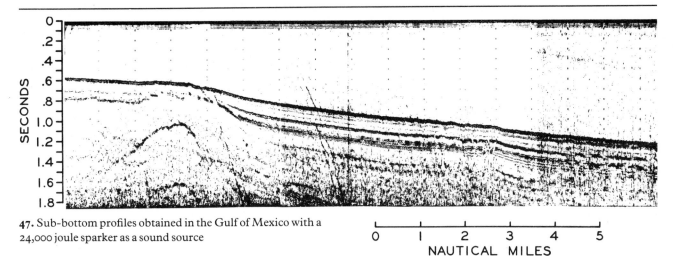

47. Sub-bottom profiles obtained in the Gulf of Mexico with a 24,000 joule sparker as a sound source

been developed for obtaining pictures of the sea-floor. These use large acoustic transducers which have a fan-shaped beam (figure 30). Each pulse of sound isonifies or 'illuminates' a thin strip of the sea-floor and the echoes received back from it depend on the angle of the bottom and on its reflectivity, a good reflector sending back a strong echo and marking the chart. Thus each transmitted pulse produces a line on the recorder which is a record of reflected sound, and each successive line builds up a picture of the sea-floor (just as a television picture is built up of lines). In this way pictures of strata and sediments on the sea-floor are obtained (figure 30). Sonars are used for detecting objects in mid-water such as fish. Commercially available models use a transducer that sends out a narrow beam which may be trained in azimuth and elevation and which can detect a single fish at 800 yards. One problem with echo-sounders and sonars is that all echoes from reflectors at the same range arrive at the receiver at the same time, and unless the transmitted beam is very narrow no information on the bearing is obtained. To overcome this, very sophisticated sector scanning sonars have been developed. In these a pulse is transmitted over a fairly wide angle but the receiving transducer is, with electronic circuitry, made highly directive, and this highly directive receiving beam angle is swung rapidly back and forth within the transmitted beam angle. In this way the direction of a reflecting body as well as its range may be ascertained. Equipment of this type has made possible the detailed study of moving underwater bodies such as trawls. Much work is in progress towards developing sonars with electronic processing of received signals to provide very long range detection and superdirectivity. Another

uses the mixing of two high frequency signals of slightly different frequencies and high intensity, to produce a highly directive low frequency signal. A form of sonar which is being widely used is the continuous transmission frequency modulated sonar (CTFM). In this, a long acoustic transmission is sent out with its frequency changing linearly with time. The range of a reflecting body is then found by comparing the frequency of the reflected signal with that being transmitted. The transmission is sent out in broad beam and picked up on a highly directional receiving transducer from which the bearing may be found and which may be scanned round to look at various reflectors within the transmitted beam. This system has a number of inherent advantages, two of which are that it provides a continuous echo from a target and that it is possible to determine the range of a target to within a few yards of the instrument.

Another widely used technique for studying the strata beneath the sea-bed is by *refraction* profiling as opposed to reflection profiling described earlier. In this system, explosive charges are set off just below the sea-surface and the pulses received up to twenty miles away. At close range the direct pulse through the water arrives first at the receiver, but at the longer range the pulses which penetrate deep below the sea-bed are refracted along layers where the velocity of sound is very much higher than that in water and thus they arrive at the receiver before the direct pulse. By studying the difference in arrival times of these pulses at various ranges, the depths of the different layers together with their characteristic acoustic velocities may be determined (figure 262). The receiver normally used is either a second ship or a buoy with a hydro-

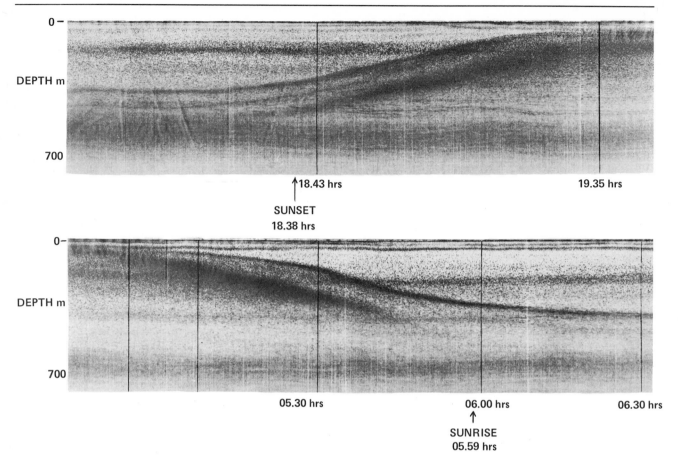

0 —

DEPTH m

700

↑18.43 hrs   19.35 hrs

SUNSET
18.38 hrs

0 —

DEPTH m

700

05.30 hrs   06.00 hrs   06.30 hrs
                                ↑
                            SUNRISE
                            05.59 hrs

48. Movements of scattering layers as recorded by the echo-sounder. Top, three distinct layers move up towards the surface at sunset; bottom, before sunrise the layers move down again. In both traces other scattering layers remain at a constant depth

phone streamed from it, both of which either record the incoming signals or radio-telemeter them back to the ship which is firing the charges. Both of these methods suffer from the environment being noisy for the hydrophone—not only is there wave-induced noise but also the hydrophone gets pulled up and down as the ship or buoy heaves. To overcome this, sea-bed receivers have been developed which are released from the ship and sit on the sea-bed whilst the refraction work goes on, recording received signals on magnetic tape or film. When the work is over, an acoustic signal is sent to them which causes the anchor weight to be released whereupon they rise to the surface and are recovered.

The rocks beneath the sea-bed are also studied by measuring the effect they have on gravity and magnetic fields. Gravimeters which are used on research ships are basically very sensitive spring balances mounted on gyro-stabilized platforms. By monitoring the change in gravity some knowledge of the density of the rocks below is obtained, since a large mass of dense rock will cause a relatively high value of gravity in its locality. The magnetic fields associated with the rocks are measured with very sensitive magnetometers. The most widely used type measures the rate of oscillation of an atomic nucleus in the ambient magnetic field. The higher the field strength, the greater is the frequency of oscillation. The detecting part of this equipment is towed in a streamlined housing some way behind the survey ship in order to get it away from the ship's magnetic field, and the oscillation rates are counted and recorded aboard the ship. By measuring the variations in the earth's magnetic field associated with the sea-bed structures, they may be correlated with the times in the past history of the earth when its field reversed its direction, and so the age of the rocks may be estimated.

Echo-sounders, as well as giving records of the sea-bed depth, give records of animals in the sea. Certain animals reflect certain frequencies strongly, but if a variety of frequencies is used it is found that all of the sea down to 1,000 m depth contains sound reflecting animals. These animals tend to arrange themselves in layers which are called scattering layers because they scatter the sound pulses (figures 48, 49). For studies of animal distribution it is necessary to know at what depth a sampling net is being towed so that it can be adjusted to fish in the layer being studied. Since towing warps containing conducting cores are very expensive it is normal to telemeter information from the net to the ship acoustically. One telemeter which is becoming widely used sends out a reference pulse every two seconds exactly and a delayed pulse, the delay being a measure of depth. If the echo-sounder recorder pen traverses the recorder once every two seconds, the pulses from the telemeter correlate on the record and plot out the depth of the net. Other parameters may be telemetered to the ship at the same time. These are temperature and flow of water past the net. The flow is measured with a propeller mounted on the end of the telemeter housing. It is useful to know the flow (i.e. speed) of water past the net, since the currents at the depth at which the net is fishing may be very different from the surface current through which the ship is steaming, and thus the ship's log indication of speed may be very misleading. Another parameter which is telemetered is the opening and closing of a sampling net. If samples are required from a discrete layer it is necessary to prevent the catch from being contaminated with samples caught whilst lowering and raising the net to and from the layer. Thus, nets have been developed which remain closed whilst lowering, and when they have reached the required depth (as indicated by the telemeter) acoustic signals are sent to a release gear on the net which opens it. When the fishing of a layer has finished, another acoustic signal is sent and the release gear closes the net which can then be hauled in (figure 53). Another method of determining the depth of a net is to fix an inverted echo-sounder to the bottom of the net. This then gives the depth below the sea-surface and also some indication of the animals entering the net. It does, however, need a separate cable to it and this becomes very difficult to handle if the net is fished at any appreciable depth. This gives better results than the acoustic telemeter in areas where strong stratification of the water results in poor acoustic propagation.

Plankton counters have been built which give an indication of the population and distribution of small animals and plants. These counters consist of tubes containing electrodes which are towed through the water. When an organism passes down the tube it changes the impedance between the electrodes and this is detected and recorded. The change of impedance is roughly proportional to the volume of the organism over a wide range of sizes and so an estimate of the volume as well as the number of organisms is obtained.

49. While many animals appear only as diffuse scattering layers, schools of fish often appear as separate strong crescentic shaped echoes, such as these at a depth of 180-280 m off the Arabian coast

A version of the 'Hardy' plankton recorder will shortly be installed in an undulating vehicle which will contain electronic equipment for measuring salinity, temperature and depth (figure 50). Another type for towing at fixed depths records temperature, depth and flow through the sampler, and winds on the sampling mesh at fixed time intervals or after a certain volume of water has flowed through the net.

Deep-sea cameras have been used for photographing animals as well as the sea-floor. These cameras use powerful electronic flash units for illumination, and may have no shutter when operating in the sunless deep waters. The film is not fogged between flashes, the film being wound on automatically after each flash. If the flash is made to trigger when a baited hook is pulled, then fish and squid may initiate their own photographs (figure 24). Cameras may also be triggered when an animal interrupts an infra-red beam in front of the camera (as in a burglar alarm) or by having an echo-sounder mounted alongside the camera which is triggered when an echo is received from an animal in the camera's field of view. When photographs are being taken of the sea-bed the camera must be the correct distance off the bottom in order to be well focussed. This may be done with a pinger, as described previously, or by having a weight hanging down below the camera which triggers the flash unit when it hits the bottom. Since light is highly attenuated by sea-water, the range obtained with underwater cameras is not very great. Underwater television also suffers from this problem and has also been rather impracticable to use at the ends of long cables for use on the sea-bed. Now, however, a telemetering television system has been built and should be very useful for studying the sea-bed. Acoustic holography (the forming of three-dimensional images using sound waves), does not at this time appear to be a practical replacement for the camera. However, with the big effort which is going into this field it will probably become a very useful tool in the future. The same cannot be said for lasers (coherent light sources) which in their present state of development, even using pulsed techniques (which reduce the backscatter), still have a very limited range.

Since light levels play an important part in where animals and plants live and grow, many instruments have been developed which measure light intensity at various depths in the sea (down to about 1,000 m where virtually no sunlight penetrates). These use very sensitive light detectors and, with a system of optical filters, measure the light intensities at various parts of the visible spectrum. Other light meters have been developed which measure the total light influx from all directions simultaneously and give an output proportional to this total flux integrated over a period of time. In order to measure the scattering of light by particles in the water, light meters have been built which send out a narrow beam of light: a part of the illuminated water column is viewed by a light detector from various angles and thus the light scattering dependence on the viewing aspect may be determined. When the light sensor is looking directly into the light source the actual attenuation of light by the water may be found. Some animals generate their own light (bio-luminescence), and instruments have been built which detect this spontaneous light generation. Since these animals tend to produce light flashes when disturbed it is necessary to look for them at a range beyond the influence of the detecting instrument. In order to do this two detectors are used which each look at a thin column of water and only give an output if a flash occurs in the volume which is common to both columns, this volume being about one metre from the instrument.

These flashes may also be used to trigger cameras for photographing their source (figure 26).

Many animals in the sea make noises. The best known are the clicking shrimp, and the porpoises and whales. Recording the sounds produced by these animals requires very quiet conditions for the hydrophone since, unless the source is very close, the background noise tends to swamp it. However, in spite of this, many excellent recordings have been made and this type of investigation offers a very promising line of exploration for the future.

GENERAL

All of the telemetering instruments described use batteries. Some large buoy projects store energy in the form of liquid gas which may be used either in a fuel cell or to power a generator which recharges batteries when they get low. Atomic power sources are only economic when there is a steady power requirement that goes on for a long time and are thus finding use in navigational beacons and buoys.

One thing that must be borne in mind when considering any instruments lowered over the ship's side, is that sooner or later they will probably be lost due to breaking wires etc. Thus devices must be built as cheaply as possible whilst retaining the highest possible reliability because too much time cannot be wasted on faulty instruments (a large oceanographic research ship may cost about £1,000 per day to run).

Electronic instruments have enormously increased the amount of data collected on any cruise and, in order to deal with it, computers are used. These are not only used at oceanographic institutions, but are being installed on board research ships where they are used to process most of the data as it is collected, to log all depths, meteorological and hydrographic data on passage, and to work out all the navigational calculations.

Sea-going Computers *M.J.R.Fasham*

A computer was first installed on an oceanographic research ship in 1962 and since that time the number of ships carrying one has grown rapidly.

The two main incentives for using shipboard computers were the requirements for more accurate navigational information and the ever-increasing complexity of oceanographic data reduction. This used to mean that, although in some cases it was possible to do a little preliminary interpretation at sea, a detailed interpretation could not usually be carried out until up to six months after the cruise. This often

resulted in further trips being necessary to study interesting areas in more detail.

However, with a computer these calculations can be carried out, quickly and accurately, as the data is acquired, thus relieving the scientist of much of the drudgery and freeing him for creative interpretation. If interesting anomalies are revealed, immediate adjustments in the cruise programme can be made, ensuring that ship's time will be used as efficiently as possible.

Most of the ship-board computers are a special type

known as process control computers. This means that it is possible to connect a piece of oceanographic equipment, such as a magnetometer, echo sounder, gravimeter or gyrocompass, directly into the computer. The computer can then be programmed to activate the instruments, read the parameter that the instrument is designed to measure, and finally store it in the computer's memory for further calculations.

A process computer also operates in what is called 'real time'. This means that it can respond to external events as they occur. For instance, if a satellite passes over the ship, the satellite navigator console will start picking up signals from it. The console then sends what is called an 'interrupt' into the computer which causes it to suspend its present activity and deal with the data being received from the satellite. Also it is possible to programme the computer to read a magnetometer at, for example, ten minutes past each hour or at any interval the scientist may require.

In most applications data are sampled by the computer at regular intervals, perhaps every two minutes. The computer then carries out any necessary calculations that are required and stores the data (with a note of the time at which it was sampled) on a backing store. This backing store may be a magnetic tape or, more often, a magnetic disk file. The advantage of the magnetic disk is that the average access time for a piece of data is faster than for magnetic tape. A modern disk file has a storage capacity of the order of half a million words of data. Thus, during a fairly routine survey sampling every two minutes, it would take about three weeks to fill up one disk.

Having stored the data, the computer is then required to communicate it to the scientists on the ship. This can be done by printing it on typewriters connected to the computer, punching it on to punched paper tape, or using a graph plotter which can give a graphical representation of the data. This last method is very useful for plotting profiles of bathymetry, magnetic and gravitational fields, or a chart of the ship's course.

As will be realised, the computer programmes required to carry out the sort of operations described above can be very complicated. However, with most computers a lot of this organisation is carried out by what is called an operating system. To use the previous example of the satellite navigator, the operating system recognizes the satellite interrupt, saves the programme that is being executed at that time, and then reads from the disk into the core store the programme that interprets the satellite data. When the satellite has passed over, the operating system then returns the computer to the programme that was in the core before the interrupt. This method of operating the computer

is called time-sharing because the computer is sharing its time between more than one activity.

One of the greatest problems in oceanographic research in the past has been knowing exactly where the ship was at any given moment in time. It is therefore worth studying in more detail how a computer helps to solve this problem. As the ship proceeds on its way the computer measures its heading, say, every second by means of a gyrocompass and its speed through the water by means of a two-component electromagnetic log. Each time it takes a sample it also records the time from an accurate digital clock. From these data the computer can calculate a dead reckoning course, which will deviate from the true course of the ship by an amount due to the surface current. However, the true position of the ship can be determined by using Decca or Loran or satellite navigation. The advantage of satellite navigation is that it can give a fix anywhere in the world and not just in an area where there happens to be a Decca or Loran network. A satellite fix is usually obtainable every one to one and a half hours and gives the position of the ship to an accuracy of a tenth of a nautical mile. The computer then compares this fix with the dead reckoning position for the same time and from the difference it can calculate a mean speed and direction of the surface current. Finally, using this current velocity it can calculate a corrected dead reckoning course in between satellite fixes. Thus as well as calculating an accurate course the computer is also yielding information about ocean currents while the ship is under way.

Let us now try and look into the future and forecast some possible lines of development. It is certain that computers themselves will become faster, smaller, and have vastly increased storage capacities. But apart from this, the use of shipborne computers will also evolve and some of the lines of development are already becoming apparent. When computers were first used on research ships it was mainly to carry out data logging and interpretations that were capable of being done, albeit much more slowly, by conventional means. This use of computers has become fairly routine and scientists are now devising experiments which could not be carried out at all without a computer. Many of these new applications will make use of high-speed data acquisition, sampling a variable up to 10,000 times a second.

As an example it is hoped to use this method to obtain more accurate results from a shipborne gravimeter. At present, although the gravimeter is installed on a stabilized platform, errors due to the ship's motion still arise. However, by sampling every second and programming the computer to carry out a mathematical filtering, it is hoped to improve the results considerably.

Other possibilities are signal processing of the output of air-gun arrays or obtaining three-dimensional pictures of the sea bed by holographic techniques.

Another line of development that has already begun is the use of telemetry to communicate between the ship's computer and apparatus floating or moored in the surrounding ocean. The research ship of the future may navigate to a predetermined point completely under computer guidance; there it will moor buoys which will continually relay information about current, temperature, salinity etc. back to the ship while it is in the area. Neutrally buoyant floats will be dropped overboard which will send signals to the ship so that the computer can continuously monitor their position and build up a synoptic pattern of the surrounding currents. If necessary, the computer could communicate via satellites with computers on land, either to obtain information or programmes that it requires, or to send back newly acquired data.

Computers and electronic instrumentation are developing at such a rate that it is impossible to predict all the uses to which ship-going computers will be put over the next few years. This makes the future an exciting prospect.

Biological Samplers *Malcolm R. Clarke*

The function of biological samplers is to catch plants or animals. What is rather less obvious is that the sampling technique must vary according to the reason for catching them. For many purposes the biologist just needs to catch the beast. For studying an organism's taxonomy, function or structure he does not always need any details of area, depth or cohabitants. For studying ecology or behaviour, however, it is essential to know more. Then, the hope is that the sample gives a miniature picture of the total population or community at the position, depth and time the sample was taken. As will be explained below, *this* kind of sampling, the kind which makes possible complete studies of migration, feeding and predator relationships, interaction of species, growth and many other ecological properties, cannot be achieved at present. The ecologist is seriously limited by his sampling techniques and he must always examine his data with these limitations in mind.

Limiting factor number one is that fine-mesh nets catch small organisms but produce very high drag when pulled through the water. Fine meshes therefore cannot be used in large nets unless towed very slowly. Small nets or large nets towed slowly do not catch large animals. Thus, clearly several sizes of net must be used to collect anything like a representative sample of the population. Every type and size of net has a different catching potential which even varies according to whether it is hauled vertically, obliquely or horizontally through the water. In addition, observations show that different species vary in their net-avoiding skill and this is just as much a function of their type of movement, activity or behaviour as of their size.

What can the ecologist do? He can limit his studies to particular species or to limited problems. He must always ask 'Have I got a truly random sample of the total population of the animals I am concerned with and how can I allow for or obtain any missing section of the population?' Many are the fine ideas which have had to be abandoned because the sampling was not good enough and the difficulties or expense of going back and sampling again were insurmountable!

The impossibility of obtaining a truly representative sample of an animal community has led to three approaches to sampling oceanic animals. The first approach was used by early expeditions and involved the fishing of as many different samplers as possible at any location. However, time clearly limited the use of much gear at any one station so that subsequent ecological studies were often hampered by sampling bias. The material obtained proved invaluable for basic taxonomy and anatomy. The second approach is to design the sampling to answer ecological questions posed by a specialist in one group or by a group of specialists tackling a single ecological problem. This type of sampling is usually very expensive in ship's time and money if specimens inadvertently caught during the sampling programme, and not required for the problem being studied, are just thrown away. The third approach is to have restricted aims in view during sampling but the extra material caught simultaneously is preserved and stored for later dispersal to specialists who wish to study aspects of biology for which the samples are suitable. Thus, for the price of storage, a great deal of additional information is obtained which often forms the columns upon which many of the more imaginative ecological arches must rest. This method is fortunately spreading in spite of ill-informed scepticism of the value of storing samples for work which may wait several years for the interested specialist and several more years for his published report.

Sampling animals in the deep sea has been likened to

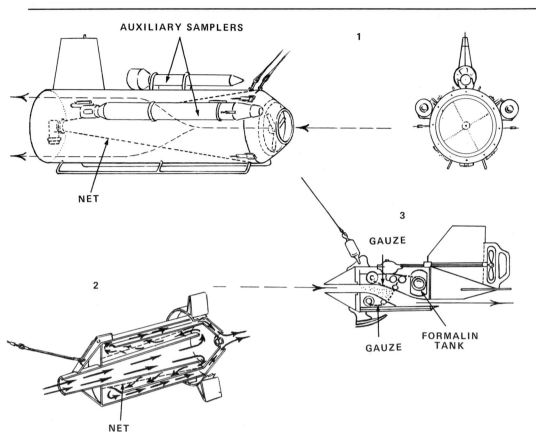

AUXILIARY SAMPLERS

1

NET

2

NET

3

GAUZE

GAUZE

FORMALIN TANK

50. 1. Gulf III high speed plankton sampler; water flows from the small mouth through a net encased in a metal housing. 2. The Jet Net in sectional view; water speed through the net is reduced by its more circuitous path than in the Gulf III. 3. Continuous plankton recorder; as the recorder passes through the water the propeller winds a continuous length of filtering gauze across the path of the entering water. Plankton is collected on the gauze and is sandwiched by a second strip of gauze before being wound into a tank of preservative (1, after Beverton and Tungate 1969; 2, after Clarke 1964; and 3, after Hardy 1936)

sampling from a helicopter, in the dark, two miles high, towing nets or hooks on or near the ground. Any samples would provide a poor picture of terrestrial life! When one remembers that the helicopter could tow far bigger nets because of the lower drag in air, and animals would not be warned of the nets' approach by luminescent organisms hitting the meshes and lighting it up, one cannot help thinking that any catch from deep water is a remarkable achievement.

Fishing methods currently in use all have particular targets, difficulties and characteristics which will now be considered by reference to the more useful and usual modern gear.

SURFACE SAMPLING

The main problem in sampling the fascinating and distinctive animal community living within a few centimetres of the ocean surface is the difficulty of avoiding water mixed up by the ship's passage. One very successful net is towed from a boom near the bow on modified water skis and all disturbed water is avoided (figure 19).

MIDWATER SAMPLING

To start with the smallest organisms, bacteria are collected in messenger operated sterilised bottles or polythene bags. The messengers are brass weights which are allowed to slide down the wire and operate opening levers on the gear. In the polythene version the vessel is automatically resealed to avoid contamination. The smallest plants and animals will pass through the finest meshes known and have to be collected in water bottles similar to those used for collecting water samples (figure 120). They are then removed from the water by filtration, sedimentation or centrifugation.

Slightly larger plants and animals are caught by using samplers towed at relatively high speed (over six knots) and this necessitates a small mouth area, to reduce drag, and robust construction. Two of the most widely used samplers are the 'Hardy continuous plankton recorder' and the 'Gulf III' (figure 50). The Hardy recorder traps small animals, entering the orifice on its nose, between two layers of gauze which is moved by a propeller as the sampler is towed through the water. This has proved of great value in studying general distribution of small animals. Its principal disadvantage is that it is towed at a set depth (10 m) but it is hoped that shortly an undulating vehicle will be adapted for the same work so that a range of depths may be sampled.

A similar device has recently been adapted for use at the cod end of a plankton net (Longhurst sampler) and this is already proving very valuable in the study of vertical and horizontal distributions of small animals.

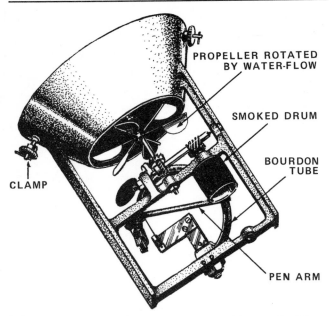

PROPELLER ROTATED
BY WATER-FLOW

SMOKED DRUM

BOURDON
TUBE

CLAMP

PEN ARM

51. This flowmeter is mounted in the centre of the mouth of the net. Rotation of the propeller drives the smoked drum round, and the pen draws a line. At the same time increasing pressure (= depth) deflects the pen up the drum. The result is a spiral line whose length is proportional to the flow through the net and whose height is proportional to the depth

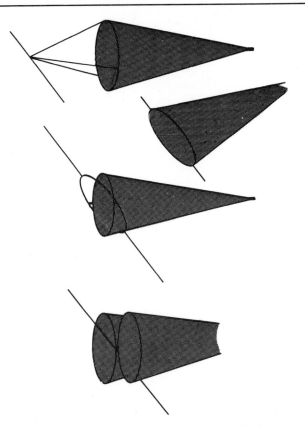

52. Several ways in which ring nets may be attached to a warp. A weight is hung on the bottom of the warp. Top, the most usual method where the warp and bridles lie in front of the net mouth. Bottom, the system used in the 'Bongo' net where the two nets balance one another. Two other methods are also shown

The 'Gulf III' is really a metal net having a small mouth and a metal covering sleeve (figure 50). It incorporates a flow meter and both fine mesh collectors and environmental sensors may be clamped on its sides.

For some purposes, it is adequate to catch small plants and very small animals with a small diameter (50 cm) ring net of very small mesh (down to 200 meshes to the inch) hauled very slowly vertically. Slightly larger animals (up to about 1 cm) are caught by a similar net with a mouth of up to one square metre in area and a larger mesh (figure 18). Two examples of this type of net are the Indian Ocean Standard Net (IOSN) which is a non-closing net designed for standard sampling from 200 m to the surface and the NF70V net which can be closed by a messenger and incorporates a flowmeter (figure 51). This net is slowly payed out open to its maximum depth and is then hauled in steadily at 1 m per sec. with the ship keeping carefully on station. The rate the messenger falls is known so that a skilful operator is able to close the net within a few metres of the depth selected. In this way a series of samples may be taken from horizons between the surface and great depths. The flowmeter in the mouth of the net indicates the amount of water filtered by the net and the depths at which the net started fishing and closed. When the net is fishing, the water passing into the mouth of the net turns the propeller of

the flowmeter and this rotates a smoked glass drum. A pointer mounted on a bourdon tube, which bends with increasing pressure, indicates depth by the extent of its movement across the drum; the number of rotations marked on the drum indicates the amount of water passing the propeller and filtering through the net. These parameters are necessary for comparisons because of the possible influence of sub-surface currents. This net has proved very valuable in ecological studies of small animals. A disadvantage is that some of the turbulent surface water is sometimes filtered when the net is lowered, and this may contaminate deep hauls.

Larger and more active animals cannot be caught by these small, vertically hauled nets. Instead, the ship must move forwards towing the net behind it. A small net can then be towed at the depth animals are expected and the catch greatly increased. However, if larger animals are to be studied a larger net must be towed. The mouths of small nets (up to 113 cm diameter) are

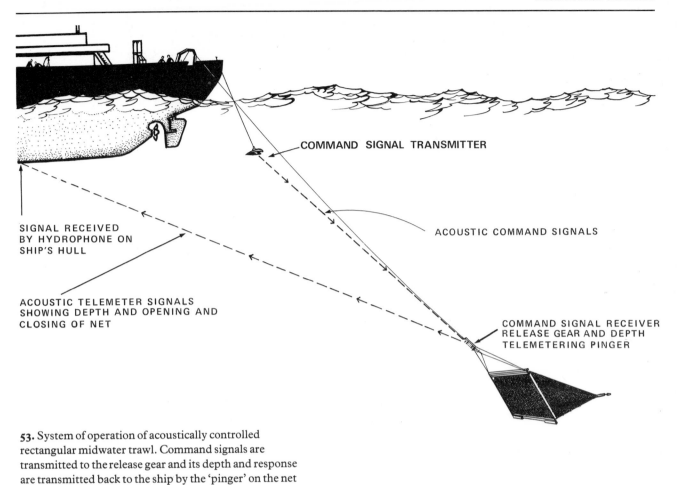

COMMAND SIGNAL TRANSMITTER

ACOUSTIC COMMAND SIGNALS

SIGNAL RECEIVED
BY HYDROPHONE ON
SHIP'S HULL

ACOUSTIC TELEMETER SIGNALS
SHOWING DEPTH AND OPENING AND
CLOSING OF NET

COMMAND SIGNAL RECEIVER
RELEASE GEAR AND DEPTH
TELEMETERING PINGER

53. System of operation of acoustically controlled
rectangular midwater trawl. Command signals are
transmitted to the release gear and its depth and response
are transmitted back to the ship by the 'pinger' on the net

usually held open by a ring of metal and such nets
become cumbersome if made much larger. These
'ring' nets also have bridles which lie in front of the
mouth and scare off an unknown proportion of the
animals encountered by the net. This bridle problem
has been overcome in several ways. The ring may be
towed from one side and a kite-like depressor or a
weight hung from the opposite side (figure 52) or two
rings may be tightly clamped with the wire in the
middle. In larger nets, however, rings have been
replaced by other devices.

The Isaacs–Kidd midwater trawl has a mouth
supported by a large V-shaped depressor plate, a bar
and bridles (figure 42). It has been popular with a six
foot or a ten foot depressor and is used at speeds of up
to six knots; the catch is rather damaged above about
two knots because the animals are pressed hard
against the meshes of the net. The depressor holds the
net very steady in the water and pulls it down so that
maximum depth is achieved with minimum wire; a

time and money saving feature. Unfortunately, the
bridles partly obstruct the mouth and the depressor
almost certainly produces noise which scares some
animals away.

Another means of holding the net open is by using a
horizontal bar at top and bottom of the mouth and a
weight or depressor to pull the mouth as vertical as
possible (figure 53). This system is successful but the
mouth is never vertical because of the drag of the net
and this distorts the net and reduces the mouth area.
These features can be allowed for by selecting an angle
of say 45° at which the mouth will be fished, shaping
the net so that it is not distorted at that angle and, by
experiment, ascertaining the speed at which it should
be towed to achieve this angle. This necessitates a
device to record the mouth angle when towing so that
the exact speed of the ship necessary for proper functioning can be determined. By this means, the net
becomes much more valuable for comparative work.
Nets constructed in this way (rectangular midwater

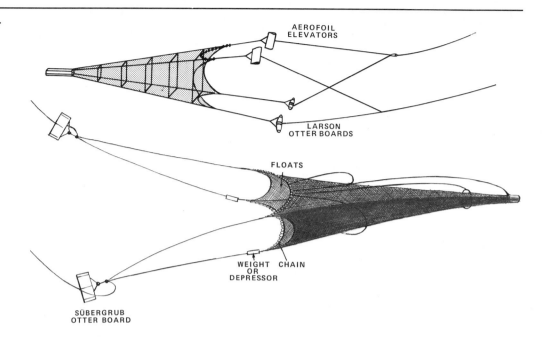

54. Two large midwater trawls; the Cobb (top) and Engels (bottom)

AEROFOIL ELEVATORS

LARSON OTTER BOARDS

FLOATS

WEIGHT OR DEPRESSOR   CHAIN

SÜBERGRUB OTTER BOARD

trawls) with fishing mouth areas of 1 sq. m, 8 sq. m and 90 sq. m have been used successfully. They have all been used with various mesh sizes and afford a very useful means of sampling a very large size range of animals with the same basic design of net.

Larger commercial midwater trawls are regularly used for biological sampling, and here the mouth is held open by otter boards, floats on the headline and large weights on the footrope. Examples are the British Columbia midwater trawl, the Engels midwater trawl and the four-door Cobb midwater trawl (figure 54). These trawls are difficult to operate from multipurpose oceanographic vessels. However, they are valuable in sampling animals up to about half a metre in length. The catch is not easy to use in comparative studies because the mouth area and shape change with speed, and gradation of the mesh size along the net makes interpretation difficult. Even with nets of this enormity with mouth areas of 2,800 sq. m and 750 sq. m (less when fishing because they close up due to drag) the larger animals, known to be present by observation, are not caught.

The necessity of knowing the depth at which animals are caught has led to the invention of several opening-closing devices from simple messenger-operated throttles to more sophisticated telemetering equipment.

Messenger operated throttles (figure 55) have the advantage of permitting the use of a string of nets on one wire simultaneously, but, in addition to other disadvantages of ring nets, part of the catch is sometimes 'spilt' at closure. The Clarke-Bumpus net is opened and closed by rotation of a flat plate but this method clearly limits the size of the net (figure 55). The

dual 'bongo' net can be opened and closed and has the advantage of producing two samples from each haul.

Difficulties with throttling or covering the mouths of larger nets led to development of 'cod end' devices to separate catches taken at different depths (figure 56). These can be operated by a mechanical or electrical pressure operated device, electrical impulses down a cable from the surface, or by a telemetering system similar to the one described above (page 57). While these cod end devices have proved extremely useful with nets up to a mouth area of about 1 sq. m, with larger nets they lead to slight mixing of deep and shallow catches. Realization of this disadvantage has led to improved devices for opening and closing the mouths of larger nets.

The Isaacs-Kidd midwater trawl has been modified by having an extra panel of netting which can slide down to open the net and, to close the net, the catch in the cod end is throttled off. This is still prone to contamination similar to that of cod end devices. The rectangular trawls described above are easily adapted for opening and closing by supporting the mouth of the net by extra bars which are held together before the net is opened. When opened, the lower of these bars, holding the lower side of the net mouth, slides down the side wires. When closed, after fishing, the upper bar holding the upper side of the net mouth is released, slides down the side wires, and lies against the lower bar (figure 57). The Tucker net employs three bars and is operated by a timing device. The rectangular trawl employs four bars and is operated by either a simple pressure activated release or, more usually, by an acoustic signal broadcast from

55. Three types of opening and closing systems: 1. One of Motoda's methods in which the first messenger releases the net from a canvas flap, and it opens, and the second messenger releases the bridles and the net closes; 2. Leavitt release method: the first messenger releases the throttling rope and opens the net, the second messenger releases the bridles and the net is throttled again; 3. Clarke-Bumpus sampler: the mouth is covered by a butterfly lid which is opened and closed by messengers. All three methods allow the use of several nets on a single wire for at each operation another messenger is released to trigger the net below (1, after Motoda 1967; 3, after Barnes 1959)

56. 1. Catch-dividing bucket fitted at the end of a net allows the catch from different depth regions to be diverted into different legs. 2. Sectional view of the bucket showing the flap that controls the flow of the catch. 3. Controls for the bucket can be operated by pressure, or electrically, or acoustically. 4. The Bé multiple sampler fishes three nets in one frame. A pressure sensitive piston release allows the successive release of three levers each forming one side of a different net. When released, elastic pulls each lever into a vertical position, thereby closing the mouth of its net and allowing the next net to fish (4, after Bé 1962)

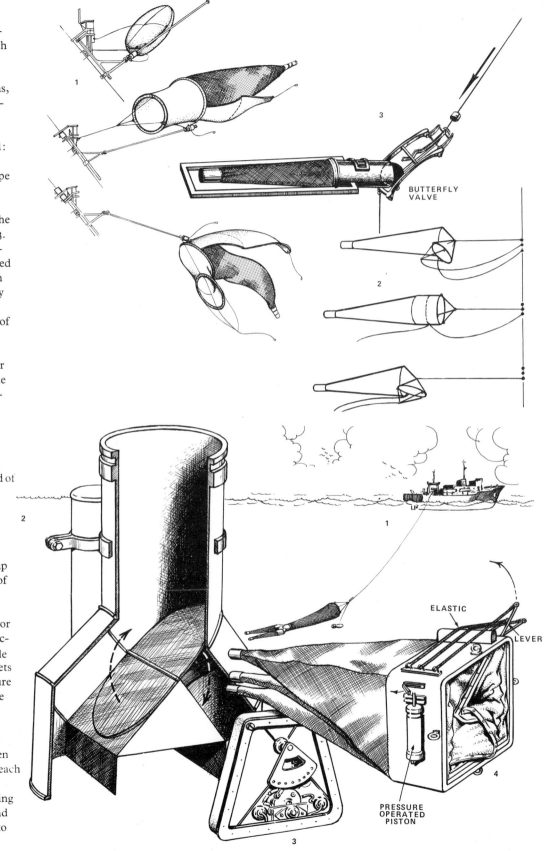

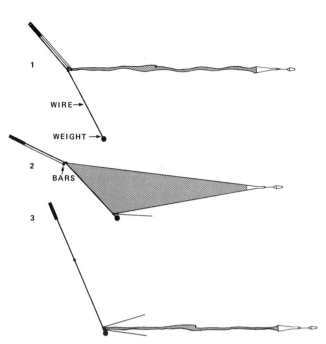

57. Operation of a rectangular opening and closing trawl: 1. net is lowered closed; 2. net in trawling position; and 3. net closed at end of trawl

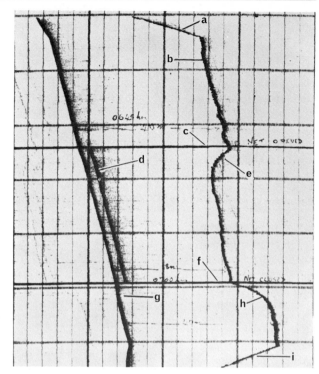

58. Record received on the echo-sounder from the 'pinger' monitoring the operation of an acoustically controlled rectangular midwater trawl. At (a) the increasing separation of the two traces indicates that the net is sinking as the wire is payed out; at (b) the winch is stopped and depth remains constant until an acoustic signal is sent to open the net (c). The opening is indicated by the appearance of the third trace (d) and the increased drag of the open mouth causes the net to rise slightly (e). A second signal closes the net at (f), the third trace disappears (g) and the decreased drag causes the net to sink (h). At (i) the net is hauled up again

the ship (figure 53). Acoustic telemetering devices give complete control over the net and full information of its performance while being towed. Some details are given elsewhere, but the way the gear is used is worth summarising here. Above the bridle attachments there is a depth-telemetering pinger unit (figure 53) which continuously sends signals to the ship; these are recorded on moving paper so that the depth of the trawl is known throughout the haul. When the closed trawl reaches the depth to be sampled it is opened by broadcasting a sound frequency from the ship. As it opens, the net changes its depth and an additional signal from the pinger registers as a third line on the paper recorder; from these records the shipborne operators know the net has opened successfully. After fishing the required time, the net is closed by another signal and then brought to the surface, both the operation and depth again being recorded (figure 58).

For quantitative work the angle of the net must be monitored, and in a recent development this is done by having a telemetering current meter above the bridles having previously calibrated speed through the water against net mouth angle. For special studies the environmental data such as temperature or oxygen concentration may be recorded by means of the same telemeter.

Large commercial trawls have not yet been opened and closed but it may prove possible to open and close the 90 sq. m rectangular trawl before long. Handling this net is relatively easy and additional bars would add little to the problems of its use.

BOTTOM SAMPLING
Commercial otter and beam trawls are used to some extent for biological sampling on the sea bottom. Their use on the continental slope is extremely hazardous

59. Dredge for obtaining rocks from the bottom being lowered over the stern of RRS *Discovery*

because of the steep contours and rock outcrops. Use on the abyssal plain is restricted to a very few vessels because of limitations of winch power and warp length.

The usual bottom nets used in deep water are relatively small and they are sometimes so constructed that they only open while on the bottom (figure 60). Such nets usually only take small or slowly moving animals.

Slow, attached or burrowing animals can be collected by dredges (figure 59) but they cannot be used for detailed ecological work because distribution of the animals on the bottom is not known. For such work, grabs are used and many have been designed but few have proved reliable quantitative samplers. To be successful a grab must be as large as possible, must not disturb the sediment when closing, and must not let the sediment escape or mix up when it is hauled in.

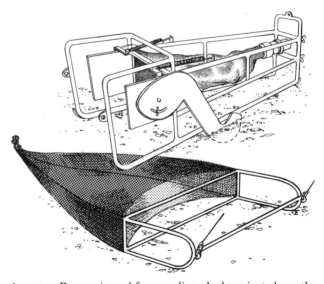

60. *top*, a Bossanyi trawl for sampling plankton just above the bottom; the spring-loaded side arms hold the doors shut until they touch the bottom. *Bottom,* an Agassiz bottom trawl (after Barnes 1959)

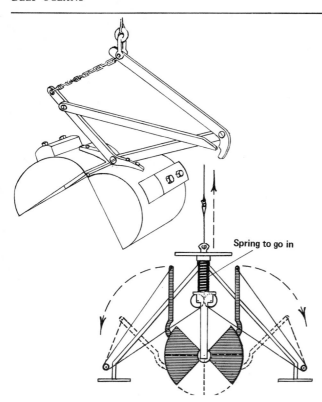

Spring to go in

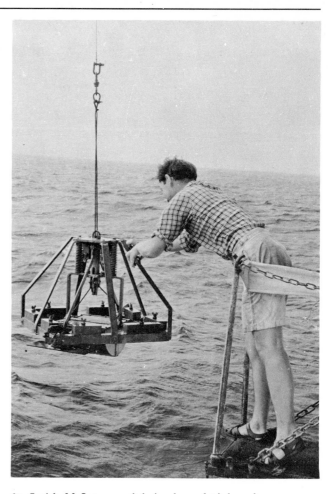

61. Two grabs; *top*, a Van Veen type in which the locking notch disengages as the grab hits the bottom, and the long arms are pulled together closing the jaws as the suspending wire is drawn up again. *Bottom*, the Smith-McIntyre grab; when the trigger plate hits the bottom the springs drive the bucket into the bottom and hauling on the wire closes the jaws (Top, after Barnes 1959, bottom, after Smith and McIntyre 1954)

62. Smith-McIntyre grab being brought inboard

Thus, a grab should sample the environment as well as the animals on it and within it. Popular grabs for deep water, in use at present, are modifications of the 'Petersen', the 'Van Veen' and the 'Smith-McIntyre' (figures 61, 62). The Petersen grab is tripped by a weight hanging below it and falls heavily into the sediment; it closes as the warp is pulled in. The Van Veen grab releases two long arms when it hits the bottom and these give good leverage to force the jaws through the sediment as it is pulled up. The Smith-McIntyre grab is spring loaded and penetrates sediments well. Principal difficulties of using even good grabs in deep water are premature release and detection of the moment the grab strikes bottom. The latter is now detected by a pinger on the grab which may prevent serious tangling of the wire. The latest grabs contain cameras which photograph animals *in situ* as the grab strikes bottom. Comparisons of grab samples with observations of skin divers suggest that considerable errors are possible in numerical computations based on grabbing.

OTHER METHODS

Many other devices displaying the full ingenuity of man are at times used for particular studies but they cannot be treated fully here. Lining is dealt with in Chapter 9; powerful pumps may be used to collect small animals from fairly near the surface; harpoons are useful for larger fish and cetacea; dolphins are also caught by special nets on the bow of a fast boat or by seines; dip nets, often of considerable size, are used with attractive lights at night; fish traps of various sizes and sometimes designed to pop-up after a delay are used in very deep water. Of all nets, the ones used on the front of submersibles must be regarded as the most attractive because they permit the first limited chasing of animals observed. Man can act as an undersea predator!

To conclude, let us not forget animal predators as samplers. Will we ever achieve their prowess? Within their stomachs may be found many species rarely collected or unknown from man-made samplers. Until we can chase and catch our prey cheaply, the predator will out-fish us.

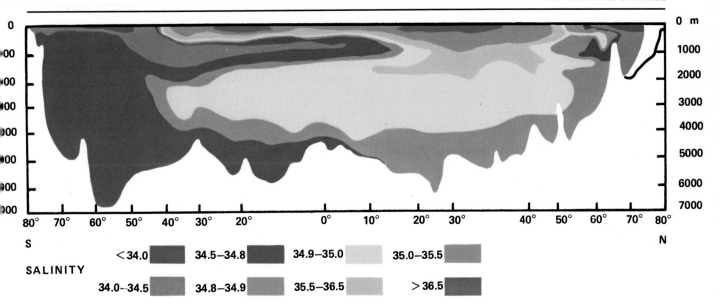

SALINITY

< 34.0	34.5–34.8	34.9–35.0	35.0–35.5
34.0–34.5	34.8–34.9	35.5–36.5	> 36.5

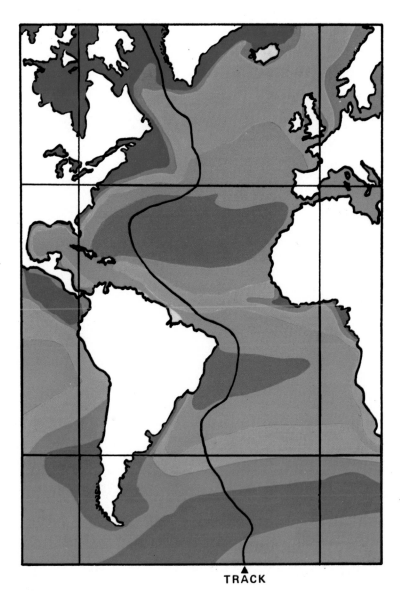

TRACK

63. Salinity distribution in the Atlantic Ocean; *top,* vertical distribution along the track shown in the lower chart, in which the distribution in the surface waters is indicated (after Wüst 1928)

Manned Submersibles *Roswell F. Busby*

INTRODUCTION

In his study of the deep ocean the oceanographer has collected data from a variety of vessels which range from the conventional surface ship to more unconventional platforms such as naval submarines, sea-based towers, aircraft, satellites, and even drifting icebergs. To this list has recently been added the manned submersible; a deep-diving, pressure-resistant capsule, capable of unrestricted manoeuvring while carrying one or more passengers to great depths and equipped with windows to allow a direct view of the sea and its inhabitants.

BATHYSPHERES AND BATHYSCAPHES

Probably the first recorded instance where man purposefully used anything other than a rock to assist him in going underwater was Alexander the Great's excursion in a diving bell, *circa* 320 BC. Following this are numerous records of diving for military and salvage purposes, but it was not until the early 1930's that diving in pressure-resistant vessels for scientific purposes was seriously conducted.

In August 1934 biologist William Beebe and engineer Otis Barton reached the unprecedented depth of 3,028 feet (923 m) in a cable-suspended, steel sphere called a bathysphere. The bathysphere received its power from a surface ship to which it was attached by cable; it carried its own oxygen and had fused quartz windows (viewports) through which the passengers observed the world around them. The restrictions on the bathysphere's manoeuvrability, owing to its dependence on the surface ship, were severe, and although the denser-than-water steel sphere was more than a match for the pressures it endured, entrusting the lives of two men to the strength of a wire rope did little to encourage this approach to marine biology, no matter how rewarding it might be.

In the early 1950's Swiss scientist Auguste Piccard began construction of the first bathyscaphe (deep boat) which, in 1960, culminated in a dive to 35,800 feet (10,910 m) in the US Navy bathyscaphe *Trieste* (figures 65, 66). By safely and comfortably carrying passengers to the deepest part of the ocean, *Trieste* demonstrated that man was technologically capable of personally exploring any part of the world's oceans, and, in the process, it effectively terminated a race to establish depth records. Instead, the competition became one of demonstrating who could build the most useful vehicle, and the result is that some 40 or more submersibles are now operating.

The bathyscaphe is, in essence, an underwater balloon and consists of a bag (called the float) and a cabin (called the sphere). The float is filled with a lighter-than-water liquid (gasoline) which is vented on the surface and automatically replaced by sea water; this provides the negative buoyancy necessary for the bathyscaphe to sink. To decrease the vehicle's weight, variable amounts of iron pellets can be dropped by the pilot. In the event of a power failure the pellets, which are electromagnetically held in ballast tanks, automatically dump. For horizontal manoeuvring *Trieste* was limited to five small electric motors which provided a range of approximately one mile. Being almost 60 feet long and weighing 50 tons (without gasoline) the bathyscaphe is awkward to manoeuvre and requires extensive support facilities. Equally as constraining is the necessity to tow it at low speed or carry it aboard a very large ship when surface transportation from point to point is required; in effect, the bathyscaphe is a very costly and cumbersome deep sea elevator.

FIRST GENERATION SUBMERSIBLES

The response to *Trieste*'s shortcomings ranged in size from the two-man, 22 feet long *Perry Cubmarine* to the 40-man, 90 feet long *Auguste Piccard,* and encompassed the diverse roles of research, surveys, salvage, engineering, search, rescue, and recreation. Although general configuration and performance vary significantly from vehicle to vehicle, there are many features common to all such as those discussed below.

Viewports: The windows of submersibles (viewports) are generally made of truncated cones of acrylic plastic. Glass has the characteristic of failing catastrophically under pressure; on the other hand, plastic gives warning well prior to failure.

Internal Atmosphere: All submersibles generally operate at atmospheric pressure and are dry. Oxygen is carried in tanks and carbon dioxide is removed by scrubber compounds such as barylime or lithium hydroxide. In the tropics and subtropics internal temperature on the surface generally exceeds 100°F and humidity commonly reaches 100 per cent. Beyond depths of 1,000 feet temperature and humidity become more tolerable.

Power: Electrical power is generally supplied by lead-acid batteries. Most batteries are mounted externally in an oil-filled container connected to an oil reservoir made of soft material such as rubber. As the sea water pressure increases with depth it automatically squeezes oil from the soft reservoir to the battery case at a rate equal to the pressure increase, thereby maintaining the pressure inside the battery case equal to that outside.

Speed: The majority cruise efficiently at one knot and can reach two to three knots in an emergency.

64. *opposite* The mouth of the Colorado River in Baia California photographed from Gemini IV (orbital height 100–175 miles). The discoloration of the sea is caused by the outflow of suspended sediment from the river

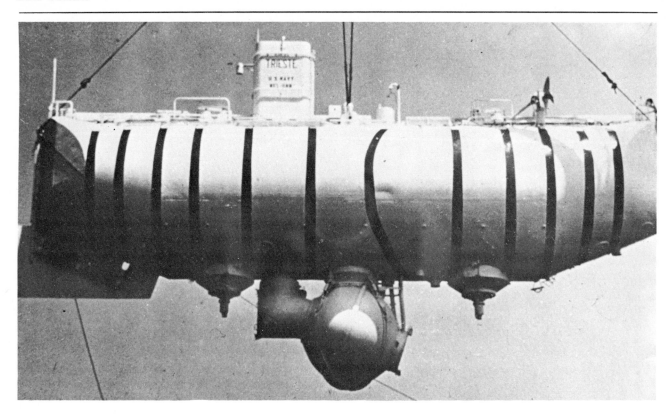

65. The bathyscaphe *Trieste*

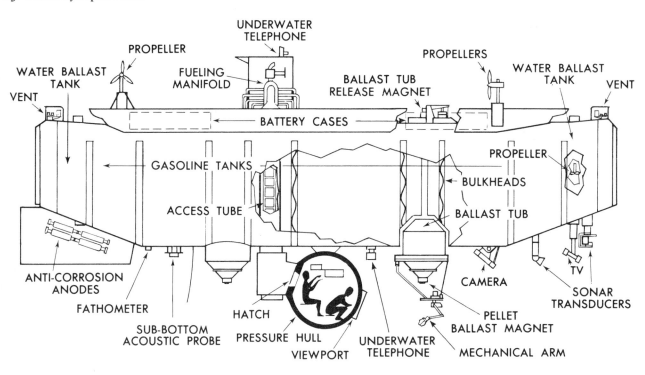

66. Internal arrangement of *Trieste*

Length: 59 1/2 Ft.   Crew: 1 Pilot 1 Scientist   Propulsion: 5 @ 2 Hp. Electric Motors   24 Hrs. (Max.) 2 Men

Beam: 11 1/2 Ft.   Speed: 1 Knot (Cruising)   Power Supply: 145 KWH Lead Acid   Buoyancy Control: Gasoline (Positive)

Weight (Air): 50 Tons (Wetlock 6 Gasoline)   Pressure Hull Dia: (I.D.): 6 Ft.   Batteries   Shot Ballast (Negative)

Payload: 12 Tons   Thickness 5 to 7 In.   Life Support: 10 Hrs. (Normal) 2 Men   Viewports: 2 @ 1/2 In. I.D. 16 In. O.D.

Duration: Limited primarily by power, most dives are from five to eight hours' duration. Life support systems can generally sustain the crew from 24 to 48 hours in an emergency.

Crew: Personnel can range from two to six, but more generally one pilot and one or two scientists can be accommodated.

Payload: Payload (the amount of equipment/ weight that can be added beyond that needed to safely run the vehicle) ranges from a few hundred pounds (*Deepstar*–4000) to several tons (*Aluminaut*).

Pressure Hull: High-yield steels constitute the majority of pressure hulls. A fail-safe device in all hulls rests in their being positively buoyant at operating depth and requiring weight (which can be jettisoned in an emergency) to hold them down.

Transportability: Most submersibles can be carried by air, truck or small ship. At sea they are launched and retrieved for each dive with a powered crane or catamaran/ramp arrangement.

Cost: Building safe, reliable and rugged submersibles is no business for the financially embarrassed. Small vehicles such as *Deep Diver* may cost up to half a million dollars; large vehicles such as *Aluminaut* may cost from three to four million dollars.

Given these general similarities let us see how three individual submersibles have solved the problems of deep submergence.

Deepstar–4000 (*Figures 43, 67*)
Deepstar–4000 (DS–4000) was launched in 1966 and almost immediately commenced a diving programme off the California coast that was to last for over two years, conduct over five hundred dives, and encompass the Atlantic as well as the Pacific.

DS–4000 is uniquely suited to perform a wide variety of research tasks: two viewports overlap in their area of coverage and allow both the pilot and scientist to see the same thing at the same time, while a smaller window between the two is used solely for

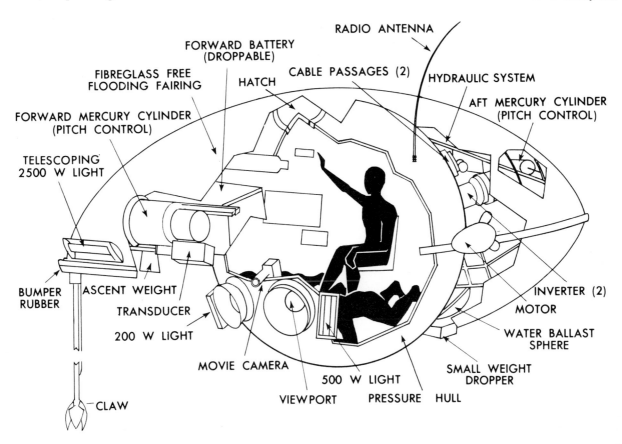

67. Internal arrangement of *Deepstar*–4000

Length: 18 Ft.	Speed Max.: 3 Knots	Located Amidships	Trim Control: Mercury Trim System
Beam: 10 Ft.	Cruise: 1 Knot	Power Supply: 3 Lead Acid Batteries, 400	Buoyancy Control: Forward Battery and
Pressure Hull Dia. (O.D.): 6.5 Ft.	Operating Depth: 11,000 Ft.	Amp.-Hr.	Mercury Dump Variable Ballast Tanks
Height: 7 Ft.	Payload: 350 Lbs.	Propulsion Endurance: 12 Hrs. @ 1 Knot	2 Weights: Descent 231 Lbs. Ascent
Weight (Air): 9.5 Tons	Hull: 1.2 Inches HY-80-Steel	Life Support Endurance: Normal 12 Hrs.	187 Lbs.
Crew: 1 Pilot 2 Obs.	Propulsion: 4.5 @ HP/C Freeflooded	(3 Crew) Max. 48 Hrs. (3 Crew)	Viewports: 2 Acrylic Plastic 1 Photographic Port

photography. DS–4000 is usually transported on a suitable offshore oil supply boat and it is launched and retrieved by a modified, articulated crane of 25 tons lift capacity. Short period waves greater than five or six feet in height generally preclude diving operations. At present, the safe, routine penetration of the air-sea interface is a much greater hazard for all submersibles than the actual dive.

DS-4000 is launched with the pilot and two passengers inside and the vehicle ready to dive. When released from the surface it descends stern-first in a helical spiral at an angle of about 22° and at a rate of about 80 feet per minute. Approximately 150 feet from the bottom, a 220 lb weight is dropped from the stern and with minor trim and ballast changes the vehicle is at almost neutral buoyancy and horizontal trim 4,000 feet below the ocean surface. Depending upon power consumption, a dive generally lasts four to five hours while the vehicle cruises at about one knot. Propulsion is provided by two 4·5 hp reversible motors. Lighting, for no sunlight penetrates to 4,000 feet, is provided by quartz-iodide lamps mounted outside the hull. Limited sampling can be conducted using DS–4,000's mechanical arm while the vehicle is bottomed.

The pilot supplies oxygen as required, and lithium hydroxide is used to absorb carbon dioxide; this arrangement can support three men for a maximum of 48 hours. An emergency breathing system, used in the event of smoke or gas contamination, can support each individual for $1\frac{1}{2}$ hours.

Conditions inside the submersible are quite comfortable, although space is not over-abundant in the seven feet diameter sphere. But at 4,000 feet depth a sweater and long pants are required to ward off the chill produced by the 40° to 50°F sea water. To the passengers there is no sensation other than that produced by the wonderment of cruising comfortably and safely 4,000 feet below the ocean's surface.

Communication is maintained by underwater telephone with the surface ship, and a small, sound-emitting pinger on the vehicle is constantly monitored and its position plotted on the surface ship to track the submersible. When the dive is finished the ascent weight is dropped, the vehicle surfaces and divers connect it to the crane for retrieval. Once aboard the surface ship the passengers disembark and *Deepstar* is immediately readied for the next dive.

Aluminaut (Figures 68, 69)
With a length of 51 feet, 73 tons in weight, and a designed operating depth of 15,000 feet, *Aluminaut* is one of the largest, most comfortable, and deepest diving submersibles in the entire field. *Aluminaut*'s unique pressure hull consists of eleven aluminium alloy (7079–

T6) rings and two hemispherical end caps bolted together to form an eight feet one inch (o.d.) cylinder $6\frac{1}{2}$ inches thick. Since *Aluminaut*'s aluminium hull is less compressible than sea water, the deeper it goes the more buoyant it becomes, and at 15,000 feet depth its length is reduced about one inch and its diameter about one tenth of an inch. The bolts used to hold the cylinder together are of larger diameter than the holes they fill, and prior to insertion they were shrunk in liquid nitrogen at −320°F and then fitted into their holes where they expanded on warming.

Aluminaut's 4,000 lb payload and great volume allows attachment of a wide variety of instruments. Consequently, several tasks can be performed on one dive which would take several dives in smaller vehicles. Four 650 ampere-hour, silver-zinc batteries located inside *Aluminaut* are complemented by the oxygen/lithium hydroxide life support system providing thirty-two hours' normal dive duration for six men.

Aluminaut is hampered, however, by the very feature which provides its assets: large size. *Aluminaut*'s weight of 73 tons prohibits it from being lifted aboard conventional small ships; consequently, it is generally towed at a slow speed of five knots. *Aluminaut*'s main ballast control is, in large part, similar to that of a bathyscaphe. Positive buoyancy is derived from its rigid, lightweight hull; negative buoyancy is attained primarily by shot ballast as with *Trieste*. In a ready-to-dive condition, a pre-determined amount of shot ballast will have been loaded and sea water is vented into the air ballast tanks to descend. Unlike DS–4000, *Aluminaut* descends in a horizontal position and can, if required, be driven downward or restrained by its reversible vertical thruster. Descending at about 60 feet per minute, the pilot begins releasing shot a few hundred feet from the bottom to slow its rate of descent and land softly in a slightly negatively buoyant condition. If progressing into shallower water, the pilot periodically dumps shot to become 'lighter'. The disadvantage with this system is that the de-ballasting is irreversible and once *Aluminaut* becomes lighter it cannot return to a much greater depth until surfacing and taking on more ballast.

If biological or geological samples are desired, *Aluminaut* is especially able to comply by using her two hydraulically-powered, bow-mounted, mechanical arms. The arms are nine feet one inch long, have six degrees of motion, can lift 200 lbs each and will only be exceeded in capability by those planned for North American's *Beaver*, which will have the added advantage of operating at variable rather than fixed speed.

Generally three scientists and three crew members are present on each dive; both the chief pilot and chief scientist share the forward hemi-head while the

68. *Aluminaut*

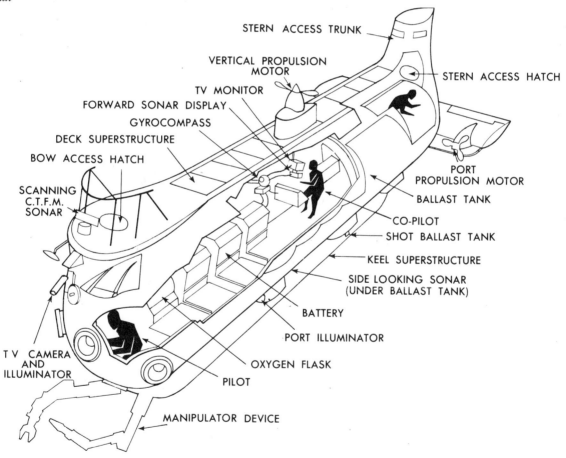

STERN ACCESS TRUNK

VERTICAL PROPULSION
MOTOR

TV MONITOR

STERN ACCESS HATCH

FORWARD SONAR DISPLAY

GYROCOMPASS

DECK SUPERSTRUCTURE

BOW ACCESS HATCH

PORT
PROPULSION MOTOR

BALLAST TANK

SCANNING
C.T.F.M.
SONAR

CO-PILOT

SHOT BALLAST TANK

KEEL SUPERSTRUCTURE

SIDE LOOKING SONAR
(UNDER BALLAST TANK)

BATTERY

PORT ILLUMINATOR

T V CAMERA
AND
ILLUMINATOR

OXYGEN FLASK

PILOT

MANIPULATOR DEVICE

69. Internal arrangement of *Aluminaut*

Length: 51 Ft.
Beam: 15 Ft.
Pressure Hull Dia. (O.D.): 8 Ft.
Height: 14 1/4 Ft.
Weight (Air): 73.2 Tons
Crew: 3 Operators 3 Obs.

Speed Max.: 3.5 Knots
  Cruise: 2.5 Knots
Operating Depth: 15,000 Ft.
Payload: 4,000 Lbs.
Hull Material: 6.5 Inches Aluminium
Propulsion: 2 @ 5 HP Stern Motors
  1 @ 5 HP Vertical Motor

Power Supply: 115 V. Silver Alkaline
  Batteries, 2600 Amp-Hr.

Propulsion Endurance: 30 Hrs. @ 2.5
  Knots
Life Support Endurance: Normal 32 Hrs.
  (6 Crew) Max. 72 Hrs. (6 Crew)

Trim Control: Rudder, Diving Planes,
  Variable H_2O Trim System, Portable
  Lead Ballast
Buoyancy Control: Droppable Steel Shot,
  Air Ballast Tanks, 4400 Lbs. Droppable
  Lead Bar
Viewports: 4 Acrylic Plastic

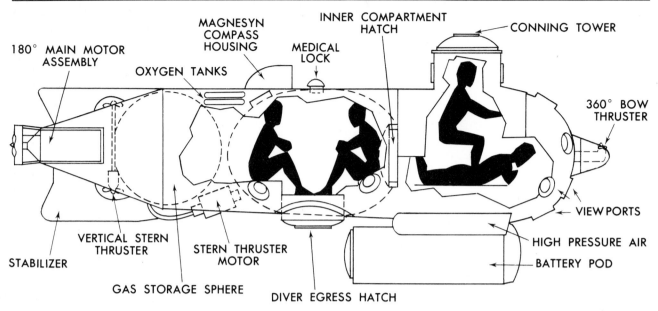

180° MAIN MOTOR ASSEMBLY

MAGNESYN COMPASS HOUSING

OXYGEN TANKS

MEDICAL LOCK

INNER COMPARTMENT HATCH

CONNING TOWER

360° BOW THRUSTER

VIEW PORTS

HIGH PRESSURE AIR

BATTERY POD

DIVER EGRESS HATCH

GAS STORAGE SPHERE

STERN THRUSTER MOTOR

VERTICAL STERN THRUSTER

STABILIZER

70. Internal arrangement of *Deep Diver*

Length: 22 Ft.	Crew: 1 Pilot 3 Passengers	Horizontal: 1-3 HP Thruster	Trim Control: Water Trim Tanks
Beam: 5 Ft.	Speed Max.: 3 Knots (30 Min.)	Power Supply: 34 KWH (AC-DC) Lead	Buoyancy Control: Water Ballast Tanks
Pressure Hull: 0.5 In. T-1 Steel	Cruise: 0.5 Knots (12 Hrs.)	Acid Batteries (In Pressure POD)	View Ports: 21 Acrylic Plastic
Weight (Air): 8.25 Tons	Propulsion: Main: 1-10 HP Single Screw	Life Support: Normal: 8 Hrs. (4 Men)	Diver Endurance: 2 @ For 2 Hrs. at Max.
Payload: 1 Ton	Vertical: 1-2 HP Thruster	Max: 24 Hrs. (4 Men)	Depth

remaining crew and scientific party operate the vehicle and oceanographic instruments.

Deep Diver (Figures 44, 70)

The submersible *Deep Diver* was selected not only as a representative of present-generation, shallow-diving (less than 2,000 feet) vehicles, but also as an example of the next generation vehicles. *Deep Diver*'s lock-out/ lock-in capability combines the payload, speed, power and endurance of the submersible with the dexterity and senses of the human, and by so doing, the designers have provided versatility of operation exceeding, by an order of magnitude, that of other shallow vehicles which lack this capability.

Basically *Deep Diver* is a two-compartment sub- mersible carrying a crew of four: a pilot and observer occupying the forward compartment and two divers occupying the connecting aft compartment. Main propulsion is supplied by a ten hp stern propeller which can, like an outboard motor, be swivelled 90° to either side of the vehicle. Delicate manoeuvring is assisted by a three hp bow thruster which can rotate 360° in the vertical plane, and a three hp, double-ended, up and down stern thruster augmented by diving planes. The arrangement provides a vehicle which can make a 360° turn within its own axis (22 feet), with the added capability of moving up, down, forward, backward and sideways. A 360° view is provided in the conning tower and 13 more viewports throughout the vehicle provide a total of 21 ports for visual observation. A unique feature of *Deep Diver*'s viewports is that they are double acting – that is, each contains two glasses which seal in opposite directions in order to withstand internal as well as external pressure.

Buoyancy is controlled similarly to that in conven- tional submarines: hard tanks containing air provide positive buoyancy, and filling these tanks with water provides negative buoyancy. Fine depth control and trim is maintained by a set of smaller trim tanks through which water is pumped either forward or aft; these are assisted by the small motor thrusters.

During a shallow (less than 130 feet) dive atmos- pheric control is similar to the other vehicles, but when diver lock-out is required at depths greater than 130 feet a mixture of oxygen and helium is generally used in the divers' compartment and normal air in the forward compartment. When divers are not required the aft compartment may be used for carrying additional equipment or non-diving passengers. With the vehicle ready to dive all personnel climb aboard, the hatches are closed, and *Deep Diver* is launched. Flooding the main ballast tank for negative buoyancy, the pilot powers down to the dive site using his diving planes and main propulsion unit. Once on site, the vehicle bottoms and the tanks are flooded to compensate for the loss of weight when the divers leave. When ready, the divers close the connecting hatch to the forward compartment and unlock the entrance/exit hatch which is still held tightly shut by the pressure of sea water. Putting on their diving equipment, the divers turn on the gas supply to pressurize their compartment. When the internal pressure equals the external pressure the bottom hatch falls open, pressurization is stopped, and the divers leave. Depending upon the job requirements, the divers may use either *Scuba* for mobility and range or, for long endurance, may use a hose attached to the breathing system in the submersible. Upon completing their tasks, the divers re-enter the submersible, close the hatch, and commence reducing the pressure inside their compartment. In the event that extensive decompression is required (gradual pressure reduction with long-period stops at pre-determined stages), the vehicle can be retrieved on board the ship and the divers fed through a 'medical' lock on top of their compartment. When another person is required to assist the divers inside, the forward compartment is also capable of being pressurized to allow him to enter. In March 1968, divers Breeze and Cook of Ocean Systems Incorporated locked out of *Deep Diver* at a record depth of 700 feet in the Bahama Islands. Although they only stayed at this ambient pressure (700 feet = 312 psi) for 20 minutes, a total of 35 hours were required to decompress, or return them to normal atmospheric pressure and gas mixture.

SCIENTIFIC UTILIZATION OF SUBMERSIBLES
As with many a new tool or concept the submersible's potential has been overstated by its proponents and understated by its detractors, the result being over-employment in some instances and underemployment in others. However, in the brief process of seeking its identity and scientific role the submersible has performed admirably.

Biological Studies
One of the major problems in the deep ocean is the behaviour and composition of the Deep Scattering Layer (DSL) (figure 48). In an effort to learn more of this itinerant 'false bottom', *Trieste, Diving Saucer,* and *Deepstar–4000* have been used. Studies off the California coast by Dr Eric Barham of the Navy Undersea Research and Development Center, San Diego, California, revealed that the primary organisms of the DSL were siphonophores, and small two to three inch long lantern fish. The presence of myctophids

was expected owing to the contents of previous net hauls, but siphonophores were torn apart by the nets and only fragments were collected, thereby explaining their apparent absence. On the other hand, studies of the DSL in the Atlantic Ocean were conducted from *Alvin* and *Deepstar–4000*, and revealed a complete dominance of the DSL by lantern fish and a virtual absence of jellyfish (figures 71, 72).

Although the DSL has received most attention, a variety of research has been conducted in other areas: the mating and spawning activities of squid were observed from *Deep Diver*; the distribution of various commercially important bottom-dwelling fish off the Virginia coast has been reported from *Alvin*; and a photographic census of bottom-dwelling forms was conducted from *Trieste* (figure 202). Dr Barham and his co-workers captured an unusual jellyfish with *Deepstar–4000* which later turned out to be a completely new species. Being of uncertain taxonomic position, the animal was appropriately named *Deepstaria enigmatica*.

Previous to the submersible sightings, the broadbill swordfish *(Xiphias gladius)* was known to range in depths to 1,000 feet; this depth range has been significantly increased by sightings from submersibles at depths as great as 2,800 feet. The broadbill's aggressive nature proved fatal to one such individual who attacked *Alvin* at 1,800 feet depth and managed to catch its bill between the vehicle's fibreglass fairing and the steel sphere. *Alvin* thereupon surfaced, the 200 lb broadbill was freed, cleaned, and provided swordfish steaks for the ship's complement.

The scallop industry was recently assisted through investigations by biologists from *Aluminaut*. Unusual results in dredging for the calico scallop *(Pecten gibbus)* off New Smyrna Beach, Florida, prompted fishery biologists to use *Aluminaut* in an attempt to explain the irregularity of dredge hauls. The reason became immediately apparent from *Aluminaut*: the scallops were lying in 100 to 300 feet wide bands separated by bare areas; when the dredge passed across the bands, the catch was small; when it passed along the band, the catch was high. Although such conditions had been previously surmised, visual observations from *Aluminaut* provided the necessary proof.

Geological Studies
One of the first groups to realise the potential of manned submersibles were the geologists, and, judging from the number of reports, the future marine geologist will rely heavily on this platform.

On the continental shelf off San Diego, *Trieste* showed the effectiveness of visual observations when sea anemones, which require a firm substratum for

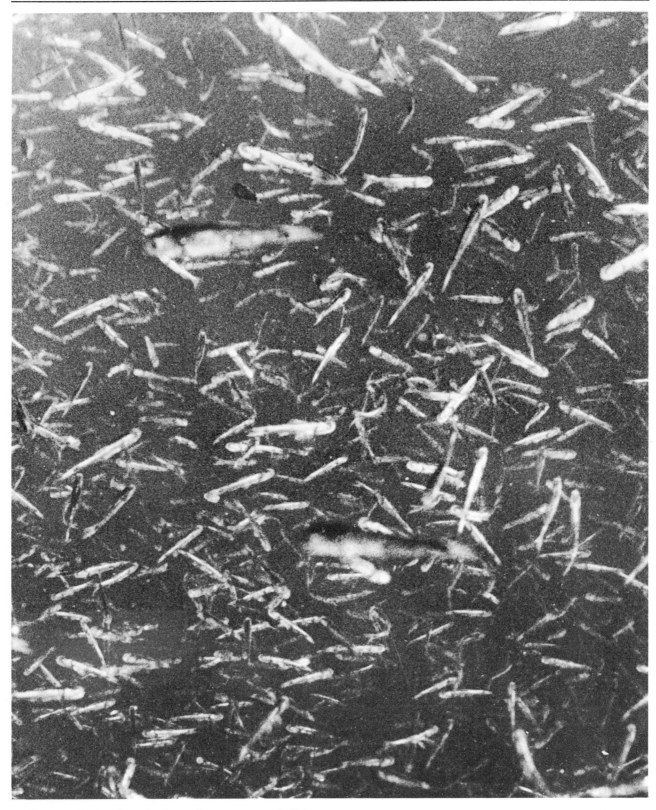

71. Photograph taken from the deep submergence vessel *Alvin* of lantern fishes *Ceratoscopelus maderensis* in the western North Atlantic

DEPTH

(metres)

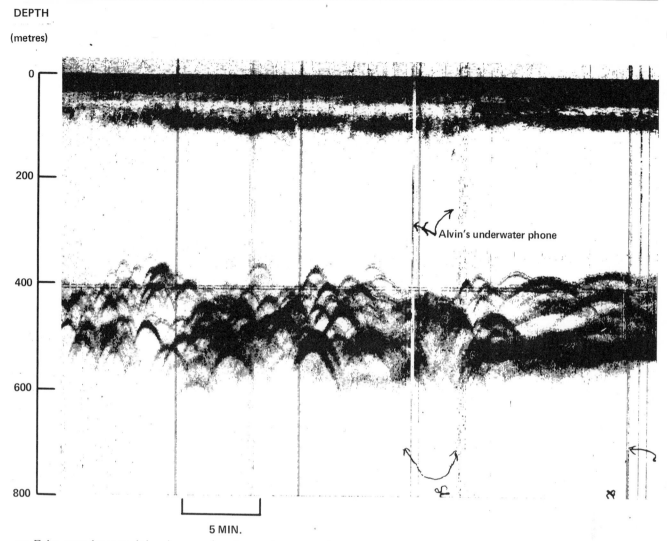

72. Echo-sounder record showing sound scatterers at 400–500 m which turned out to be the fishes photographed in Figure 71

attachment, were observed growing on sandy sediments. Although surface-obtained samples showed this to be a sandy area, the presence of shallowly buried bedrock went undetected until the anemones were observed from *Trieste*.

In other geological dives aboard *Trieste* direct observations of outcrops in the La Jolla Fan Valley assisted significantly in determining the method of migration of undersea channels and in revealing bottom features undetected by other techniques.

In subsequent studies, portions of the La Jolla and Scripps submarine canyons were so narrow that the three metre diameter *Diving Saucer* had to be employed. Vertical overhanging walls were observed which had gone unnoticed from previous echo-and-

wire-sounding surveys. The results of these operations went towards producing the most detailed and informative treatise of submarine canyons to date.

Further evidence of the effectiveness of the submersible was presented by Dr R. F. Dill in a discussion of narrow, step-like, rock terraces observed from *Deepstar*-4000. Owing to the small relief of only ten feet, these features had been missed in previous echosounder surveys because the wide-angle sound cones usually used give poor resolution.

One of the first reported studies from present submersibles in the Atlantic was a geological reconnaissance of the Bahamas' marginal escarpment in the *Perry Cubmarine*. Although primarily an engineering task, these dives produced a picture of the escarpment entirely different to that indicated by previous data.

Following its success in finding a lost H-bomb off Palomares, Spain, the submersible *Alvin* conducted a series of dives in the Bahamas for a variety of scientific purposes. *Alvin* performed another series of dives on the Blake Plateau and, among many other significant observations, it was found that what heretofore had been considered as coral banks were actually a series of ridges and troughs with coral merely colonizing the surface. In October 1968, while finishing another successful diving season south of Cape Cod, the first serious accident happened: in the process of being launched a cable parted and *Alvin* was dropped. Because of its open hatch, the vehicle sank almost immediately to the bottom at 5,500 feet. Fortunately no crew members were lost, and *Alvin* was later retrieved successfully.

Physical studies

Meaningful water current measurements from submersibles are difficult because the period of the observations is generally so short that they cannot be considered as reflecting average conditions. None the less, the submersible has shown itself to be useful in verifying other measurements and measuring in selected locations.

Using nothing more sophisticated than nylon yarn affixed to a grid external to *Trieste*, currents were measured in the San Diego Trough as slow as 0.02 to 0.06 knots. In another instance, extreme variability was observed in current speeds, from less than 0.05 knots at the top of a cliff 60 feet high to over 0.4 knots at its base. The presence of swift currents may also be detected by the ripple marks the current produces in soft sediments; such observations have been made from many submersibles.

A requirement necessary for measurement of all currents, even while bottomed, is for more power or greater negative buoyancy. This requirement was exemplified by General Dynamic's *Star III* while attempting to bottom in the Gulf Stream off Key West, Florida. *Star III*, capable of five knots maximum, was swept along by the two to three knot current as if it were a leaf in a windstorm.

Other physical data collected by submersibles have been in the field of deep ocean sunlight penetration. Beginning with the bathysphere, Beebe reported light penetration to 1,900 feet depth in the clear waters south of Bermuda. *Trieste* discovered unusually low transparency in deep layers (300 m) of the Mediterranean off Capri. Light penetration has been reported from a variety of vehicles to as much as 2,300 feet, and evidence of visibility range exceeding 200 feet laterally at 600 feet depth under ambient light has also been presented.

In other areas of ocean research, magnetic intensity profiles were collected from *Aluminaut* off St Croix in the Virgin Islands; gravity intensity profiles were made by *Aluminaut* off Vieques Islands, Puerto Rico; and the two-man submersible *Asherah* was used in studying and salvaging the remains of Byzantine shipwrecks off the coast of Turkey.

If the submersible's brief operating history of essentially only five years is any indication of its potential, the next decade should see literally an invasion by oceanographers of the ocean depths. These technological advances present a remarkable opportunity for almost unimaginable progress in the knowledge of the oceans.

FUTURE SUBMERSIBLES

The next phase of submersible development will undoubtedly be towards specialization of vehicle design and development of built-for-the-purpose instrumentation. Although the initial surge in submersible development produced a wealth of vehicles, the design philosophy was to offer a vehicle capable of the widest possible applications. Considering the millions of dollars invested by the builder in a tool for a still unpredictable market, this philosophy is understandable. None the less, experience has shown that one multi-purpose submersible will not suffice, and like aircraft, ships, and motor vehicles, special tasks require special designs. Already the trend towards specialization can be seen: the US Navy will soon launch the Deep Submergence Rescue Vehicle which will be used to rescue personnel from stricken submarines; the North American Aircraft Corporation launched *Roughneck* in 1968, which is the first of its *Beaver* class submersibles designed for use in offshore oil fields. Lockheed Missiles and Space Corporation recently won the design competition for construction of a US Navy Deep Submergence Search Vehicle to be used for search and identification of various objects to depths of 20,000 feet.

There is little doubt that future shallow (1,500 feet) submersibles will include diver lock-out/lock-in. This capability will be greatly enhanced by the advances being made in the field of ambient pressure diving (man exposed directly to sea water pressure) where dives to 1,000 feet are becoming common.

Materials for submersibles will probably come primarily from the high-yield steels where much is known of their composition, the methods of fabrication and the predictability of their performance under pressure. However, there are other materials which may offer even greater advantage as pressure capsules: titanium with its light weight and very high strength is quite promising, and one of the more unusual materials being considered for submersible hulls is glass. In spite of a few shortcomings, glass becomes stronger with pressure, and offers extremely high positive buoyancy; its greatest advantage, however, lies in the prospect of the pressure hull being one huge viewport. The US Naval Ordnance Test Station is experimenting with the submersible *Deep View*, which combines a cylindrical steel body with a glass hemi-head. The US Navy Civil Engineering Laboratory recently completed a 500-foot test dive in *Nemo*, a prototype vehicle with a $5\frac{1}{2}$-foot diameter pressure hull composed entirely of acrylic plastic.

The January 1969 launching of the US Navy's NR-1 points the way towards significantly increasing dive duration. The new vehicle is 140 feet long; it carries a crew of six and, using a small nuclear reactor, can cruise and remain submerged for thirty days or more. NR-1's long staying power and virtually unlimited electrical power will in themselves open new areas and methods of investigation. Other sources of power for longer missions may by supplied through fuel cells such as are used in spacecraft. The fuel cell, it is believed, will offer longer duration with much less size and weight than do nuclear sources.

With the air-sea interface providing the greatest single obstacle to routine submersible operations, it is not unlikely that future developments will see underwater launch and retrieval. Such tending platforms as a large mother submarine or a ship such as *Flip* which can be ballasted into an upright position and allow underwater entrance/exit of the submersible are very appealing. Although costly, such launch/retrieval systems will put submersible operations on an equal footing with more conventional platforms by eliminating the present complete dependency upon weather.

Exposure to and actual use by a large number of scientists and engineers will demonstrate the tremendous potential offered by this unique platform, which in turn will provide the catalyst towards creating greater usage and more specialized vehicles and support instrumentation.

The manned submersible offers the ocean scientist a unique opportunity. For centuries he was forced to stand on the rim of his test tube and drop a variety of instruments within. Consequently our entire concept of the ocean is inferred and is based on the results of remote sampling. Many theories derived from such data have been shown to be correct; others are questionable. The time has now arrived when inference can no longer be justified; the oceanographer, unlike the chemist, need not stand outside his test tube, but may with safety and comfort become a party to the experiment itself.

3 Physics of the Ocean

H. Charnock

Theoretical oceanographers, mainly applied mathematicians, study the ocean by deductive methods. They apply known mathematical and physical laws not to the real ocean, for that is too complex a system to be completely specified, but to conceptual models. Such models often simulate some attributes of the real ocean and suggest generalisations of great significance.

More commonly the ocean, as with many systems with many interlocking components, is studied by inductive methods. Observations are made, the structure and behaviour of the whole and of its parts are described and attempts are made to relate them in a consistent way. The ultimate test of consistency is prediction and the reward, control.

Many of the readers of this book will have interests in marine biology: the 'deep ocean' of the title represents a unique and stable environment which is of great evolutionary significance. For this reason this chapter on the physics of the deep ocean will have the inductive approach which is forced on biologists by the intrinsic complexity of the organisms they study. We shall regard the ocean as an analogy of a biological organism and attempt to summarise what is known of its superficial appearance, size and shape; its substance, the characteristics of the complex fluid we call simply sea-water; its anatomy, the observed distribution of temperature, salt and ocean currents, and its physiology, how its structure is maintained. .

There are many general problems – how is the energy of the system supplied and transformed? How are heat and salt circulated? How does the system interact with its surroundings, the solid earth and the atmosphere? Such general questions in their turn raise other more specific ones; the maintenance and the role of a particular ocean current system is an example of the general problem of interaction between components of different scale. The final question, whether the main characteristics of the ocean are likely to be changed by man's activity, either deliberately or inadvertently, has an obvious, and important, relevance to problems of biological conservation.

As there is only one ocean in the solar system biologists may well feel that to treat it, by analogy, as a biological organism is unjustified, like considering the characteristics of a whole population from the examination of a single individual. If, however, we consider not just the solar system but our galaxy, a reasonable estimate is that more than 600 million out of the 135,000 million or so stars are likely to have an ocean basically similar to our own. As there are probably more than 100,000 million galaxies in the universe the potential number of oceans is very large indeed.

Estimates such as these have been made by scientists of the RAND Corporation in an attempt to find how many stars are likely to have planets which men could live on: an ocean, or large bodies of open water, is essential. Any planet which man could live on would have to be about the size of Earth and have had a similar evolution so as to provide a similar atmosphere and a similar ocean. It would probably be rotating round a similar sun so that most of the meteorological and oceanographic phenomena would have their counterparts. Seasons; blue skies; winds, clouds and rain; deserts and ice caps: all would be similar to our own present experience.

Curiously, however, the living things would (will?) be found to differ widely in detail from those on Earth. There would be organisms which carry on photosynthesis and life forms that use others for food, the fishes would be fish-shaped, the animals legged and the birds winged. But the actual phyla of plants or animals would not be the same. Every planet would have its own particular classification of organisms – some might have more than one, if the ocean covered most of the area and the few small continents were isolated from each other.

But these are speculations, introduced here merely to defend a brief account of the ocean which attempts to illustrate general principles rather than geographic particulars.

THE GEOMETRY OF THE OCEAN

Any planet habitable by man must have large areas of open water, and Earth is amply provided. Countless

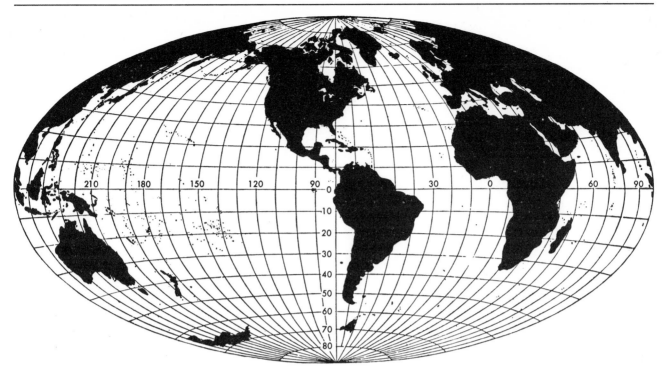

73. Land and sea over an equal area projection of the earth
(from *U.S.C.G.S. Special Publications* 1968, 5th edition)

74. Distribution of water and land areas with latitude
(from McLellan 1965)

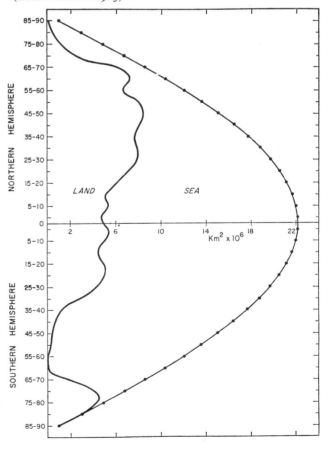

popular articles on oceanography have stressed the large fraction of the earth's surface which is covered by the ocean, and the distribution is shown in figure 73, which is an equal-area projection, unlike those of the many atlases whose maps are designed to display non-polar continents. Figure 74 shows the asymmetric distribution with latitude; near the North Pole and at about 60°S there is little land, but in general there is no extensive latitude belt which can be considered a zonal ocean.

That the ocean is wide seems obvious enough to anyone who has studied a globe or marvelled at pictures of the earth from space. Nevertheless, a long voyage across the ocean, by sea or by air, still impresses one with a concept of distance somehow more immediate than is gained by looking at the much more distant stars. Flying from the land out to sea, by day or by night, one can understand the poet's reference to the Pacific as the eyeball of the world; it does, after all, cover a third of the earth's surface.

Whether one regards the ocean as deep depends on one's point of view. The breadth of the ocean is comparable with the radius of the earth; on this scale the ocean is shallow, with a maximum depth of about 11 km, and not like an eyeball at all. The facts are displayed in figures 75 and 76, from which it can be seen that under much of the ocean the bottom is about 4 km below the sea surface. The depth to width ratio

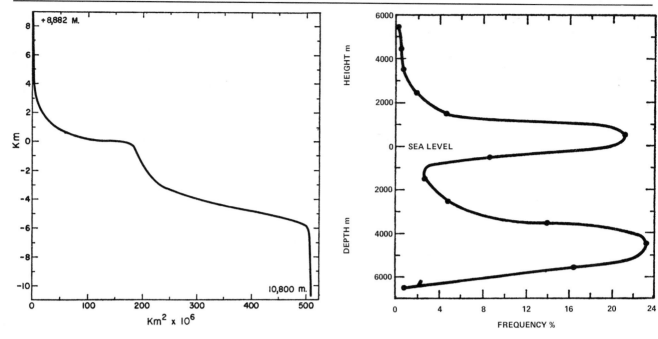

75. Area of earth below given depths
(from McLellan 1965)

76. Frequency distribution of height and depth
(from McLellan 1965)

is therefore of the order 1:1000, about the same as the diameter of the thickness of the varnish on a small globe. Several authors have pointed out that a scale model of the Pacific would have the thickness/breadth characteristics of a sheet of airmail paper; and one, Bascom, has remarked that a model which reproduced the earth on the scale of a basket-ball would be found to be slightly damp.

When oceanographic observations are displayed as cross-sections, the vertical exaggeration is usually so large as possibly to mislead. One must always try to allow for the distortions introduced. One must also realise that although the ocean is shallow, relative to its width, it is nevertheless deep relative to other important lengths. It is deep relative to the height of a man, for example, and much deeper than skin-divers, or even nuclear submarines, can safely go.

Although the total depth nowhere exceeds 11 km, within this limit is relief enough to interest many geologists and geophysicists. In general terms one can distinguish three distinct regions. The nearest to the coastline, extending perhaps 100 to 200 km out into the ocean, is the continental shelf where the water is relatively shallow, under 200 m deep, and where the bottom usually undulates gently (slope about 1:500), sometimes rising to the surface in offshore banks or islands. This is the region of most concern to mariners: here landfalls and departures are made and the risk of

collision or of going aground is much more than usual.

It is a region accessible to observation and so best known, but it is also the most complex and least homogeneous so that general statements are hard to make. Certainly it seems to be of great economic importance; here the first people may live on the sea-bed, here ocean engineering will first develop on a large scale. Already much of the world's fish is trawled from it and much of the world's sewage debouches on to it. The prospects for legal oceanography seem excellent.

Beyond the continental shelf the sea floor falls away to the abyss, the deep sea proper. The transition is abrupt, the overall slope being 1:10. The slope is rugged and frequently corrugated by submarine canyons which are probably even more dramatic than those on land. It is this continental slope which marks the lateral boundary of the deep ocean with which we are mainly concerned. Although the mean depth of the ocean is about 4,000 m there are many sorts of sea-bed, ranging from abyssal plains which are very flat, with slopes less than 1:10,000, to what is perhaps the biggest topographic feature, the long mid-ocean ridges. Many of these are described in Chapter 10 together with their important implications for our ideas about the formation of the ocean basins. For our present purpose we take the bottom of the ocean as a complicated but fixed lower boundary to the deep

TABLE 4
Certain physical properties of distilled water

Property	Compared with other substances	Importance in physical biological environment
Heat capacity	Highest of all solids and liquids except liquid NH_3	Prevents extreme ranges in temperature Heat transfer by water movements is very large Tends to maintain uniform body temperatures
Latent heat of fusion	Highest except NH_3	Tends to maintain temperature near melting point due to release or absorption of energy
Latent heat of evaporation	Highest of all substances	Large latent heat of evaporation extremely important in heat and water transfer of atmosphere Evaporation reduces body temperature
Conduction of heat	Highest of all liquids	Although this plays a small part in nature it is important in organisms
Thermal expansion	Maximum density at 4°	Important in vertical circulation and temperature distribution in lakes
Surface tension	Highest of all liquids	Important in physiology of the cell Controls certain surface phenomena and drop formation and behaviour
Dissolving power	In general, dissolves more substances and in greater quantities than any other liquid	Obvious implications in both physical and biological phenomena
Dielectric constant	Highest of all liquids	Of utmost importance in behaviour of inorganic dissolved substances due to resulting high dissociation
Electrolytic dissociation	Very small	A neutral substance, yet contains both H^+ and OH^- ion
Transparency	Relatively great	Absorbs all visible wave lengths of light almost equally Important in both physical and biological phenomena

ocean itself, assuming that the changes due to continental drift are so slow as to have no measurable effect on the ocean during the lifetime of this book.

THE PROPERTIES OF SEA-WATER

After this brief mention of the geometry of the ocean basins we turn to the fluid with which they are filled. We take it for granted, but water is a remarkable substance. It is one of the few naturally occurring inorganic liquids, and moreover is one of the few substances which occur naturally in all three phases, as a gas, a liquid and a solid. Some of the properties of water which are important to oceanography, and indeed to life itself, are listed in Table 4.

Perhaps most important are its large specific heat and latent heat. These ensure that large amounts of heat are needed to raise the temperature and to melt or vaporize water. The result is that water is an excellent fluid for air conditioning because of its high capacity for storing heat. Man uses it to cool himself by sweating, and the same properties, used in a more complicated manner, control the temperature of the earth. Everyone knows of the continental climates, with the hot summers and bitter winters typical of Siberia, and of the oceanic climates of islands and coastal regions where the ocean's heat storage moderates the extremes to produce a small annual temperature range.

Ice is lighter than water, floats on it and insulates it

against further heat loss. In spite of seasonal change, a nearly constant temperature environment is thus preserved in which living things can more readily survive.

There are many other properties of water – its solvent power, its dielectric constant, its surface tension – which ensure that all the reactions essential to life proceed rapidly; most of them are not much affected by the dissolved salts which distinguish sea-water from fresh.

Sea-water is a complicated solution which probably contains all the elements; present analytical techniques have identified about half of them. Some of the more abundant are listed in Table 6 (page 121). The others are present in concentrations of about one part per million or less.

Samples of sea-water from almost everywhere in the open ocean have been found to contain the major constituents of Table 6 in very nearly constant proportions. Over geological time the various chemical components have been very well mixed, and there is no effective process which can separate them from each other. The result is that just as we can treat the uniform mixture of atmospheric gases as 'air', so can we treat sea-water as a mixture of fresh-water with a so-called 'sea-salt' which consists of a uniform mixture of the elements of Table 6. The total amount of dissolved material in sea-water is called the salinity, measured in

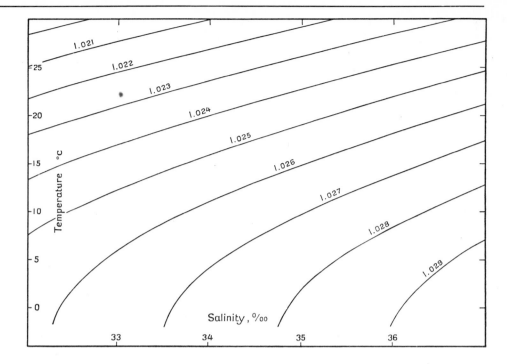

77. Graph to show how surface density is determined (from Sverdrup 1954, copyright by the University of Chicago Press)

grams per kilogram of sea-water and usually written as S‰

Many of the important properties of sea-water depend on the salinity, so it is important to be able to measure it accurately. Because of the near-constancy of composition, any one of the major components can be measured and the salinity deduced. The amount of chloride (or chlorinity) was commonly used and the salinity was usually estimated by titration with silver nitrate. Nowadays, the salinity is usually estimated by measuring the electrical conductivity of the sea-water sample at a known temperature and using a known (measured) relation between salinity and conductivity. The salinity can be estimated, by this method, to +0·003‰; this represents an accuracy of about one part in ten thousand of the salt.

The properties of fresh water depend on its temperature and pressure. Those of sea-water not only depend on the temperature and pressure but on the salinity also. Perhaps the most important property of sea-water is the density: this is because the ocean, like any other fluid, tends to move so that the densest water is at the bottom and the lightest at the top. The relation between the density and the pressure, temperature and salinity is called the equation of state. It is a complicated relation which has not yet been deduced theoretically but has been found by observation. The effect of pressure, or the compressibility, is not so large as may be suspected. Sea-water is nearly incompressible, the density increasing only by a few per cent at pressures corresponding to the deepest depths in the ocean.

This near-incompressibility of the ocean means that in general a solid body will either float at the top or sink to the bottom. It is possible, by careful design, to construct a body which will float at an intermediate depth (see Chapter 2), but the stories one hears of drowned men or sunken ships drifting disconsolate at mid-depth have no basis in fact.

For many applications in oceanography the effect of pressure can be avoided by comparing densities at the same pressure, usually at the surface where the pressure is one atmosphere. The empirical relation between the temperature, the salinity and the density (at one atmosphere) is shown in figure 77.

The shape of the density curves as the temperature and salinity vary can be seen in this diagram. Taking a particular density-line, one can see that it varies in slope as it goes from higher to lower temperature. This means that at low (polar) temperatures it is change of salinity, not temperature, which changes the density. At high (tropical) temperatures the situation is reversed.

We have already mentioned that fresh-water is densest at 4°C. As the salinity increases the temperature of maximum density decreases. So does the freezing point but, as shown in figure 80, the result is that water whose salinity is greater than 24.69‰ goes on getting heavier until it freezes. Common sea-water has a salinity of about 35‰ so must cool to nearly −2°C before it freezes.

Another most important physical property of sea-water is its capacity for absorbing electromagnetic

78. In 1963 Surtsey Island rose out of the sea off Iceland.
Similar undersea volcanic activity probably provides considerable material for solution in the waters of the oceans

radiation (figure 136 Chapter 5). The most penetrating component is blue-green light, with a wavelength of about 0.5 μ. Even in pure water this is strongly absorbed; its energy is reduced by half after about 35 m of travel. In ocean water the penetration is much less, the energy being reduced by about half after 5 m of clear water and after only 0.5 m in the more turbid coastal water. The reduction in penetration is due to suspended particles as well as to the dissolved material.

In clear clean water, as we have said, the blue-green light penetrates furthest. Shorter wavelengths (towards ultraviolet) and longer (red and infra-red) are more strongly absorbed. This is observed by aqualung divers and illustrated by underwater colour photographs taken without artificial light. Red or yellow things look blacker as one goes down, while blues and greens retain their colour. Both deepwater fishes and humans (whose ancestors were marine) have eyes most sensitive at low intensities to wavelengths near 0.5 μ, which penetrate deepest into clear ocean water.

Since most of the energy from the sun reaches the earth's surface as visible radiation (99 per cent in wavelengths less than 4 μ) it does not penetrate very far. Even in the clearest water less than one per cent of the incoming sunshine reaches 100 m down; in turbid coastal water it would be almost completely absorbed in the top 10 m. From this point of view the ocean is certainly deep, since the sun affects only a thin layer at the top. We shall see later that this has important effects on ocean temperature and currents, but here we shall draw attention to the obvious, important, yet often forgotten fact that the ocean is dark. Apart from the energetically trivial flashes produced by bioluminescent organisms there is no ambient light. This has been known for many years, since oceanographers long ago found that photographic plates could be exposed for many hours at mid-depths without becoming fogged.

The situation is in contrast with that of the atmosphere in which we live, and in which our whole lives are linked with the assurance that day follows night. We rely so much on vision as the most useful of our senses that it is hard to conceive of life in continual darkness. This is probably one of the main reasons why we still know so much less about the ocean than we do about the atmosphere, why the dark and secret abyss is the last great pit of ignorance in our natural environment. Meteorologists get clues about the winds and weather from seeing clouds, birds, smoke and so on; astronomers would hardly exist if the atmosphere were opaque; but oceanographers are restricted to indirect observation of a fluid through which they cannot see.

The opacity of the ocean to electromagnetic radiation is not confined to visible light alone. Shorter wavelengths (x-rays and ultra-violet) and longer (infra-red,

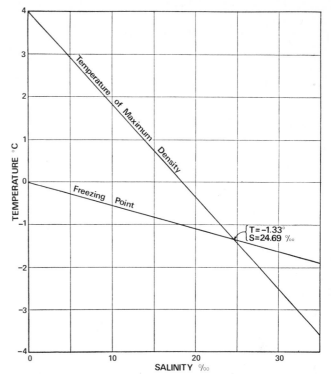

80. Dependence of freezing temperature and temperature of maximum density on salinity (from McLellan 1965)

radar, microwaves) are all even more strongly absorbed. Very long radio waves can penetrate to some extent (to about half a wavelength) but the most effective way of signalling is to use sound-waves.

For sound-waves the situation is different; they can be transmitted with much less loss in the ocean than in the atmosphere, so man and marine creatures use sound to get information. Most people are familiar with echo sounding, in which the depth is calculated from the time a pulse of sound takes to go to the bottom and be reflected back again. A similar principle is used in asdic and sonar, but here the beam is not necessarily vertical and the range is measured not to the bottom but to a submarine, a shoal of fish or some other sound-reflecting object. In early tests, men actually used their ears to detect the returning sound pulse, but this is not an efficient process since sound loses a lot of energy on going through an air-water interface. Now microphones or hydrophones are used to convert the sound energy into an electrical signal which can be amplified and displayed on a chart. Conditions in the sea are not uniform enough to allow clear pictures of the reflecting targets to be produced. This is partly due to the fact that the sound sources are about the same size as the wavelength (10 kHz sound has a wavelength of about

79. *opposite* An aerial view of the 'chalky' water of a Norwegian fjord discoloured by a 'bloom' of the coccolithophore *Coccolithus huxleyi*. The normal water colour is seen in the fjord in the foreground

81. Photographs of clouds at various levels above the Indian Ocean taken from Gemini XI at a height of 280 miles and looking north. At the top are Iran, Pakistan and India, to the left Muscat and Oman and the Arabian Sea

15 cm), but is also due to the variation in the velocity of sound with temperature, salinity and pressure. The direction of travel of the beam of sound waves is altered by the small inhomogeneities in the ocean, and the beam is confused so that a clear image cannot be formed. Similar fluctuations, produced by the effect of atmospheric inhomogeneities on light beams, produce the twinkling of stars.

Another effect is that the variation of sound velocity produced mainly by the temperature structure can refract the sound path into a curve rather than a straight line. The atmospheric equivalent here is a mirage; it is obvious that the distortion due to the near surface distribution of the velocity of sound (itself strongly linked to the temperature) is of great importance in location and range-finding in submarine warfare. Much effort is expended by military oceanographers in attempting to forecast the thermal structure of the upper layers of the ocean.

There are many books devoted to the physical properties of liquids and gases, some to air and some to water. Here we have touched only briefly on those properties of sea-water which most affect the structure and behaviour of the ocean. The basic physical properties of flesh and bone have a big effect on the anatomy of animals; their size, their shape, how they move and so on. Similarly, we shall see that the properties of sea-water determine, to a large extent, the anatomy of the ocean.

THE ANATOMY OF THE OCEAN

The superficial appearance of the ocean is well known. From the air (or from space) one distinguishes the land, the ice, and the huge area of ocean covered with the moving undulations which we call waves. An observer, carefully studying the coastlines, might distinguish the daily and half-daily sloshing about of whole ocean basins that constitute the tides. These are phenomena of motion, to which we shall return.

Photographed in black and white from space the ocean looks black but clouds look white (figure 81). This is because the clouds reflect a lot of the visible radiation from the sun whereas the sea absorbs most of the beam and reflects only a small fraction. In colour, and from lower heights, the sea has a narrow range of colour but including deep blue, greens and greenish yellow (figure 241). Colours are difficult to measure and express numerically so there are no climatic charts to show it, but in general the deep or indigo blue colour which one associates with the deep blue sea is usually found in tropical seas and in regions like the Caribbean or the Mediterranean where there is little biological production (blue is the desert colour of the sea). At higher latitudes the colour is typically the green-blue more typical of aquamarine. Polar and coastal waters are typically greenish.

The explanation for the various colours is not straightforward. The main reason for the blue colour is the selective scattering of sunlight by the molecules: the sea looks blue for the same reason that the sky looks blue. When there are particles in the water, especially phytoplankton, there is less transmission, more scattering, and probably some absorption by the soluble organic substances the plankton produce. Apart from the colour of the sea itself, reflections are important. The colours tend to be changing all the time, due to sun-glitter on breaking waves and ripples, the play of cloud shadows and reflections, and the subtle differences produced by the rippled and slick regions of the surface. Where there is a thin film of oil on the surface, due to a tanker cleaning her tanks or possibly to some planktonic source, the small ripples on the surface are damped out. The smooth areas look lighter and reflect the sky almost perfectly. Where there is no surface film the surface looks darker, partly because the front of each ripple reflects a higher (and therefore a darker

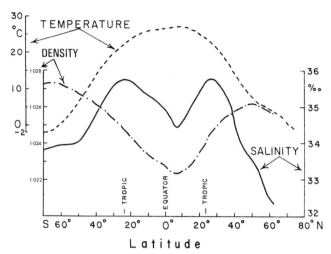

82. Average variation of salinity, temperature and density at the sea surface (from Pickard 1964)

blue) piece of the sky, partly because the reflection is less complete. Since the sea surface is usually rippled there is almost always a colour contrast between the sea and the sky at the horizon.

Beautiful though they are, the colours of the sea are hard to measure and little scientific use has been made of them, though some interesting studies have used discolorations, due to outflow from rivers, plankton blooms, and oil slicks, as indicators of water movements. Such observations are of course, confined to surface layers: for more scientific studies, instruments are needed, and these enable measurements at various wavelengths to be made at all depths.

Many of the present techniques of observation are described in Chapter 2. In this chapter only the results, not the methods, are described. Most of our information comes from the near surface layers, which can be observed using relatively simple instruments. An overall view is provided by figure 82, which shows the variation with latitude of the surface temperature, salinity and density, averaged for all oceans.

The open ocean surface temperature decreases from values of about 30°C. near the equator to about −2° near the ice of high latitudes. The temperature largely determines the density, which is lowest near the Equator and increases North and South. The salinity distribution is more interesting: it has maximum values at latitudes of about 25° N. and S. with an equatorial minimum. The reason is indicated in figure 83, which shows the similarity between the salinity distribution and the difference between the evaporation and the precipitation. Accurate rainfall measurements over the sea are nearly impossible, but there is a clear indication that the low equatorial salinity is associated with the high tropical rainfall, and

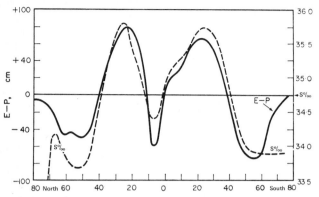

83. Surface salinity (S) compared with the difference between evaporation and precipitation (E − P) (from Groen 1967, after Wüst 1957)

the twin maxima are associated with the low rainfall of the subtropical anticyclones.

The distribution of properties near the surface of the open ocean is approximately zonal, with the various contours running mainly east-west. There are anomalies near coasts, associated with currents and with a phenomenon known as upwelling, which occurs on the eastern boundaries of the ocean and in which water from a depth of up to 500 m or so comes to the surface.

It is impossible, in a short article, to give charts showing the distribution of all the more interesting properties. These include the mean surface temperature (in winter) and its annual variation, the mean surface salinity, the major surface currents and surface winds, the limits of polar ice, the mean variation of air temperature and the mean annual precipitation. These are the elements considered by Elliott in an attempt to delimit typical oceanic regions. His useful classification is shown in figure 84 and some typical values of the elements in Table 5.

84. Climatic regions of the ocean (after Elliott 1960)

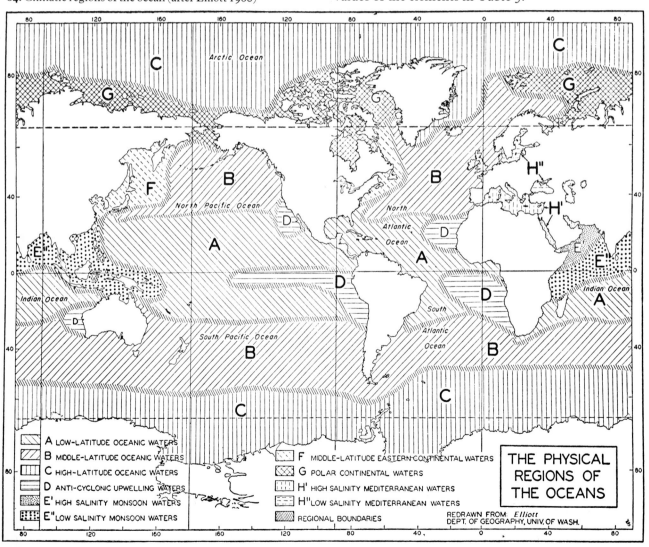

OCEANIC REGIONS

The regions shown as A, B, C on figure 84 are truly oceanic: they have the surface characteristics which one would expect if the earth had no land-masses but were completely covered by ocean.

Type A occupies the large area between the Tropics of Cancer and Capricorn. Most of the region is influenced by the Tradewinds, so the surface currents generally have a component toward the west. Surface temperatures are high and steady, and the high evaporation and low rainfall produce high surface salinities. Only in the Doldrums are some of these characteristics altered – there one finds the Equatorial currents and counter-current and the high rainfall which decreases the surface salinity.

Water with the surface characteristics of *Type B* is found in middle latitudes, between the Tropics and the polar pack ice. It extends furthest toward the north in the North Atlantic because of the broad connection with the North Polar Basin. These are the regions of prevailing Westerlies, like the Roaring Forties, with travelling depressions which produce much rain and snow. In these regions the surface temperatures and salinities decrease poleward but there is considerable variability, both regionally, from place to place, and seasonally, from time to time. The major surface currents are from the west, but with strong boundary currents like the Gulf Stream system of the eastern coasts of the continents.

The region designated as *Type C* is regarded as oceanic even though it is ice covered for most of the year; there is no continental influence except perhaps near Antarctica. Air temperatures are usually sub-zero but with large seasonal variations, while sea-surface temperatures are usually near the freezing point of sea-water and do not vary much. There is little precipitation but it exceeds the evaporation so that the salinities are low. Both winds and currents circulate clockwise round the poles.

The surface characteristics of the other regions are

TABLE 5
Characteristics of the world's oceans

THE OCEANIC REGIONS

Type	A Low-lat. oceanic waters	B Mid-lat. oceanic waters	C High-lat. oceanic waters	D Anticyclonic upwelling waters
Winter temperature	20°C to 25°C	5°C to 20°C	About −2°C	15°C to 20°C
Annual variation of temperature	Less than 5°C	About 10°C	Less than 5°C	About 8°C
Average salinity	35‰ to 37‰	About 35‰	28‰ to 32‰	About 35‰
Annual variation of air temperature	Less than 5°C	About 10°C	Up to 40°C	About 5°C
Precipitation–evaporation balance	E exceeds P	P exceeds E	P exceeds E	E exceeds P

THE CONTINENTAL REGIONS

Type	E Monsoon waters	F Mid-lat. continental waters	G Polar continental waters	H Mediterranean waters H	H¹	H¹¹
Winter temperature	20°C to 25°C	About −1°C to 5°C	About −1°C	0°C–25°C	15°C–25°C	0°C–5°C
Annual variation of temperature	Less than 5°C	15°C to 25°C	Up to 7°C	8°C–20°C	About 10°C	15°C–20°C
Average salinity	30‰ to 37‰	30‰ to 33‰	From brackish to 28‰	Brackish to 41‰	37‰ to 41‰	Brackish to 33‰
Annual variation of air temperature	Up to 10°C	15°C to 35°C	About 10°C	10°C to 20°C		
Precipitation–evaporation–run off balance		P exceeds E	P and R exceed E		E exceeds P	P and R exceed E

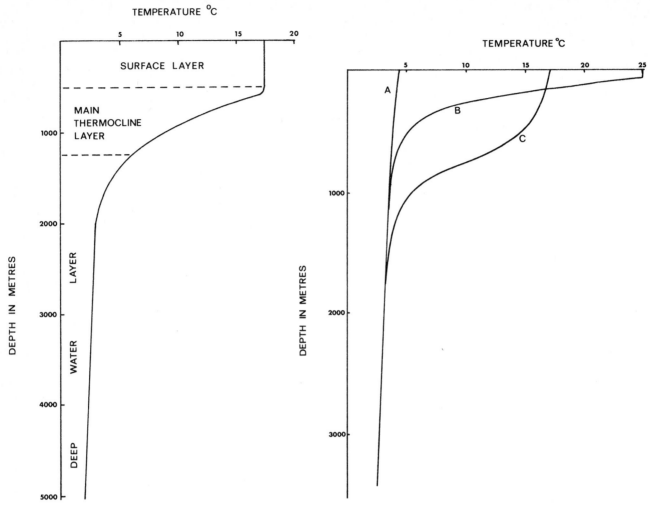

85. Schematic temperature variation with depth in the deep ocean

86. Typical temperature profiles in winter: A, high latitudes; B, near the equator; C, middle latitudes

produced by the influence of continents on the oceanic types. *Type D*, for example, is produced from the low-latitude oceanic waters of Type A by the water which rises from intermediate depths off the western coasts of continents. This up-welling water is relatively cold and only moderately salty, producing surface temperatures and salinities lower than those of corresponding latitudes in the open ocean. There is more cloud, fog and rain: the rainfall is less than the evaporation but the subsurface origin of the water keeps the surface salinity low.

Another modification of the low-latitude oceanic water (Type A) is produced by the strong seasonal variations of the monsoons. These affect particularly the regions shown as *Type E* in the Indian Ocean and the West Pacific. Here the surface characteristics show conspicuous and complicated seasonal variations. The

details vary with the water balance: in some regions dry winds produce strong evaporation and high salinity, but in others the monsoon rains and high river flow reduce the salinity. The currents also vary seasonally, usually going to the north-west with the summer monsoon and south-east with the winter monsoon.

In the middle latitudes of the Northern Hemisphere the near-surface characteristics of Type B are much modified by strong outbreaks of cold dry continental air in winter. These winds off the land produce the low winter temperatures and big temperature variations found in *Type F*. Here salinities are kept low by high rainfall and river runoff, and there are strong currents which transport cold water toward the south.

The main continental influence in the high latitudes of the Northern Hemisphere is due to fast ice, or sea ice frozen to the coast in winter. Thus *Type G* differs from Type C in that it is usually open in summer and ice-covered in winter. The main effect is to produce water of very low salinity, especially off the mouths of large rivers.

Continental influence is most obvious in the Mediterranean waters *(Type H)*. These are confined to basins which are connected to the open ocean through straits with relatively shallow sills. Their exchange of water with the world ocean is slow – the bottom water of the Mediterranean is probably renewed only every 75 years, and that of the Black Sea only every 2,500 years.

The common features of the Mediterranean type waters arise from their being almost land-locked, which produces a wide range and variation of temperature and salinity and a large annual variation of precipitation. Surface currents usually circulate anti-clockwise. There is a wide variation of characteristics but Mediterranean waters divide mainly into the high salinity type (Mediterranean, Red Sea, Persian Gulf) produced by an excess of evaporation over rainfall, and the low salinity type (Black Sea, Baltic) produced by an evaporation deficit. The high salinity type is of special interest to oceanographers since the Mediterranean for example can be regarded as a small scale model of the deep ocean, reproducing some of the physical processes important in a localised form which is easier to observe and study.

OCEAN STRUCTURE

Observations at depths below the surface are less numerous but we have a good knowledge of the mean distribution of temperature, salinity, oxygen, and somewhat less complete information on other constituents.

By far the best known is the temperature structure of the ocean. The range of temperature is from $-2°$ to 30°C. (just about the range of temperature in which man can live) but there is much more cold water than warm: the mean temperature is 3.5°C. All the water warmer than about 5°C is confined to a relatively thin layer between about 50°N and 50°S.

Between these latitudes the typical structure is of a layer of nearly isothermal water near the surface, a layer of relatively rapid temperature variation at a greater depth, and a thick cold layer extending to the bottom. This is illustrated in figure 85: the region where the temperature changes most rapidly with depth is known as the *main thermocline* of the ocean.

The way in which the typical temperature distribution changes with latitude is shown in figure 86. At latitudes greater than about 50° the temperature varies little with depth. As one goes toward the Equator the surface temperature increases and the depth of the main thermocline decreases, to become a very sharp, shallow change of temperature in the tropical and equatorial oceans.

Figure 87 shows a typical north-south section of the temperature distribution; all the water warmer than 5°C. is confined to the upper layer. The rest of the ocean is cold, even under the tropics and the Equator.

This structure to some extent is explicable in terms of the properties of sea-water we have already mentioned. The sunlight does not penetrate far and all the processes which tend to change the temperature operate within metres of the surface. After a body of water leaves the surface its temperature can be changed only by mixing, and this is not very effective. Also, as we have seen, the colder the water, in general, the heavier. It is therefore to be expected that the heaviest (coldest) will generally sink to fill the deep ocean basins. The coldest water is found at the surface in polar regions

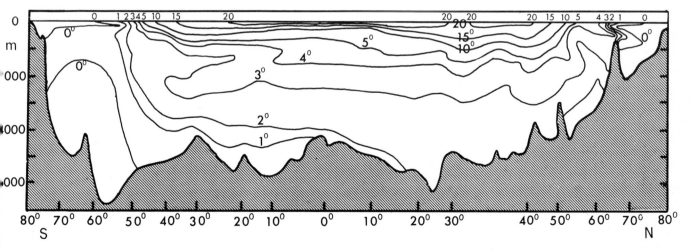

87. North-South temperature (°C) cross-section of the Atlantic Ocean, along the track shown in Figure 63 (from Wüst 1928)

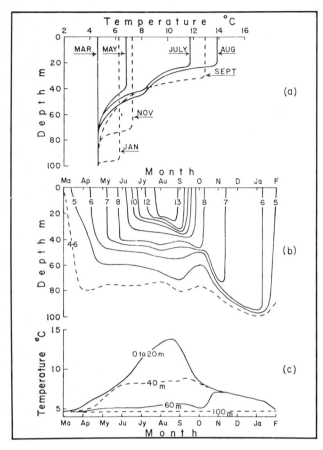

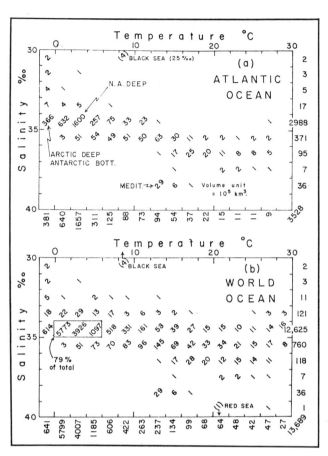

88. Growth and decay of the seasonal thermocline in the eastern North Pacific (50°N 145°W) illustrated by (a) temperature profiles (b) isotherm changes and (c) temperature changes at specific depths (from Pickard 1964)

89. Diagram to show the volume of water of given temperature and salinity in the Atlantic and the world ocean. Volume units are 10^5km^3 (after Montgomery 1958)

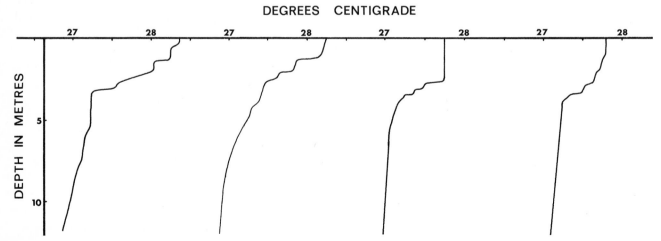

90. Four profiles showing the diurnal thermocline on one afternoon, showing differences over a short time due to varying wind and sun

in winter, after its heat has been radiated away into the long Polar night.

The observations on which figure 86 is based were all made in winter. This season was chosen because in summer the temperature distribution is complicated by the appearance of the *seasonal thermocline*. A typical summer temperature profile is shown in figure 88. The surface temperature (13°C.) is higher than in winter; again there is an almost isothermal layer near the surface, but in contrast to the winter situation this goes only to 25 m before the temperature begins to decrease rapidly as the depth increases, reaching 5°C. at a depth of 70 m. At greater depths there is no seasonal variation and the distribution is the same all the year round.

Some examples of the development and decay of seasonal thermoclines are also given in figure 88. It is the normal feature of the extra-tropical upper ocean in summer. Nearer the Equator the seasons are less pronounced, the main thermocline is shallow (see figure 86) and the seasonal thermocline cannot be clearly distinguished.

The development of the seasonal thermocline is the ocean's response to the annual cycle of solar radiation. There is a corresponding response, in suitable conditions, to the daily cycle. On calm days the sun heats the uppermost few metres of the ocean and often produces a thermal structure similar to that of figure 90 for which the term *diurnal thermocline* seems quite appropriate. In stronger winds, however, the heat is mixed down to greater depths and no temperature variation is measurable. The situation is thus controlled by local weather changes and is variable and sporadic.

The temperature structure of the ocean varies from place to place and, near the surface, from time to time. Nevertheless, it can usually be described in terms of the basic distributions we have outlined. This is possible only because the density of the water is mainly determined by its temperature and because the ocean is basically stable. That is to say, it arranges itself stably in the sense that the heaviest water is at the bottom. Clearly an arrangement in which lighter layers were below heavier layers would be unstable in the sense that a small disturbance would produce a re-arrangement of the layers, the lighter water rising and the heavier sinking to take its place.

In a stable situation the layers can only be re-arranged, or mixed, by some sort of stirring or similar process which requires energy. The more stable the layering the more work must be done against gravity to make the fluid uniform. The various thermoclines are strongly stable since the temperature decreases (and therefore the density increases) strongly with depth. Vertical motions which can produce mixing are largely

absent and the thermocline acts as a barrier: it is difficult for anything below the thermocline to be transferred to the water above and for anything above it to be transferred below. The world's fisheries (as explained in Chapter 9) occur where there are processes effective in destabilising the thermoclines, so that the nutrients which are normally below can be brought into the sunlit layers above.

It is not possible to describe the salinity structure of the ocean by giving typical profiles as for temperature. This is because the salinity affects the density to a lesser extent than does the temperature, so there is no necessity for the salinity to increase as the depth increases. Of course, the processes which affect the salinity (rainfall to dilute the water and evaporation to concentrate it) occur at the surface, just as do those which affect the temperature. Once the water leaves the surface its temperature and salinity can only be altered by mixing with other water masses. Most of the mixing processes treat heat and salt in the same way, and it is found that the water mass tends to retain a recognizable temperature/salinity (T/S) characteristic.

This can be illustrated using the temperature/salinity diagram (figure 93), which is a simple graph with temperature on one axis and salinity on the other. Any sample of sea-water can be represented by a point on the diagram whose co-ordinates correspond to the measured temperature and salinity.

The temperature/salinity diagram of figure 89 shows how much water there is in the ocean with particular values of temperature and salinity. It will be seen that these properties are by no means uniformly distributed. In the Atlantic, for example, 47 per cent of the water has a temperature between 2° and 4°C. and 76 per cent between −2° and 4°C. About 85 per cent of the water has a salinity between 34 and 35‰.

It will be seen that a particular water mass can sometimes be identified by its Temperature/Salinity characteristic; the Mediterranean, the Red Sea and the Black Sea waters are obvious examples on figure 89. The reason, as we have seen, is that the Temperature/Salinity characteristic is acquired at the surface and depends on the local climate at the place of formation. In subsequent sinking the properties are modified by mixing but their distribution is constant over large areas.

Some idea of mixing processes can also be obtained by using a characteristic diagram such as the T/S diagram. For a simple calculation will show that if two water masses, each represented by a point on a T/S diagram, are mixed, then the point representing the resultant mixture will lie on the line joining them, and at a distance determined by the proportions in which they are mixed.

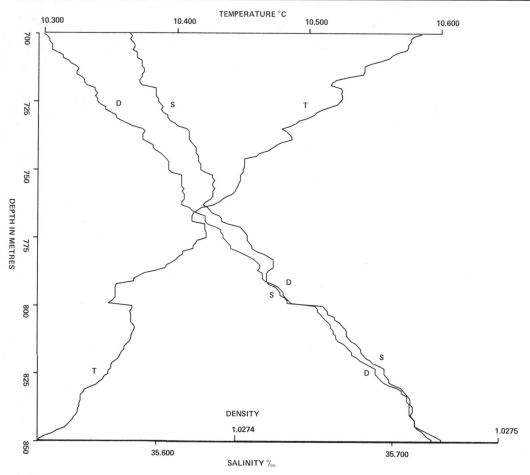

91. An example of fine structure of temperature (T), salinity (S) and density (D) in the deep ocean

The processes which mix water masses in the deep ocean present a central problem in modern physical oceanography. The basic stability tends to damp out the fluctuations which usually mix homogeneous fluids, so that wind mixing is not effective at depths well below the surface.

Recent observations of the fine structure of temperature and salinity, using the sensitive TSD probes described in Chapter 2, have shown that temperature and salinity do not always vary smoothly with depth. Sometimes they change almost discontinuously, in a series of small steps. Figure 91 is a fairly typical result.

It is not yet known how such steps are produced, how far they extend horizontally, or how long they persist before they decay. They may well play an important part in the mixing process and certainly illustrate the sheer complexity of the ocean. As in living creatures increasing magnification seems always to show structure on ever smaller scales, which never-theless has a strong influence on the overall behaviour (figure 92).

Temperature and salinity are the most important tracers for indicating the source regions of water masses. This is because they are what is called conser-vative: there is no process which puts in or takes out salt or heat except at the surface, so in the deeper layers the temperature and the salinity are conserved. The various source regions produce water masses with particular temperature-salinity characteristics which can be traced over thousands of miles of travel, being gradually modified by slow mixing with other water masses. Most of the world's water masses are indicated schematically on figure 93 and give perhaps the most compact representation of the anatomy of the ocean.

There are also, however, other tracers which are interesting, even though they are not conservative, because they can be used to measure time. Examples are oxygen, which is used as the life support of marine creatures and to decompose detritus, and ^{14}C, a radioactive isotope of carbon whose radioactivity

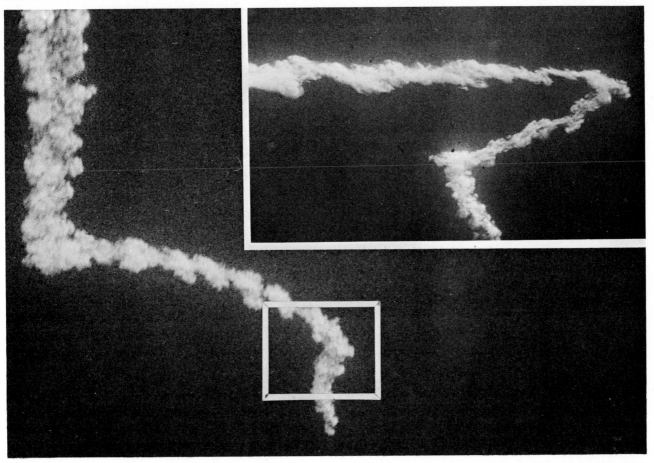

92. An example of small-scale structure in the ocean is illustrated here by a strong current stream concentrated at a density step about 70 cm thick made visible by a plume of dye in the water. The transition from weak shear in the layers above and below the sheet occurs within less than 1 cm. The inset shows the jet at the lower edge of the sheet about 30 seconds later when the dye has been pulled out farther

decays at a slow known rate (by half every 5,570 years). The water at the sea surface is usually saturated (or even super-saturated) with all the gases in the atmosphere. When the water leaves the surface it takes its oxygen with it, and the gradual reduction of the oxygen by marine life and detritus leads to a diminution of the oxygen content, which gives some indication of the time since the water was at the surface. In some regions, where the water is stagnant, all the oxygen is used up and hydrogen sulphide is present instead. The Black Sea is a classic example—it is said to have got its name as a result of the sulphides which blackened metal objects lowered into it.

Unfortunately, the rate at which oxygen is used cannot be estimated precisely (how many fish are there in the sea?), so the information is qualitative. The general impression from the temperature/salinity structure is that one has a good idea where the water originated but little of the route by which it travelled and the time it took.

This is particularly obvious from ^{14}C studies: the measurements are technically difficult but indicate that periods of a few hundred years have elapsed since the water of the deep ocean was last at the surface. This interpretation is challenged by some oceanographers who feel that these estimates are too high. We shall see that the direct measurements of ocean currents are not, as yet, able to provide a definitive answer to the question.

MOTION IN THE OCEAN

Perhaps the most obvious characteristic of the ocean, to a casual observer, is that it is moving about. It is convenient to distinguish between wave motions, which are basically oscillatory, and ocean currents, which are less so. This is a matter of convenience only, since the categories overlap and the processes which produce and maintain the various motions interact.

We deal first with the waves, which are most obvious to a casual observer. They are perhaps the best-known

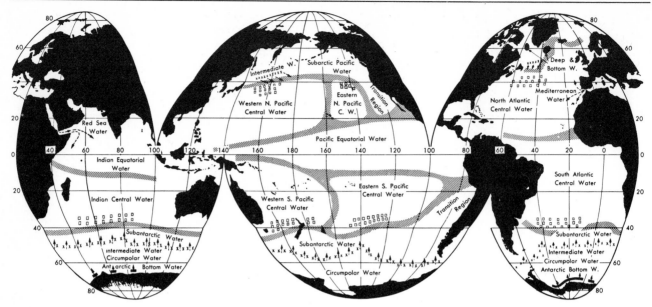

93. Water masses of the ocean and their temperature-salinity
characteristics (from Sverdrup, Johnson & Fleming 1942)

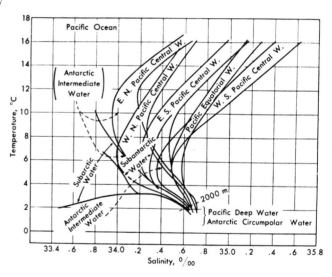

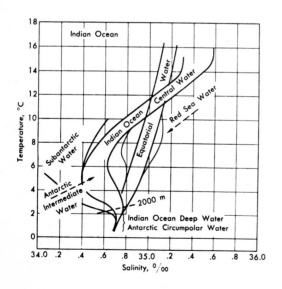

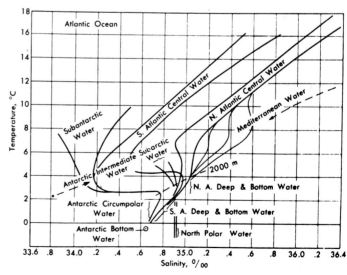

oceanic phenomenon and, until relatively recently, the least understood. When the wind blows over water, waves are formed which have a wide variety of size, shape and speed. It is difficult to relate them to the regular, smooth waves of the mathematician or laboratory worker, which have a well defined period, length and speed (figure 98).

Nevertheless, it is possible to estimate values of these parameters, even without instruments, and typical measured values agree with theoretical expectations. The longer the *interval* between succeeding wave-crests (the period), the greater is the *distance* between succeeding wave-crests (the wavelength) and the greater is the wave speed (the speed is proportional to the period, and the length to the square of the period).

The size of the waves is controlled mainly by the strength of the wind, the time it has been blowing, and the fetch, or length of water over which the wind is blowing. In the open sea, where the fetch is long, the strong winds of a depression generate a complicated sea which can usefully be regarded as a mixture of waves of all periods and speeds. These travel in a variety of directions (but mainly in the same direction as the wind) until they leave the region of strong winds (the generating area) and thereafter travel as swell, which is the term used for a wave which is no longer under the influence of the wind.

Useful results can be obtained by treating the mixture of waves as the sum of a large number of simple waves, each travelling independently and not reacting in any way with any of the others, as illustrated in figure 94. There are mathematical and physical techniques for sorting out a complicated wave structure into its component parts and both involve the use of a filter. A filter, in general, lets through one fraction of a mixture and retains the rest. In electronic use, a filter lets through a signal of a chosen period and suppresses the others.

Suppose one has an instrument which provides an electrical output proportional to the height of the waves at a fixed point and one makes a recording of it. If the electrical signal is played back through a filter which lets through only oscillations with a period between, say, 29 and 30 seconds, the mean energy of the output is a fair estimate of the energy of the 29 to 30 second waves. The process could be repeated with 28 to 29 second waves, 27 to 28 and so on down to 0 to 1 second waves. A plot of the energy in each wave-band relative to the period is called a wave spectrum. Such a graph is now the main tool in wave analysis.

Idealised spectra for waves in equilibrium with different winds are shown in figure 95. As the waves travel out from the generating area the long period, long wavelength components travel fastest and so

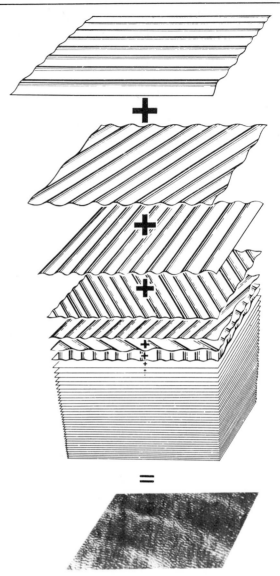

94. Diagram to illustrate how simple waves can add up to a confused sea (from Pierson, Neumann & James 1955)

arrive first at a distant recording station: the speed of waves with different periods is known, so by monitoring the arrival of different components the distance to the generating area can be estimated. It is found that waves can travel over long paths – they have been detected over paths of several thousand miles.

In spite of the developments of these methods for describing waves, we do not yet have an agreed theory of how they are generated, though there has been much progress in recent years. It seems that the small waves, or ripples, play an important part in that they determine the friction between the air and the sea. This is the most important single link between the atmosphere and the ocean: it affects how the wind drives the ocean

95. Wave spectra to show how energy is distributed with wave period in A, slight winds, short fetches; B, intermediate conditions; C, strong winds and long fetches. Note how the energy scale is gradually stretched because of the large variation in energy with windspeed and fetch

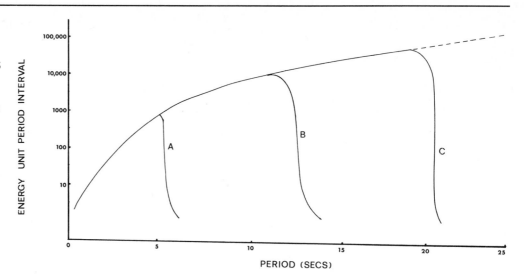

currents and determines how water is evaporated and heat transferred from the sea surface. The part played by the larger waves in other air-sea interaction problems is obscure – sometimes it seems as if they exist just to make the oceanographers' task harder.

The surface waves with which we have so far been concerned are short waves in the sense that their length is not more than the depth of water. More picturesquely, they do not feel the bottom.

Long waves (wavelength considerably greater than the water depth) behave differently. Their speed is no longer dependent on their wavelength or period but only on the depth of water. (They travel at a velocity given by the square root of the product of water depth and the acceleration of gravity).

Good examples are provided by tsunami, which are long waves generated by earthquakes and which travel at great speeds (typically 200 m/sec.) across the ocean (figure 97). One dreadful example was the tsunami of 1 April 1946 which was generated by an earthquake near the Aleutians. It reached Hawaii (3,700 km away) after travelling for 4 hours 40 minutes, doing great damage and killing over 150 people. Its wavelength was about 200 km and its height in mid-ocean was estimated at little more than 50 cm, but this increased to over 10 m when the wave ran into shallow water. There is now a tsunami warning service which gives as much notice as possible to islands which are likely to be affected.

Even longer ocean waves are, of course, the tides. They are driven by the differential attraction of the sun and moon, but the theory is much more difficult than most elementary books indicate.

We can get some insight into the mechanism by considering, for simplicity, only the tides generated by the moon. Those generated by the sun are quite similar, the resulting tides being the sum of the lunar and solar tides which behave almost independently of each other. The distance from the centre of the earth to the centre of the moon (or of any other earth satellite) is determined by a balance between two forces: the gravitational attraction tends to bring them together while the centrifugal force due to the rotation tends to make them separate. Only when these forces are in balance can a satellite remain in orbit, since the attraction decreases and the centrifugal force increases as the earth-moon distance increases.

The tidal forces are very small. They arise because of the relatively slight differences in the distance between the moon and various points on the earth. Water in the ocean nearest to the moon is not quite so far from it as the centre of the earth; it is slightly more strongly attracted to it by gravity and slightly less strongly repelled by the rotational (centrifugal) forces. The result is a slight bulge of the ocean toward the moon. Conversely, water in the ocean furthest from the moon is slightly more distant, has less gravitational

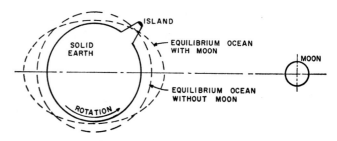

96. Diagram to show the equilibrium tide due to the moon (from McLellan 1965)

97. Wreckage of a house at Keaukaha, Hawaii, moved 100 feet inland by the tsunami of 1 April, 1946

98. Waves on the North Sea

99. The relative energy per unit frequency of various surface waves (after Munk 1950)

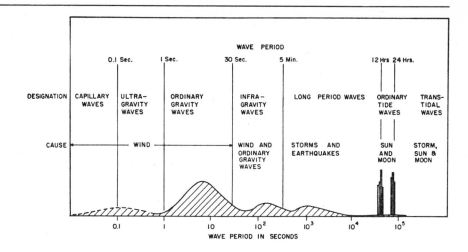

attraction and more centrifugal force: it will therefore have a slight bulge away from the moon.

This is the basis for the 'equilibrium theory' of tides as illustrated in figure 96. It gives some insight into the generating forces and a qualitative explanation of some tidal features, but it does not predict either the amplitude or the phase of the observed tides. There are several reasons: the ocean has continents as boundaries; the tides are long waves and restricted to a speed depending on the water depth; the rotation of the earth has an important effect; the various ocean basins have natural frequencies of oscillation which affect their response to the tide generating forces.

A proper theory of the tides is possible but requires long computation on modern large electronic computers. Practical tidal predictions are still done using empirical methods, as they have been with considerable accuracy for many years. Such predictions can only be made for places where long series of observations are available.

Between the more common short ocean waves and the long tidal waves there are many others of interest. The spectral distribution of some of them is illustrated in figure 99 and ranges from waves of period less than 0·1 sec. to those with periods measured in days. With slow variations, whose 'periods' are measured in years and decades, one is concerned about the slow changes in mean sea level which are affected by both climatic and geological changes.

All the ocean waves we have so far mentioned are surface waves in which the air-sea interface moves up and down. There is another class of waves – internal waves – which hardly affect the surface but have their largest motions at some depth below it. They are made obvious by repeated measurements of the temperature structure, as illustrated in figure 103. The layers near the main or seasonal thermocline are found to be oscillating in a more or less random but wavelike manner.

Less is known about internal waves than about surface waves; they tend to be of larger amplitude and longer period than the typical surface waves but they travel more slowly. They are being actively studied both for their practical importance in underwater acoustics and because they may bring about oceanic mixing by processes which bear some resemblance to the breaking of surface waves.

OCEAN CURRENTS

The ocean currents near the surface affect ships and have been known to mariners from time immemorial. Even now, most of our information on them is based on the collection of ships' drift data reported by seafarers. A ship steering a constant course at a steady speed does not travel at that course and speed relative to the ground but is displaced from it, or set, by the ocean currents. Given reasonable navigation a simple calculation will provide an estimate of the effective speed and direction of the near surface current. For many years, national authorities have collected such estimates and produced average values of them in the form of charts. We have by this means a considerable knowledge of the surface circulation of the ocean, though most of the observations are concentrated on shipping routes and there are few observations from many of the more remote parts of the world.

The general features are shown in figure 104; although the Pacific, Atlantic and Indian Oceans are quite different in shape they have a rather similar current pattern, or general circulation. There is a huge clockwise circulation (or gyre) in their northern parts and a counter-clockwise one in the south. This is conspicuously asymmetric, the currents being much stronger in a narrow region near the western boundary of the North Pacific and the North Atlantic (the situation in the Indian Ocean is complicated by the

100. One of a series of plastic water-sampling bottles is here taken off the wire after being brought back up to the surface. The sample is enclosed by the two spring-loaded end caps, and the metal holders for the two thermometers can be seen on the right of the bottle

101. *top* The prevailing wind has caused a 'bloom' of the alga *Trichodesmium* to accumulate in lines or windrows. The Red Sea is so-named because very heavy blooms sometimes turn the surface a red-brown colour

102. *below* The absorption of wavelengths of light by principal photosynthetic organisms. The absorption bands are from pigments which absorb the light for photosynthesis. These pigments are the green chlorophylls of which there are three types, yellow carotenoids of many types and other pigments which absorb green light. The curve for brown algae includes diatoms and dinoflagellates (from Yentsch 1966)

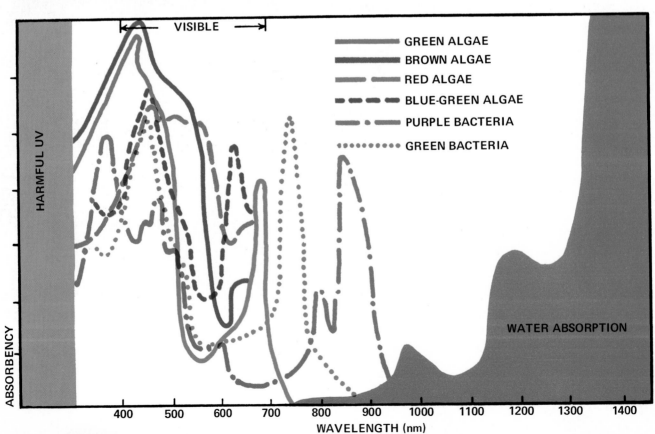

103. Internal waves from repeated temperature sounding (from Lafond 1962)

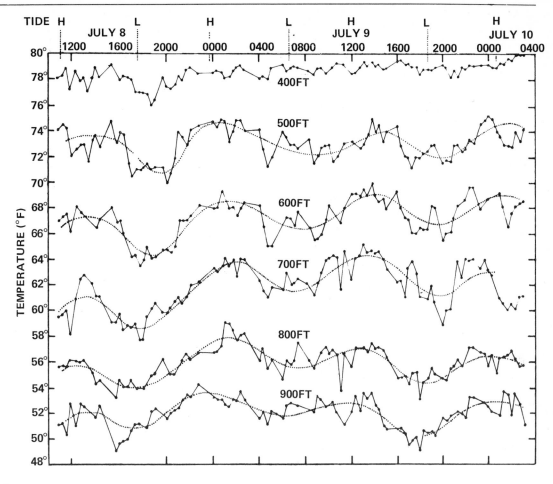

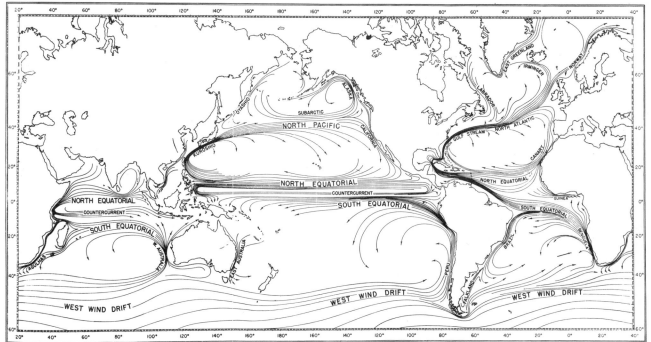

104. Climatological average of the ocean currents of the world (from *NRDC Summary Technical Report* Division 6 Volume 6A 1946)

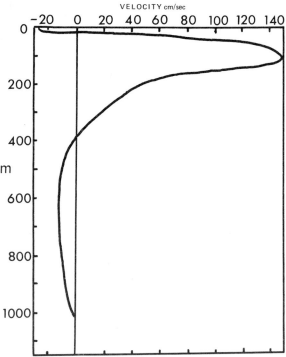

105. Vertical profile through the equatorial undercurrent of the Pacific, the Cromwell current (from Knauss 1960)

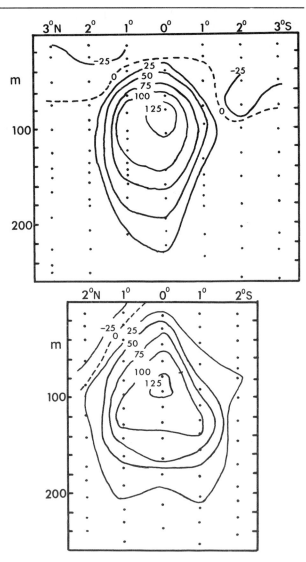

106. North-South cross-section through the Cromwell current at two different longitudes; velocities in cm/sec (from Knauss 1960)

seasonal variation due to the monsoon). These strong boundary currents (the Atlantic Gulf Stream and the Pacific Kuroshio) are the best known currents of the ocean.

Near the Equator in all three oceans there are two west-flowing Equatorial Currents. The South Equatorial Current lies at or south of the Equator and the North Equatorial Current to the north of it. In the Pacific and Indian Oceans, and in part of the Atlantic, the two west-going Equatorial Currents are separated by an east-going Equatorial Counter Current.

In the Southern Ocean, surrounding the Antarctic, there is no continental barrier (though the relatively narrow Drake Passage may have a similar effect), and the main surface current flows round the earth as an east-going flow referred to as the Circumpolar Current or as the West Wind Drift.

It must be emphasised that the charts of ocean currents are climatological; that is, they are based on averages of observations made over a long time. On any particular occasion a ship may find a current very different from the average current portrayed on the Pilot Chart. This is especially noticeable in the region of a fast western boundary current, such as the Gulf

Stream, which meanders and changes the position of its axis in an unpredictable way.

In such a case the climatological chart can be misleading, for the observations at a particular place are averaged over a long period, irrespective of whether the current is present or not. It can be seen that a strong narrow current, which varies its position, will be represented on a climatological chart as a broader but slower current. In this way, what one may call the 'climatological Gulf Stream' (as represented on a time-averaged climatological chart) is perhaps ten times wider, and considerably weaker, than the Gulf Stream on any particular occasion.

The Pilot Charts, nevertheless, give a good and useful representation of the currents in the top 10 m of the ocean. At deeper levels the position is much less

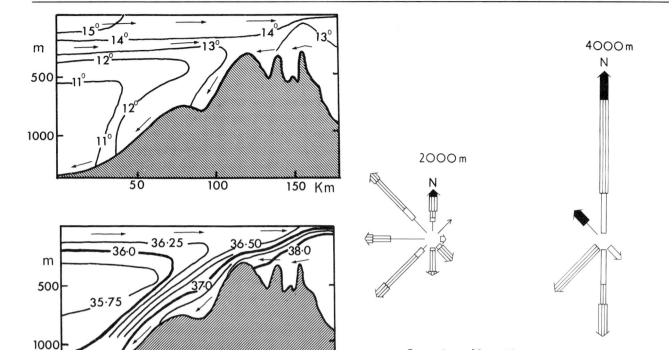

107. The inflow of Atlantic (on left) and outflow of Mediterranean water at the Strait of Gibraltar shown by temperature and salinity profiles (from Sverdrup, Johnson & Fleming 1942)

108. Diagram illustrating the variable ocean currents observed in the deep ocean near Bermuda (from Swallow 1961)

well known, mainly due to the difficulty of observation. The surface currents, though they vary with wind and weather, are almost always present in recognizable form. There are few sub-surface currents of which one can say the same.

Perhaps the most interesting of the steady sub-surface currents are the Equatorial Undercurrents. These are found in the Pacific and the Atlantic (and sporadically in the Indian Ocean) as a swift narrow current flowing from the west at a depth of about 100 m exactly along the Equator.

The Equatorial Undercurrent in the Atlantic was adumbrated by Buchanan over eighty years ago, but it was the one in the Pacific which was rediscovered (in 1952) and thoroughly investigated (1958) by US oceanographers.

The recent work started with the observation (by Cromwell) that long-line fishing gear, set near the Equator to catch tuna of various kinds, tended to drift to the east, against the surface flow. This implied a sub-surface current, which was called the Equatorial Undercurrent. It was thoroughly studied by Knauss (see figures 105 and 106), who found that it was a narrow (300 km), thin (0.2 km), fast (up to 150 cm/sec.)

current extending at least from 150°W to 92°W (6,500 km). The core of the current lies along the Equator at a depth which rises slowly from 100 m in the west to about 40 m around the Galapagos Islands. It transports about 40 million tons of water per second – about the same as the Florida Current, which is the fastest part of the Gulf Stream.

The existence of a similar current in the Atlantic was confirmed in about 1960 and attempts to find it in the Indian Ocean show that it is sometimes present, in spring, but not in all years and not in such a marked form as in the Atlantic and Pacific. The difference is due to the variability of the monsoon winds over the Indian Ocean.

Other semi-permanent sub-surface currents are known which take the form of overflow currents. These are found whenever dense water is formed in a basin with a sill: the dense water overflows the sill as a current into the ocean basin outside. Typical examples are the flow of the heavy deep water of the Mediterranean into the Atlantic at the Strait of Gibraltar (figure 107), and the flow of dense water from the Arctic into the Atlantic at various sills in the ridge which connects Greenland, Iceland and Scotland. Our

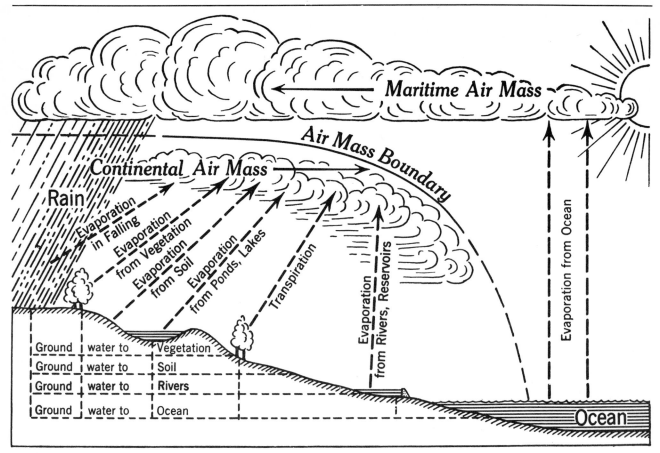

109. The hydrological balance of the earth (from Trewartha 1954)

knowledge of these overflows is still relatively scanty, although they play an important part in the dynamics of the deep ocean.

Apart from these few semi-permanent flows our knowledge of sub-surface currents is very sparse. It was once thought that they would be very slow, perhaps less than 1 mm/sec. (100 m/day), but recent observations using neutrally buoyant floats have shown that much bigger values are common. Some impression of the currents in the deep ocean near Bermuda is given in figure 108. Both speed and direction are highly variable and one cannot distinguish a prevailing current in the same sense as one can a prevailing wind. Instead, the impression is one of unsteady currents, with speeds of the order of 10 cm per second, which vary randomly in speed and direction, with a dominant period of a few weeks. Such a variable flow is difficult to distinguish from a complicated wave motion and may indeed be produced by some complex sort of wave.

Superimposed on these relatively slow variations are faster ones, some of tidal period, commonly of rather more than 12 hours, and others of inertial period. The inertial period varies with the reciprocal of the sine of the latitude, being 12 hours at the poles, 24 hours at latitude 30° N or S, and so large as to be meaningless at the Equator.

There is, then, a sort of weather in the ocean. Anyone who continued to live for long periods at the high pressures of the oceanic depths would find it rather monotonously cold, wet, salt and dark. The currents would seem variable, small in speed relative to the winds of the atmosphere but capable of exerting more force because of the high density of water relative to air. Visibility would be limited to a few metres, even in the clearest water, and in general marine creatures would be uncommon, unless our hypothetical hydronaut provided food for them or his lights aroused their curiosity.

PHYSIOLOGY

Having summarised what may be thought of as the anatomy of the ocean we turn to the more difficult question of how it is maintained.

We can start with the relatively simple question of the hydrological balance, as illustrated in figure 109.

110. Iceberg and small floes in the Antarctic

The ocean contains about 98.5 per cent of the earth's water, most of the rest being locked up in continental ice (figure 110) and only relatively small amounts as fresh water. An even smaller amount, equivalent to a layer of water only 25 cm thick over the earth's surface, exists as water vapour in the atmosphere.

The 1·5 per cent as ice and fresh water may seem a small amount. If the ice were to melt, however, mean sea level would rise by 50 m or so with immense consequences for civilisation. Fortunately the climatic variations which melt and re-form glaciers are slow, so there should be adequate warning for our descendants, who may in any case number weather-control among their skills.

About 1 m of water is evaporated from the ocean in an average year, being subsequently condensed as rain or snow and reprecipitated. The evaporation and the precipitation balance over the earth as a whole but not, as figure 83 illustrates, at individual places or even in latitude bands. It is the local differences in evaporation and precipitation which bring about variations in surface salinity and indeed which make the Atlantic slightly saltier than the Pacific.

Near coasts and in landlocked seas, such as the Mediterranean and the Red Sea or the Baltic, the run-off from rivers is important, and the important balance is between evaporation on the one hand and precipitation and run-off on the other.

The salt balance reached a near-equilibrium eons ago. The evaporation does not transfer any salt from ocean to atmosphere but a small amount is transferred when waves break to form bubbles at the sea surface and when the droplets of sea spray evaporate. The tiny salt particles which remain are carried by the wind, and may play an important part in the water cycle by acting as nuclei on which cloud particles can form. Most of them fall back into the ocean in rain but some fall on the land and are returned to the ocean via the river flow.

The water of rivers is fresh, so it comes as a shock to find that they contribute 4,000 million tons of dissolved solids to the ocean annually. The total quantity of solids dissolved in the ocean is much bigger (10^{16} tons), so even if the transfer of salt from ocean to atmosphere is neglected it would still take a few centuries before an increase in salinity could be measured.

THE HEAT BALANCE

Just as we can assume that the total amount of salt in the ocean is constant, so we take it that its total heat content stays the same. The evidence is not conclusive: repeated measurements of temperature in the deep ocean agree within experimental error, but the heat capacity of sea water is so large that a vast quantity of heat is needed to change its temperature by a measurable amount. From indirect evidence, however, it seems likely that the heat balance of the earth is changing only slowly; the heat balance is harder to calculate than the water or salt balance because a major part of the transfer is from outside the system by radiation. Apart from tides (which involve relatively small amounts of energy), all the motions of the atmosphere and ocean—all the winds, currents, evaporation, clouds – are powered by energy from the sun.

The earth is a satellite of a star which is of the right size and at the right state of development to produce energy by thermonuclear reactions. It is converting mass to energy at the rate of 4,000 million tons/second. This produces an amount of energy so large as to be beyond description. Fortunately the earth intercepts only a tiny fraction of the total; most goes on to space. Even so, the earth receives energy equivalent to about 50 megaton bombs per second.

Most of the energy, as we have said, reaches the earth as shortwave radiation, mainly in the visible region at a wavelength of about 0·5μ. The atmosphere absorbs some and some is scattered back to space, but about 40 per cent reaches the earth's surface (about 25 per cent directly and 15 per cent scattered downward from the atmosphere). Some of this is reflected but most is absorbed in the upper layers of the ocean.

To balance the heat budget, this energy has to be accounted for; there are various ways in which it is used.

1. Heat is lost as radiation to the atmosphere and to space; since the ocean is much colder than the sun it radiates at a longer wavelength, in the infrared. As we have said, the ocean looks black from space because most of the short wave (visible) radiation is absorbed and re-radiated in the long wave (infrared). Clouds are better reflectors and look white.

2. Heat is used to evaporate water from the sea surface. Since the latent heat of water is so large, much heat is used in this. The heat is transferred to the atmosphere, at a different place and later time, when the water vapour condenses. This is the most important component of the interaction between the atmosphere and the ocean.

3. Heat is transferred from the ocean to the atmosphere when the sea is warmer than the air above it. On average, the sea surface is slightly warmer than the air above it and there is a net transfer from ocean to atmosphere.

Taking the whole ocean as a unit, and averaging over a long time, the heat input by radiation amounts to about 320 calories/sq cm/day and is balanced by these three terms, the back radiation (130), the heat used in evaporation (170) and the heat transfer (20).

If we consider smaller areas and shorter times the situation is more complicated; heat can be stored in the upper layers of the ocean as a diurnal or seasonal thermocline and heat can be transferred from place to place by ocean currents or by small scale mixing processes. Perhaps the most important point, however, is that the ocean is heated at the top (the interior of the earth is extremely hot but the heatflow through the ocean floor is energetically trivial). One of the more obvious thermodynamical principles, which can be verified in any kitchen, is that heating a fluid from above is an inefficient process. The ocean is no exception – very little of the energy from the sun is converted into the kinetic energy of oceanic motion. Only in polar regions, where the ocean surface can be strongly *cooled* at the top, are sinking motions produced; the water is cooled until it is dense enough to sink and gradually to fill the deep ocean basins.

This mechanism is too inefficient to account for the ocean currents we observe – how then are they produced? The answer to this question is well known—they are driven by the wind–and the obvious implication is that for any profitable discussion the general circulation of the ocean must be linked with the general circulation of the atmosphere. It is being increasingly recognised that oceanography and meteorology are not distinct branches of earth science and cannot be understood in isolation. In most of the advanced countries they are already studied in the same institutions and by the same methods.

To put it in a rather formal way, the atmosphere and the ocean, regarded as one system, are kept on the earth by gravity and irradiated from outside (by the sun). The atmosphere is relatively transparent to the incoming radiation (visible) and the ocean almost opaque to it. So the atmosphere is heated from below and is thermodynamically active; vertical instability, overturning and violent motions are characteristic. The ocean, as we have seen, is relatively sluggish and slow-moving and its properties are only modified near the surface. The ocean is much wider than it is deep, so that a scale model might well resemble a sheet of airmail paper. But, as Ewing points out, all the important messages are written on the top surface.

The energy used to maintain ocean currents comes from the sun but by a complicated route. First, through

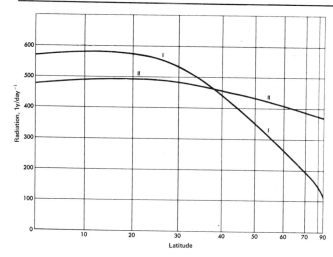

111. The radiation balance of the earth: 1. incoming solar radiation absorbed by earth and atmosphere; 11. outgoing long-wave radiation (from Houghton 1954)

the atmosphere to the sea surface, where it is absorbed and transferred in a different form to the atmosphere. There some of it is converted, by complicated processes, to depressions and anticyclones whose winds then use their energy to generate waves and drive ocean currents. It is a complex system, and its many interlocking components cannot be considered in isolation; we have compared it with a living organism, or indeed with a whole food-mesh, and the analogy seems not inapt.

It seems clear that the first stage in trying to understand the circulation of the ocean is to get a clear idea of how the winds of the world and the general circulation of the atmosphere are maintained. Again, tides apart, the energy is supplied by the sun: the motions arise because the lion's share of the solar radiation falls on low latitudes and maintains a temperature difference between the tropics and the poles.

The tropical atmosphere gets heat from the sun, as well as by evaporation and heat transfer; it has a surplus even after providing the long-wave radiation outwards. The polar atmosphere, on the other hand, has a deficit, as figure 111 shows; it gets less than do the tropical regions and provides about the same outwards. If the atmosphere and ocean were still, the tropics would get hotter and the polar regions colder. They are maintained at their present temperatures only because there is a transfer of heat from low to high latitudes. This is brought about by the winds and, to a smaller extent, by ocean currents.

In this way the atmosphere can be thought of as a gigantic – and very inefficient – heat engine, absorbing heat in the hot equatorial belt and losing it nearer the poles. Since much of the heat transfer is due to evaporation and condensation of water vapour, one can think of the mechanism as a steam engine and use such expressive terms as 'the equatorial firebox' and regard the giant cumulo-nimbus cloud towers (figure 112) as the cylinders in which the steam is converted to liquid water drops. The engine is very inefficient, but it is certainly large; the heat which is being transferred from Equator to Pole crosses a particular latitude circle at the rate of about 10^{15} cal/sec, a large figure which represents nearly a million times the power consumption of all the world's population.

How does the atmosphere bring about this huge heat transfer from the tropics to the polar regions? There are two main mechanisms, one which operates between the equator and about 30°N or S and another which operates at higher latitudes.

In the mechanism which operates at lower latitudes the essential feature is that the air rises at the thermal equator and then spreads northward and southward toward the poles. Most of it sinks again at about 30° N or S and returns towards the Equator, acquiring an easterly component due to the rotation of the earth, and forming the Trade Winds. These carry less energy than the poleward winds aloft, and so there is a net transport of energy from low latitudes to high.

The essential features of this circulation are the two closed wind patterns north and south of the Equator; they were described by Hadley as long ago as 1735 and are known as the Hadley cells. Their existence provides a rational explanation for the equatorial rain belts (ascending air motion) and the subtropical anticyclones around 30° N or S. These are linked with descending air motions, little rainfall, and hence the desert areas on land and the deep blue and high salinity water at sea. The associated horizontal winds are the Trades, at the surface, and the subtropical jet stream aloft. The subtropical jet stream is a band of strong westerly winds in latitudes 30° to 40° at a height of about 12 km. It arises because the air, which has risen from the Equator, tends to retain its angular velocity as it moves poleward. When it reaches 30° or so it is moving much faster towards the east than the earth beneath it.

The Hadley cells provide a good explanation for many observed phenomena, and the mechanism of the general atmospheric circulation in low latitudes is relatively well understood. In both hemispheres, poleward of 30° N or S, the heat transfer is not brought about by motions of the Hadley cell type but by disturbances on a smaller scale. Here the prevailing winds are westerly at all levels, unlike the tropical regions where the Trade Winds at the surface have an easterly component.

Embedded in the westerlies of middle latitudes are

112. Towering cumulo-nimbus clouds in the Atlantic

the travelling depressions which provide the typically unsettled weather. They have a complicated wind and temperature structure, such that air which is warmer than average tends to be going poleward and air which is colder than average tends to be going towards the Equator. It will be seen that on average the heat will be transferred in the required sense – from Equator to pole. The disturbed westerly regime and the complex travelling depressions are essential features of the heat transfer.

The situation is obviously complicated but the overall features of the general circulation of the atmosphere as indicated in figure 113 are clear. Dynamical meteorologists are using computers to simulate the behaviour of the atmosphere, with results sufficiently lifelike to make one feel that the essential physics of the situation is correct. Once the winds of the world are understood the way is open for a study of the ocean currents they produce.

The atmosphere and the ocean have certain basic similarities – both are vast bodies of fluid on a rotating earth – but we must recognise their differences. We have emphasised the differences in their physical properties, especially those controlling the transmission of radiant energy, but we must recognise that their geometry also plays an important role. There are no barriers in the atmosphere which correspond to the continental barriers to the oceans.

From our account of the current structure of the ocean it will be clear that these boundaries play an essential part; water obviously cannot flow through them but has to flow parallel to the coast. The shapes of the Atlantic and the Pacific ocean basins are very different, yet the major currents, the Gulf Stream and the Kuroshio, are very similar. Clearly the 'Western Boundary' has a decisive effect and presents an important problem – that of westward intensification.

Among the major features of the atmospheric circulation are the subtropical anticyclones, those semi-permanent high-pressure areas centred about 30° N and S around which the winds circulate steadily. These winds produce ocean currents which circulate in the same sense. Yet the wind systems are symmetrical, while the ocean currents are not – they have an intense boundary current flowing along the western boundary.

The reasons for this are difficult to understand and even more difficult to explain. All the large-scale flow in the atmosphere and ocean is affected by the rotation of the earth. The relatively small-scale flows which we meet in bathroom, kitchen or laboratory are dominated by other forces, so the effect of the rotation of the earth is beyond our experience. The flow patterns it produces in the atmosphere and ocean sometimes seem bizarre.

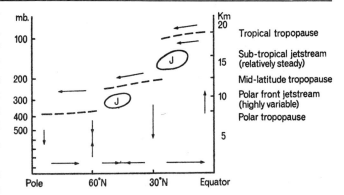

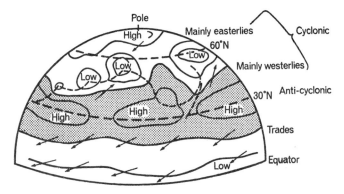

113. Schematic representation of the general circulation of the atmosphere on an idealised earth, ignoring the effects of continents and oceans and the seasons. Lower diagram shows the surface pressure pattern as it might be on one day. The long period average would show simple zones of low and high pressure. The upper diagram represents a vertical cross-section from pole to equator (J: jetstream; mb: millibars pressure) (from Sutcliffe 1966)

The basic notion is one of the spin which anything on a rotating globe must have. If one imagines a man standing astride the North Pole, for example, it is obvious that he will be rotating, or spinning, about his own axis, at the same rate as the earth – once per day. If the same man now stands astride the Equator, the earth continues to rotate about its axis but he no longer rotates about his. His local rate of rotation or spin is zero. So the spin which affects anything on the earth is zero at the Equator and increases to one revolution per day at the Poles.

In these examples, our hypothetical man – we could equally have considered a parcel of fluid – was at rest relative to the earth. He – or the fluid – could also have been rotating on his own axis relative to the earth. The total spin is obviously made up of two components – the spin relative to the earth and that due to the rotation of the earth beneath it.

It is a difficult but important concept – important because, in relatively shallow fluids like the atmosphere and the ocean, a body of water moves in such a way

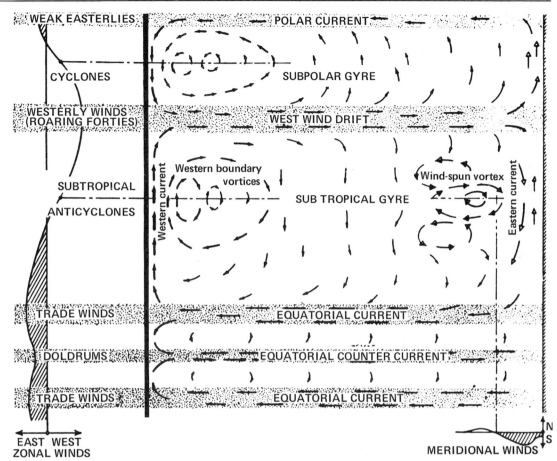

114. Computed currents in a rectangular ocean (*right*), due to effect of observed mean wind distribution (*left*) (from Munk 1950)

that its spin stays constant.

As an illustration of the way in which this can be applied to an ocean circulation problem we consider, following Stommel and Munk, an ocean in the form of a rectangular dish – since we observe that the shape of an ocean makes little difference to the general distribution of currents in it, we can choose any convenient shape. We suppose that our rectangular dish represents one of the N. Hemisphere oceans, say the North Atlantic, and we know, from observation, that the winds over the northern half are Westerlies while those over the southern half are replaced by the Trade Winds, with an easterly component.

We consider a particular parcel of water which in the absence of wind will have an anti-clockwise spin due to the earth's rotation. The wind distribution will tend to give it a clockwise spin relative to the earth and so will reduce the anti-clockwise spin. To keep its spin constant the parcel of water must move to where the anti-clockwise spin due to the earth's rotation is less, that is, toward the Equator.

In general then, the flow over the whole of the hypothetical rectangular ocean will be to the South; any compensating flow to the North must be accompanied by an increase in the anticyclonic spin. One process which can produce the required effect is that of friction with the boundary. Only if there is a strong western boundary current can a parcel of water maintain a constant spin and so complete its circuit of the basin.

This is a necessarily over-simplified account of the process which it is thought produces strong western boundary currents, such as the Gulf Stream, in oceans with North/South boundaries. Perhaps the result can be illustrated by figure 114, which shows the results of detailed calculations by Munk. The ocean is taken, schematically, as a rectangular dish. The wind field over it is taken from climatological observations.

It will be seen that in the region between the Westerlies and the Trades the flow over almost all the basin is to the South. All the water returns to the North as a single strong current near the Western

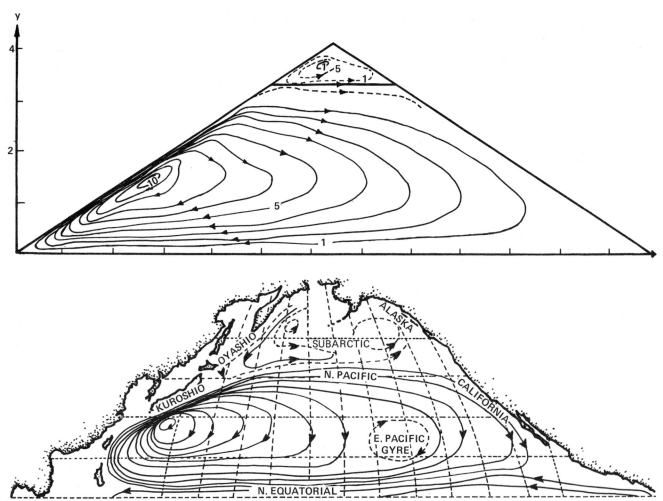

115. Computed currents in a triangular ocean (*top*), due to effect of observed mean wind distribution, compared with the average currents of the North Pacific (*bottom*); arrows indicate the approximate direction of transport of water, and the areas between adjacent contours enclose equal amounts of water transported (after Munk & Carrier 1950)

Boundary. Although the rectangular basin does not resemble any particular ocean the general pattern of currents which the calculations indicate have similarities with the currents in most of the real ocean basins. Comparison with the North Atlantic, for example, shows currents analogous to the Labrador Current, to the Gulf Stream and to the Equatorial Currents. There is a calm area which resembles the Sargasso Sea and, on the Eastern Boundary, a current rather like the Canary Current.

Another example of the same type of calculation is shown in figure 115. The Pacific can be represented approximately as a triangular ocean, and it turns out that the currents calculated using the observed winds over an ocean in a triangular dish are very similar to those observed in the Pacific.

For oceans with roughly North-South boundaries,

like those of the Northern Hemisphere, one can fairly claim that the main features of the current structure are reproduced and that a first order understanding of their generation has been achieved. The Southern Ocean which has no meridional boundaries, but only a constriction at the Drake Passage, has not been satisfactorily modelled. Many other problems remain; our notions of the deep currents are still untested, we do not know exactly how the wind moves the water or the effect of changes of the wind in time and space.

There are also difficult problems associated with Eastern Boundaries which have great importance for fisheries. One of the bizarre effects of the earth's rotation is that a wind blowing over the sea produces a transport of water at right angles to the wind direction (to the right in the Northern hemisphere). If a wind blows from the North along an Eastern Boundary the water is transported away from the coast and sub-surface water upwells from depths of a few hundred metres to take its place. The situation is fairly common in the Trade Wind belts, and the process is important because the upwelling water has properties different from the usual surface water. It is, of course, colder but also it contains nutrients which can strongly affect the biological productivity. The great fisheries off the coasts of California and Peru are supported by upwelling, and there is said to be a potentially huge fishery, as yet unexploited, off the S. Arabian coast where upwelling is known to occur in summer (figure 144).

The phenomenon is well known but badly understood. It occurs sporadically, in localised patches, and can easily be confused with Eastern Boundary currents which are independent of the upwelling mechanism.

As regards the currents which are not driven by the wind, but by the cooling and sinking of water in high latitudes, our ideas are still largely hypothetical. These currents are probably less strong but may be equally important in transporting heat and salt, to maintain the oceans' contribution to the transport of heat from tropics to poles to offset the radiation surplus. On average, warm currents flow poleward and cold currents equatorward so as to transfer heat in the appropriate sense.

For example, there is increasing evidence for a cold current, originating from cold water sinking in the North Atlantic, which flows to the South, at depth, off the coast of North America and down to the Southern hemisphere.

Our knowledge of the physical processes which maintain the general circulation of the atmosphere is now such that vast programmes are being undertaken to increase the observations available and to acquire bigger and faster computers to analyse them. The hope is that, given fast enough telecommunications, long-period weather forecasting will be possible. Such forecasts will also need more knowledge about, and observation from, the ocean; and there is every reason to hope that our dream of a conceptual model of the ocean and the atmosphere, considered as a single system, will at last become a reality.

MODIFYING THE OCEAN

One of the things that distinguishes man from other creatures is that he can control his environment. Not just physically, by wearing clothes, and living in houses – or by wearing a wetsuit and breathing helium and oxygen – but also biologically, by breeding cattle and plants, and by the development of frightening hormone weedkillers, and so on. There is little evidence of attempts to breed people selectively, as yet, though the simultaneous expansion in organ transplants and in computer marriage-bureaux must give one pause.

Oceanographers and others who study our natural environment probably do so, consciously or not, in order to feel more secure within it. With understanding comes a lessening of our fear of the unknown, the abolition of strange gods with specialised functions and the gradual vanishing of the dragons and fierce sea creatures from the unknown seas of the early charts.

Given our present knowledge of the observed characteristics of our environment and our general idea of how they are maintained, it is natural to ask whether we can make large-scale modifications to it, for the benefit of mankind. This is by no means a new idea – was it Mark Twain who first grumbled that everybody talks about the weather but nobody does anything about it? – but it is more often raised now that large amounts of energy have become available from controlled nuclear reactions.

We know from earlier sections that the atmosphere and the ocean together form a vast and inefficient heat engine which is powered by radiation from the sun. To modify such a huge thermonuclear reactor at such a large distance is surely beyond our present technology. We shall not be able to alter the amount of solar radiation reaching the earth but only its distribution.

Since an appreciable fraction (about a third) is reflected from the earth unused, an attempt could be made to change the reflectivity. It has been suggested that the snow and ice of high latitudes might be blackened by a suitable substance such as soot, or that the Sahara might be covered with a film of black polythene. The ocean absorbs most of the solar radiation incident upon it but it might be made more reflective if it were covered by some suitable white, highly-reflecting foam. This might have the effect of reducing the rate of evaporation, which takes about half the incoming solar energy.

Experiments have been made using thin films of hexadecanol (a fatty substance derived from the heads of sperm whales). On confined bodies of water substantial reductions of evaporation have been achieved, and the technique is certainly useful in the important task of conserving fresh water. But there are technical problems in maintaining such a film intact on large bodies of water affected by strong winds.

In this method, and in all methods which involve the deposition of films over large areas, the costs are very large. As an example, suppose we need a surface film only 0·001 mm thick over an area 1,000 km square. The weight of substance might well be a million tons and the cost perhaps $2,000 million.

The same sort of calculation indicates the impossibly high cost of forming artificial clouds high in the atmosphere to cut down the incoming radiation. Here we have some idea of possible effects because huge volcanic eruptions put vast quantities of dust into the upper atmosphere, and the succeeding weather seems to differ very little from what it was before, except that the sunsets are more picturesque.

To change the cloudiness is a notion that has been popular for a long time. Modern rainmakers seed clouds using silver iodide, in the hope that it will change the microphysical processes which form raindrops. Their efforts have not, so far, made much difference to the observed rainfall and there is no plan to use them on a global scale.

The air-sea heat engine is inefficient and converts but little of the solar energy to motion. Nevertheless, much energy is used in overcoming friction and a difference could be made by modifying the bottom boundary. One could consider, for example, removing selected mountains, a process for which one would need great faith or a few million megaton bombs. It has even been suggested that the friction between the air and the ground could be increased by erecting many hoardings on the prairies and steppes, the cost to be met by the advertisers.

More sophisticated suggestions involve the ocean, and a number have been suggested. The blocking of the Straits of Gibraltar would certainly have oceanographic and meteorological consequences as well as political ones. The climate of the Mediterranean countries would be much affected and the circulation of the North Atlantic and the World Ocean would probably be modified by the absence of the warm and salty water from the Mediterranean basins.

To isolate the Arctic by blocking the Bering Strait is another possibility which has been canvassed. Here the idea is that the melting and freezing of the Arctic Ocean is associated with the bigger climatic changes such as Ice Ages.

Why do all these suggested ways of modifying the environment seem like science fiction? Not, one feels, because they involve such huge amounts of energy, or because they would cost so much, but because we cannot, from our present knowledge of the working of the atmosphere and the ocean, predict the consequences of our brute force and ignorance. Much of the harm done by science stems from attempts to do good; especially in complicated situations where the consequences of particular actions cannot be foreseen. Most people are familiar with what is known as Murphy's Rule – that if anything can possibly go wrong, it will –

and since our lives are so bound up with our present climate and weather there is little room for error.

For some time yet we must be content to feel at one with our environment, because we can describe its present state and have some understanding of the processes involved. Not until we know enough to simulate its behaviour, using computer models of increasing complexity, shall we have the confidence to start environmental engineering on the global scale. We do not know enough to make large-scale experiments without involving innocent bystanders; but fools rush in, and some way of assessing and controlling proposed experiments would not be premature.

We have already mentioned the probability that there are many planets with oceans within our galaxy. But the nearest one is many light years away, so unlikely to be visited by other than the most intrepid astronauts for many generations. In the meantime mankind is living on a spaceship; it is an ironic fact that although we cannot do much deliberately to control our environment we may well be polluting it, inadvertently, yet irreversibly.

There are many examples. Carbon dioxide, important for photosynthesis and an absorber of radiation, is increasing in concentration. Within a few centuries we shall put back into the atmosphere and ocean the organic carbon stored in sedimentary rocks over millions of years. Will this melt the Arctic ice and submerge the world's cities? Will the pollution from the world's sprawling megalopolis gradually obscure the sun and the sky?

More dangerous is the possible pollution of the ocean by the synthetic materials which have been invented to benefit mankind: lead from gasoline, detergents and pesticides. There seems little point in attempting to make a comprehensive list, since technology is advancing so rapidly. What is needed is the realisation that we have perhaps a few generations left in which to develop some understanding of the interaction between our biological and physical resources. For centuries, man has protected himself against the environment – now we must protect our environment against men.

The ocean and the atmosphere are expected to last, in more or less their present form, for hundreds of millions of years. Will man's world end with a nuclear bang or an overpopulated whimper? Or can we learn enough of our environment to ensure that the new technology will not extinguish either life itself or man's joy in it?

4 The Constituents of Ocean Waters

F. Culkin

Speculation about events which occurred millions of years ago inevitably leads to differences of opinion. An early theory that the earth was formed from a large body of gaseous material, which broke away from the sun and gradually cooled, gave way to the current 'cold' theory in which the present earth was formed by agglomeration of a large cloud of dust particles. Contraction of this cloud and radioactive decay of elements such as uranium, thorium and potassium caused the temperature of the core to increase initially, and the volatile constituents such as water, hydrogen chloride, ammonia, sulphur dioxide and carbon dioxide were driven to the surface. Some of these volatiles probably reacted with the molten rocks and some escaped to form an atmosphere. As the new earth cooled down, the water vapour in the atmosphere condensed and fell as rain to form ocean waters, which were probably acidic at first but which were neutralized fairly rapidly by reaction with the basic rocks in the earth's crust.

The volume of the early ocean, formed some three thousand million years ago, was probably much less than that of the present ocean. Expulsion of water and other volatiles from the interior of the earth has continued, and the oceans now occupy a volume of 300 million cubic miles and cover about three-quarters of the surface of the earth. Much water escapes into the atmosphere by evaporation from this vast surface and is later precipitated as rain or snow, mainly over the continents. This water is then returned to the sea by rivers, which supply 35,600 cubic km (8,440 cubic miles), carrying about 13,000 million tons of suspended material and 4,000 million tons of dissolved material every year from the rocks over which they run. Continental borders are also being continually eroded by the action of tides (figure 116), thus making a further contribution to the dissolved and suspended matter of the sea. Another source of material is underwater volcanic activity. Much of this probably takes place unnoticed on the deep ocean floor, but occasionally sufficient molten material erupts to form a new island such as the recently formed Surtsey Island in the

TABLE 6

Concentrations of the major dissolved constituents of a typical sea-water

Constituent	g/kg of sea-water
Chloride	19·353
Sodium	10·760
Sulphate	2.712
Magnesium	1.294
Calcium	0·413
Potassium	0·387
Bicarbonate	0·142
Bromide	0·067
Strontium	0·008
Boron	0·004
Fluoride	0·001

These constituents are dissolved in *ca* 965 g water which consists of 857·8 g oxygen and 107·2 g hydrogen.

North Atlantic Ocean. This island rose through water 130 m deep to a height of 180 m above sea level (figure 78).

Despite the fact that all of these processes have been going on for so long, the concentration of dissolved matter in sea-water is today only three to four per cent and has probably never varied very much from these limits. This is because there are many other processes such as biological activity and sediment formation which effectively remove material and keep pace with its accumulation. The net result is a solution containing probably all the elements which occur naturally in the earth's crust, ranging from nearly two per cent of chloride to minute traces of rare elements such as gold and uranium.

More than 99 per cent of the dissolved matter in sea-water is made up of the eleven major constituents, which are listed in Table 6 together with their concentrations in a typical sample of sea-water. The chemical analysis of these constituents has interested chemists since the early years of the nineteenth century and the principal methods used in their analysis (gravimetric and volumetric) will be familiar to most

116. Erosion of a rocky coastline in Portugal has led to the formation of these pillars and arches

students of elementary chemistry. Sulphate, for instance, can be determined by precipitating it as barium sulphate, filtering and weighing the precipitate. The weight of sulphate in the sample taken can then be calculated from the weight of barium sulphate obtained. Chloride, on the other hand, is best determined by a volumetric titration in which a solution of silver nitrate is added slowly from a burette until all the chloride has been precipitated as silver chloride, as shown by indicators which change colour at this point, or as shown by a change in the electrical potential between two electrodes immersed in the sample. Despite efforts to increase the accuracy and precision of analytical methods, so as to be able to detect small differences in composition between samples, the

interesting fact has emerged that the major constituents occur in almost constant proportions to one another. There is evidence that deep waters contain slightly more calcium, relative to the other constituents, than do surface waters, probably because of biological removal of calcium in the surface layers and re-solution of calcareous shells in the deeper waters. Generally, however, the composition of the major salts dissolved in sea water is almost constant, though the degree of dilution may vary from place to place.

A solution such as sea-water, of which there is almost an unlimited supply and which has a fairly constant composition, has many advantages over land deposits as a source of chemicals. Sodium chloride (common salt) was first extracted from sea-water over

117. *opposite* The deep-water medusa *Periphylla* (5 cm diameter), here consuming a lantern fish, has a rather variable shape; some specimens have a tall conical top, others a low rounded one

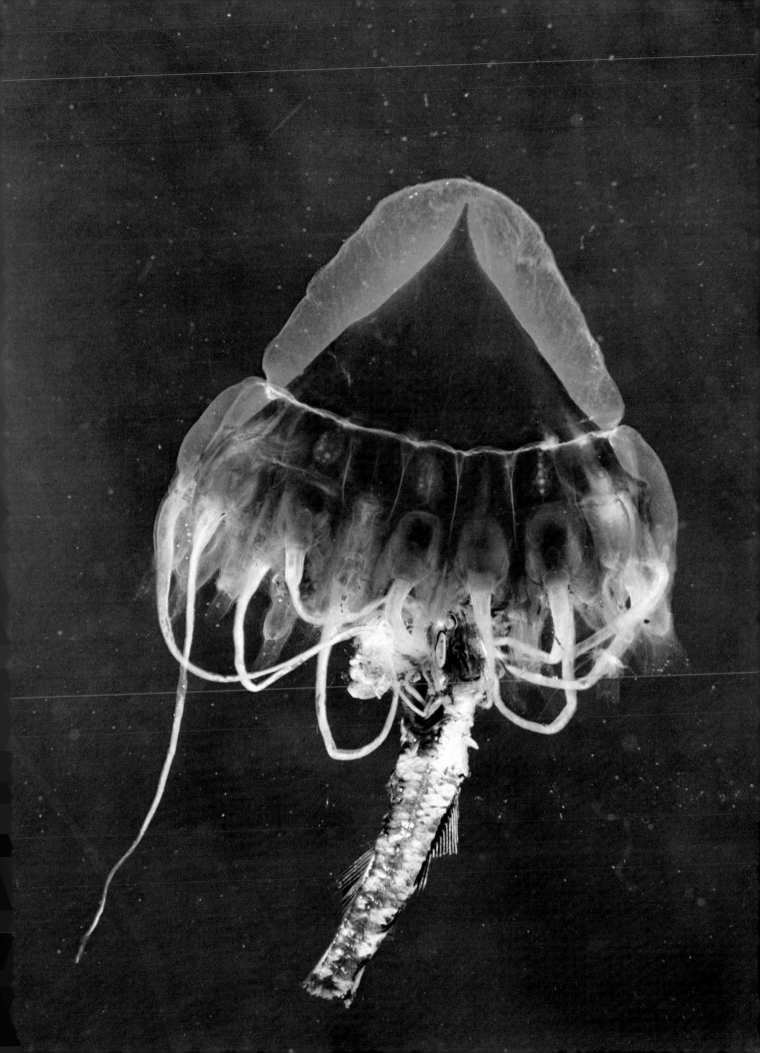

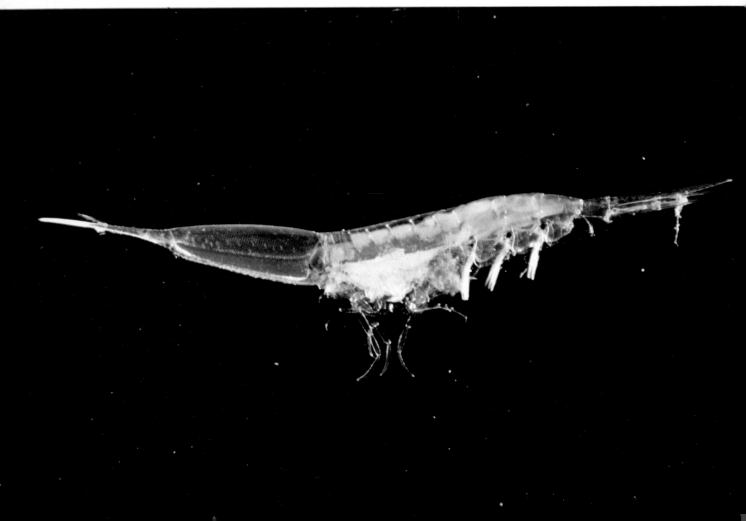

4,000 years ago by the Chinese, and today over six million tons are produced annually throughout the world from this source. In hot dry climates, sea-water is simply allowed to evaporate to the stage where salt begins to crystallize from solution, whereas in cold climates an initial concentration is effected by freezing out some of the water as ice, followed by evaporation.

The concentration of magnesium in land deposits is several hundred times greater than in sea-water, but the sea is now the major source of this important metal. The basic process, in which magnesium hydroxide is precipitated from sea-water by addition of milk of lime manufactured by calcining oyster shells, is interesting in that one element from a marine source is used to extract another.

Despite the fact that the concentration of bromine in sea-water is only 0.007 per cent, the oceans contain over 99 per cent of the world's naturally occurring bromine. The development of anti-knock additives for motor-car fuels has led to a big increase in the demand for bromine in recent years, and today the world's supply of this element is derived almost entirely from sea-water.

Another very valuable constituent of the oceans is, of course, water itself. De-salination plants involving distillation are in operation in many parts of the world, mainly to supplement existing supplies of fresh water. The most widely used process is distillation, but electrodialysis, involving the use of membranes through which dissolved salts pass more readily than do water molecules, reverse osmosis in which the membranes permit passage of water molecules but not dissolved salts, and ion exchange resins are other promising methods. Production of pure water from sea-water is still costly, however, even when there is available a cheap source of power such as solar energy or a nuclear power station.

The presence of these dissolved salts, the total concentration of which is called salinity, has an appreciable effect on the physical properties such as freezing point, electrical conductivity, refractive index and density of water. To the physical oceanographer the most useful of these is density. A knowledge of the distribution of density with depth is important in studies of the movement of water masses, stability of water columns etc., but as density cannot be measured *in situ* it has to be calculated from measurements of temperature, salinity and depth. Samples of water are collected from different depths by means of reversing bottles of the types shown in figures 100 and 120. The bottle is attached to a wire and lowered to the required depth. A brass weight (messenger) is then slid down the wire and when it reaches the bottle it trips a mechanism which closes the two ends of the bottle, thereby

118. *opposite above* The colourful amphipod *Pegohyperia* (3 cm) has a very striking yellow and black eye

119. *below* In the curious amphipod *Streetsia* (3 cm) the head, together with the eye, is lengthened forward into a narrow cylinder

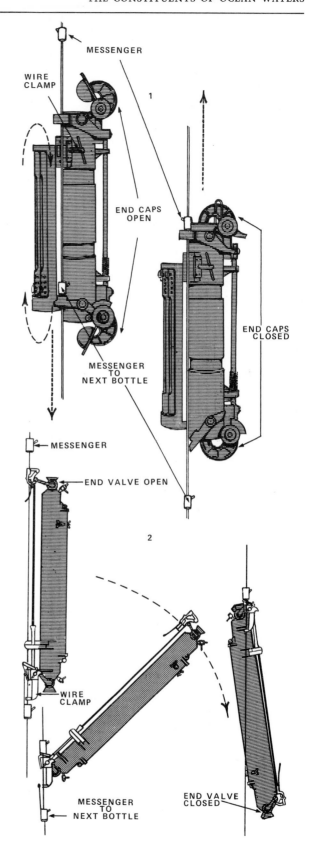

120. Two types of water sampling bottle. 1. An N.I.O. water bottle with the end caps open (*left*) before being triggered by a messenger. When triggered the caps close and the thermometer frame turns through 180° (*right*). 2. Nansen bottle operation in which the whole bottle turns upside down, closing the end valves at the same time trapping a sample of water. At the same time either the thermometers are inverted with the water bottle (Nansen type) or a frame carrying two reversing thermometers is rotated through 180° and the thermometers register the temperature of the water. Two types of thermometer are used together. In the 'unprotected' thermometer the hydrostatic pressure of the water acts on the bulb, causing a high reading (the pressure increases by one atmosphere for every 10 m depth). The other type is the 'protected' reversing thermometer in which the thermometer is protected from the pressure of the water by a partially evacuated glass tube, and so registers the true temperature of the water. The difference between the readings of the two types of thermometer is used to calculate the depth at which the sample was taken (a difference of 1°C corresponds to a depth of approximately 100 m). In this way, samples of water are obtained from known depths for analysis. For many years it was the practice to determine the chlorinity (Cl‰), i.e. the chloride plus the bromide, of the sample by titration with silver nitrate and obtain the salinity (S‰) or total dissolved salts from the relationship. $S‰ = 1.805 Cl‰ + 0.03$.

In recent years this method has been replaced to a large extent by the use of electric salinometers which compare the electrical conductivity of the sample with that of a standard sea-water of known salinity. This standard (Copenhagen Standard Sea-Water) is a natural sea-water collected in the North Sea, whose chlorinity (and hence salinity) has been determined very accurately. It is distributed in sealed glass ampoules by the Standard Sea-Water Service, Copenhagen, to oceanographic laboratories all over the world and has been used for many years by oceanographers for standardizing silver nitrate solutions in the chlorinity determination and, more recently, for calibrating salinometers.

By use of these methods it is possible to obtain a picture of the distribution of salinity, temperature and density with depth by taking samples at selected depths. Such sampling techniques will probably be superseded in the future by the use of the recently developed Temperature-Salinity-Depth probes, described in Chapter 2, by means of which continuous profiles (i.e. at all depths, not just selected depths) can be obtained with an accuracy comparable to that obtained by the older methods and with much greater speed.

In addition to the major constituents, sea-water contains very small concentrations (down to a few parts per million million) of a large number of other elements and compounds (Table 7). They include some of the most abundant elements in the earth's crust, such as aluminium, iron and silicon, which, although they are present in river waters in much higher concentra-

TABLE 7

Minor constituents of sea-water

Element	Concentration (mg/kg)	Element	Concentration (mg/kg)
Silicon up to	3.0	Vanadium	0.002
Nitrogen	0.5	Titanium	0.001
Phosphorus	0.07	Tin	0.0008
Iodine	0.06	Antimony	0.0005
Barium	0.03	Cobalt	0.0001
Iron	0.01	Chromium	0.00005
Aluminium	0.01	Thorium	0.00005
Molybdenum	0.01	Silver	0.00004
Zinc	0.01	Lead	0.00003
Copper	0.003	Mercury	0.00003
Arsenic	0.003	Bismuth	0.00002
Uranium	0.003	Tungsten	0.00001
Manganese	0.002	Gold	0.000004
Nickel	0.002	Beryllium	0.0000006

TABLE 8

Enrichment factors for the concentration of various elements in marine organisms (from Mero 1965)

Element	Concentration of element		Enrichment factor	Marine organism
	In sea-water (mg/kg)	In organism[1] (mg/kg)		
Titanium	0.001	40	40,000	Algae
Vanadium	0.002	560	280,000	Tunicates
Cobalt	0.0005	1	2,000	Algae
Nickel	0.002	5	2,500	Algae
Molybdenum	0.01	60	6,000	—
Iron	0.01	1,000	100,000	Algae
Lead	0.00003	700	20,000,000	Fish Bones[2]
Tin	0.003	1,000	330,000	Fish Bones
Zinc	0.01	10,000	1,000,000	Fish Bones
Chromium	0.00005	2	40,000	Algae
Silver	0.0003	7	21,000	—
Rubidium	0.12	150	1,000	Algae
Lithium	0.17	6	30	Algae
Strontium	8.0	3,000	400	Algae
Barium	0.03	100	3,300	Algae
Manganese	0.002	120	60,000	Algae
Copper	0.003	3,000	1,000,000	Fish Bones
Gold	0.000004	0.0014	1,400	—
Germanium	0.00007	0.5	7,600	—
Iodine	0.06	50	30,000	Algae

[1]Concentration in the dry ashes of the organisms.
[2]The element was not necessarily concentrated in the bones while the fish was living.

tions, are very reactive in the more alkaline waters of the sea and are rapidly removed from solution by sedimentation processes.

The concentrations of some of the elements in Table 7 may seem almost negligible, but the total weight in all the oceans of an element such as gold amounts to about nine million tons, which at current market prices is worth about five million million pounds sterling. When gold was first detected in sea-water in 1886 it was reported that the concentration in the English Channel was 65 mg/ton. In 1900 another investigation revealed that this figure was much too high and the true concentration was reported to be about 6 mg/ton. After the 1914–18 War, the German chemist and Nobel Prize winner, Dr Fritz Haber, suggested, on the basis of these figures, that the German war debt might be paid by extracting gold from the sea. Unfortunately, the analytical methods used in the earlier determinations of gold in sea-water were grossly inaccurate, and Haber's search of the oceans failed to reveal gold concentrations much greater than 0.001 mg/ton. There have since been many other attempts to extract gold from sea-water but all have proved unprofitable.

The concentration of uranium in sea-water is one thousand times that of gold, and the increasing demand for uranium as a fuel for nuclear reactors has led in recent years to an interest in extracting it from the sea. A process in which sea-water is passed through a bed of titanium dioxide, which adsorbs uranium, has been developed but cannot yet compete commercially with extraction from land ores. This type of process could become important, however, if prices of uranium increase, especially if it could be combined with the simultaneous extraction of other elements.

The failure, so far, of man to profit from the vast quantities of these and other rare elements in sea-water is due largely to their very low concentrations. A number of elements are, however, selectively concentrated by natural processes occurring in the sea. Large areas of the deep ocean floor are littered with black manganese dioxide nodules (figure 258). In addition to being very rich in manganese these nodules have in the course of their formation scavenged some of the trace elements from sea-water. These manganese nodules could become an important source of manganese, cobalt and copper if the problems of mining them from the deep ocean floor can be solved.

The concentrations of many elements in marine plants and animals are considerably higher than their concentrations in sea-water (Table 8). Some of these elements are known to have specific biological functions, e.g. copper is an essential constituent of haemocyanin, the respiratory pigment of some invertebrates. Less is known about the role of others and the mechanisms by which they are accumulated. Smaller organisms extract some elements directly from sea-water but larger ones, in addition, derive some from their food. Although the enrichment factors shown in Table 8 are very high, in most cases the actual concentration of the element in the organism is still low. With the exception of seaweeds, which have been successfully exploited for their iodine, potash and alginic acid contents, marine organisms are still far more valuable as food than as sources of chemicals.

Because they have a shorter life cycle than most animals and have to get their food directly from the sea-water, marine plants (phytoplankton) have a much more marked effect on the composition of the water than do the animals. Phytoplankton grow only in the upper layers of the ocean exposed to sunlight. During the day the process of photosynthesis takes place, in which the plants remove carbon dioxide from the water and at the same time discharge a similar volume of oxygen. At night the reverse process takes place and oxygen is consumed while carbon dioxide is produced. In addition to oxygen and carbon dioxide, of which there is an ample supply in the surface layers exposed to the atmosphere, phytoplankton also require other nutrients such as phosphate, silicate and nitrogen compounds. During a bloom of phytoplankton these are removed from the sea-water so efficiently that they may disappear almost completely from solution. When growth stops as a result of nutrient depletion and the plants die, these nutrients are gradually returned to the sea-water by decomposition processes and the whole cycle can be repeated when conditions are favourable again (figure 133). Studies of seasonal variations of nutrients in sea-water give the biologist information on these cycles.

Below the euphotic zone the concentrations of the nutrients are much less variable. From near saturation values at the surface, the concentration of dissolved oxygen decreases with increasing depth to a minimum at intermediate depths. This oxygen minimum layer is due to consumption of oxygen by bacteria in decaying organic matter falling from the upper layers. At greater depths the dissolved oxygen content may increase again or level off to a constant value. Circulatory processes in the deep water replenish oxygen used up in respiration and decomposition of organic matter. It is only in certain areas such as the Black Sea and parts of the Baltic, where circulation is restricted or non-existent, that the water is completely devoid of dissolved oxygen.

The decomposition processes which lead to the formation of the oxygen minimum layer also liberate nutrients, the concentrations of which increase at intermediate depths. The deep waters of all the major oceans originate in high latitudes where cold, dense, surface waters, rich in nutrients, sink and spread towards the equator. These water masses can be identified and traced by means of their high silicate and phosphate contents as well as by their characteristic temperature and salinity (figure 63).

For the study of nutrient distributions it is desirable to carry out the analyses on board ship as soon as possible after sampling, in order to avoid changes which might take place in a sample during storage. Despite the obvious difficulties of carrying out such work at sea, suitable methods have been developed which make this possible. Dissolved oxygen is determined by a volumetric method, but gravimetric and volumetric methods generally are not sufficiently sensitive for the determination of the other nutrients. Instead, these are determined by fairly simple colorimetric methods in which the nutrient is converted to a coloured compound. The amount of colour produced, which depends on the amount of nutrient taken, is then determined by means of a spectrophotometer or by comparison with standards. Much of the tedium of such analyses has been eliminated in recent years by the development of automatic systems of analysis in which the sample of sea-water is fed into an instrument which then adds the necessary reagents to produce the coloured derivative of the nutrient. The resultant solution then passes through a photometer which automatically measures and records the amount of colour.

CONTAMINATION OF THE OCEANS

As well as all the naturally occurring elements which are being continually added to the oceans by rivers, a number of less desirable additions are also being made, both directly and indirectly, as a result of man's technological progress.

In the past 20–30 years, increasing quantities of synthetic herbicides and pesticides have been used in agriculture. Some of these are fairly quickly broken down in the soil, the rivers and the sea to harmless products; but others, such as the chlorinated hydrocarbons (of which DDT is probably the best known), are extremely stable and persistent. Their presence in the tissues of land animals and birds is not surprising in view of their widespread use. However, the finding that they are present in penguins and seals in the remote Antarctic has caused some concern. The route by which these pesticides can travel from the continents of the Northern Hemisphere, where they are most widely used, across thousands of miles of sea, is not clear. They are fairly volatile and could get into the atmosphere and be transported by winds before being precipitated with rain or snow. This process is only likely to make a minor contribution, however, as there is comparatively little precipitation over the Antarctic continent and pesticides have not yet been detected in Antarctic snows. They could also be transported directly via the sea-water, but this does not fit in very well with the known distribution of the ocean currents. Indeed, it has been calculated that it would take about 20 years for North Atlantic water to reach the Antarctic. A third, though again not entirely satisfactory, explanation is that the chlorinated hydrocarbons could be transported along the marine food chain. The stomach contents of the contaminated Antarctic penguins consisted almost entirely of krill, a species of euphausiid which is very abundant in the region and which forms the main food of the birds. Analysis of these stomach contents revealed the same pesticides as were found in the penguins, though in lower concentrations. Further evidence that marine food chains are involved, is the fact that chlorinated pesticides have been found in many species of sea birds. A particularly alarming case is that of the Bermuda petrel, which feeds only far out at sea, mainly on squids, and whose eggs and chicks have been found to be seriously contaminated with DDT residues. Reproduction of this bird has declined over the past ten years at a rate of over three per cent per year, and at this rate will fail completely by 1978. This is probably an extreme case, however. There is no evidence that pesticide levels in the marine environment generally are reaching levels which are dangerous to man, and increasing legislation to prevent their indiscriminate use should cause levels to fall in the future.

Another example of a pollutant which is added to the oceans indirectly as a result of man's activities is lead, the concentration of which in the surface waters of the northern oceans has increased five- to ten-fold during the past 50 years. Some of this comes from industrial and agricultural sources but most of it is attributable to the increased use of the anti-knock compound lead tetra-ethyl in motor-car fuels during the last 30 years. It is estimated that 350,000 tons of lead are introduced into the atmosphere every year and much of this finds its way into the oceans by way of rain and rivers. In the surface waters the element is ingested by marine organisms which, after death, fall to the ocean floor, thus removing about half the annual input of lead to the sediments fairly rapidly. At present, therefore, the presence of lead in the oceans is a much less serious problem than its presence in the air.

Radioactive contamination is one of the penalties of the nuclear age. A radioactive atom (or isotope) is identical chemically to a non-radioactive atom of the same element. It differs in that it emits a radiation which can be harmful to most forms of life exposed to it. Consequently, any plant or animal which accumulates, for example, strontium from sea-water will also accumulate radioactive strontium (strontium 90) if any is present, because it does not distinguish between the two. Because the radioactivity can be harmful, it is desirable that marine organisms should not be exposed to high concentrations of radioactive material.

Radioactivity enters the oceans from a number of sources, the principal one of which is fall-out from nuclear bomb tests. Some of the products of nuclear fission are short-lived and rapidly decay to harmless isotopes. Others retain their radioactivity for much longer periods, but, although these may be present in' significant amounts in surface waters of the ocean immediately after an explosion, they are fairly rapidly dispersed. The pool of radioactivity released by the explosion of a nuclear device beneath the sea surface in 1956, for instance, had been effectively dispersed within four weeks of the explosion.

The method of disposal of radioactive wastes from hospitals and laboratories which use radioactive isotopes and from nuclear reactors depends on the nature of the material. Each radioactive isotope decays at its own characteristic rate and there is no known chemical or physical process for accelerating the decay. Strontium 90 and caesium 137, for example, both lose half their residual activity every 28 years (which is called the 'half-life'), whereas carbon 14 takes 5,570 years. Large volumes of low-activity waste can be discharged safely into estuaries or directly into the sea where they will be diluted rapidly by natural mixing processes. Small volumes of high-activity waste are usually encased in metal or concrete for long-term storage at the bottom of the deep ocean in sites where there is little danger of accidental recovery or damage to underwater cables. All radioactive discharges are subject to strict control, and areas round disposal sites are carefully monitored to ensure that radioactivity does not return to man, either directly or as a result of commercial fishing. Although one case has been reported of a whale which became radioactive as a result of feeding on plankton near a nuclear waste disposal site, it has been calculated that humans who eat mainly sea food are exposed to considerably less radioactivity (as a result of their diet) than those whose diet consists mainly of food from land sources.

Oil pollution of the oceans started to become a serious problem soon after the First World War when oil began to replace coal as a fuel and large tankers were used to transport crude oil to refineries in Europe and America. Discharge at sea of the sludge which settles in the tanks and of the sea-water which is used both to wash out the tanks and to partly fill the tanks as ballast, during the return journey, has since led to widespread pollution of coastlines. Numerous attempts by international bodies to introduce legal control have resulted in a ban on the discharge of persistent oil or oily mixtures from ships within a zone which extends more than 1000 miles into the Atlantic Ocean. A procedure known as 'Load-on-Top', in which only relatively clean water, carefully separated from tank washings and ballast, is discharged at sea has been adopted by about 80 per cent of the world's tankers, so oil pollution caused by deliberate discharge is being dealt with in a responsible manner by the oil companies. Accidental pollution, however, raises tremendous problems because of the much larger quantities of oil involved. Between 1954 and 1964 there were 91 groundings and 238 collisions involving oil tankers, resulting in 39 cases of cargo spillage. When such accidents, involving hundreds of thousands of tons of crude oil, occur near land, the first objective is to prevent the oil from reaching the coast, either by removing it from the damaged ship or by disposing of it at sea. Setting fire to the oil is not very satisfactory as it results in rapid loss of the more volatile constituents and leaves semi-solid tarry residues. Emulsification of the oil by application of detergents and subsequent dispersion in the sea-water is more effective but expensive. It also suffers from the disadvantage that the solvents and detergents are extremely harmful to marine life, more so than the oil itself. Sprinkling a high density powder, such as chalk, over the surface has been successfully used to sink the oil, and this method is neither toxic nor expensive, but the dangers of blanketing bottom-living fauna, fouling fishing gear, tainting of marine organisms, and the return of the oil to the surface have all been raised as objections. The real damage from oil pollution, however, is not so much at sea, where it starts, but on the coastlines when it reaches land. A survey by the Royal Society for the Protection of Birds has revealed that 100,000 birds die every year as a result of oil pollution on the shores of Britain alone. Other forms of wild-life, both flora and fauna, suffer drastically from the measures adopted by local authorities, justifiably alarmed at the threatened loss of valuable tourist trade, to clean up their beaches with detergents which are toxic, even in very low concentrations. The most acceptable solution to the problem of coastal oil pollution, for both the naturalist and the local authority, appears to be the future development of less toxic detergents.

In the future, man will undoubtedly take more from the sea in the form of food and valuable minerals. It is probably reasonable to put material back into the sea in the form of wastes which cannot be conveniently or safely disposed of on land. It is essential, however, before they are dumped in the sea, that due regard should be paid to the possible ultimate fate of such wastes.

5 The Fields of the Sea

The Ocean Flora *Theodore J. Smayda*

INTRODUCTION

It has been said that the beginnings of botany are in the sea; that land plants are dried-off seaweeds given a waxy cuticle to prevent desiccation, an internal plumbing system for flow of food and water, an anchorage system of roots, and a flexible rigidity to permit growth towards the sun. Some plants, such as eel grass, *Zostera,* and the turtle grass of tropical waters, *Thalassia,* have returned to the sea where they co-mingle with seaweeds to form communities of variable organization and abundance in response to the prevailing light, temperature, substrate, and surf and tidal characteristics. Like the seaweeds, sea grasses require light for photosynthesis and a suitable substrate for attachment. While seaweeds have relatively simple holdfasts which permit attachment to rocks and shells, an ancestral adaptation important for the evolution of land plants, the sea grasses have a rootlike rhizome system usually anchored in sedimentary material. It remains uncertain whether these plants assimilate nutrients primarily through this rhizoid network or, as in the seaweeds, by direct diffusion into the thallus.

The requirements for light and attachment to the bottom (*benthic* environment) have greatly influenced distribution of these plants. The sea grasses are restricted to the very fringes of the sea, whereas the seaweeds are confined to the littoral zone or the upper levels of coral reefs. In common with land plants, these plants literally grow towards the sun. And just as an upper limit on maximum elongation of about 100 m characterizes land plants, total length is also restricted in attached sea plants, but for different reasons. Elongation is limited on land by water transport problems, and in the sea by light, which is equivalent to saying that vertical distribution is light-limited in the latter environment. The depth limit appears to be the level to which only one per cent of the light incident on the sea surface penetrates – the depth (compensation depth) at which algal respiration equals photosynthesis. Since coastal waters vary in their transparency, regional differences occur in vertical distribution of algae and sea grasses. They are generally found at greater depths in the tropics than elsewhere, and at shallower depths in regions influenced by significant, turbid river run-off. If, as has been suggested, some Arctic seaweeds can utilize dissolved organic substances for their energy source (*heterotrophy*) in place of photosynthesis, then the depth of penetration of such forms would be independent of light transmission, unless light-dependent uptake (*photo-assimilation*) of these compounds occurs. Notwithstanding this, the maximum depth of occurrence of viable seaweeds is around 400 m in the Mediterranean, about 10 – 100 times the maximum depths of occurrence of the various sea grasses. The seaweeds and sea grasses are thus extremely restricted in their distribution; in fact, they occur in only one to two per cent of the ocean's illuminated area.

An awareness that seaweeds grow along coasts in the littoral zone partly triggered the mutiny of Columbus's crew during their westward voyage in 1492. The extensive patches of the floating seaweed, *Sargassum* (figure 121), encountered in the Sargasso Sea were falsely interpreted to indicate the presence of dangerous submarine rocks on nearby banks, with imminent risk of running aground or, in more imaginary visions, becoming hopelessly embedded in the weed. However, they were in the open sea. In an immense eddy of the Sargasso Sea, at temperatures above 18°C, extensive patches of this drifting 'gulfweed' often collect in enormous quantities over an area comprising $4 \times 10^6 \text{ km}^2$ or more. Some believe that the *Sargassum* weed is recruited from specimens torn loose from the coasts of Florida and Central America. Once within the Sargasso Sea they appear to be capable of vegetative propagation only, unable to form new reproductive organs unless re-attachment occurs – a scant possibility in the open ocean. Growing while adrift, the *Sargassum* weed continues to elongate and form numerous branches, but fails to elaborate a proportionate number of gas bladders vital for suspension. The density of the gulf-weed gradually increases with growth and this causes it to sink lower into the water column until it eventually disappears beneath the illuminated layer (*euphotic zone*) and, no longer capable of photosynthesis, perishes in deep water. Thus not even *Sargassum* is really adapted for life in the open sea, even though it represents the most successful attempt of the seaweeds to do so. This partially successful effort suggests that the existence of an indigenous high sea flora requires that it be capable of both suspension within the euphotic zone as well as sexual reproduction. Otherwise, occurrence is dependent on recruitment from other areas.

PHYTOPLANKTON – PLANT WANDERERS

In 1847 the Danish scientist A. S. Ørsted, during a voyage from Denmark to the West Indies, noticed a water discoloration in aquaria. Examination revealed *microscopic* algae to be responsible, prompting him to collect numerous sea-water samples from open tropical seas for microscopical examination. His startling observations revealed that, contrary to prevailing opinion, the open ocean was not devoid of plant life, aside from the drifting *Sargassum*. A luxuriant, floating microscopic plant assemblage also existed! Further, he concluded that these microscopic algae must represent the primary food source upon which the vast majority of animal life in the sea is dependent. His brief report,

121. Branch of a *Sargassum* species, natural size (from Winge 1923)

written in Danish, was generally overlooked by the scientific community and the tremendous importance of his theory not fully comprehended until the late nineteenth century, when renewed interest in these microscopic plants – the *phytoplankton* – arose.

Several decades after Ørsted's report, the notion prevailed that marine and terrestrial plant life showed similar abundance patterns – sparse in polar regions, but luxuriant in the tropics. This view made it difficult to reconcile the unquestionable animal richness of polar waters, evident from fish and whale catches, with the notion of low plant abundance there. It was understood that terrestrial animal life was ultimately dependent on plant life, and a similar dependence on plants for food and oxygen seemed a likely characteristic of marine food chains as well.

Aware of this dilemma, German scientists, notably Hensen, initiated a census of marine organisms, with one of the aims the determination of the causes of their seasonal and regional variations in abundance. These studies quickly revealed the astonishing diversity and abundance of unicellular, microscopic algae previously encountered by Ørsted and others wherever water samples were collected in illuminated layers. Without the holdfasts of the higher algae, these unicellular forms float, reproducing while in suspension. This pelagic mode of life makes them vulnerable to translocation to less favourable regions by currents, or removal from the euphotic zone and sufficient light for photosynthesis. Hensen aptly termed this assemblage of plants *phytoplankton*, or plant wanderers, derived from the Greek stems *phytos* (plant) and *plankton* (to wander). He also established, counter to the prevalent notion, that plant life in the form of *phytoplankton* was extremely abundant in colder waters and, in fact, exceeded tropical populations. Regional differences in phytoplankton abundance and its causes are still being elucidated.

Hensen's investigations also confirmed that phytoplankton are indeed at the base of the marine food chain, i.e. that they are the 'grass of the sea'. Their importance is readily seen from a comparison of their annual production of new plant material, as determined by recent measurements, with that of land plants. Expressed as carbon, the annual phytoplankton production of 10^{10} tons is at least equivalent to that of land plants, and probably greater by two- to three-fold. Thus, not only are phytoplankton the most abundant form of plant life in the sea, but these microscopic, unicellular, marine algae are also the most abundant form of plant life on earth since they inhabit about 70 per cent of its surface and account for at least one-half of the total global plant production – all while adrift in the ocean's surface layers.

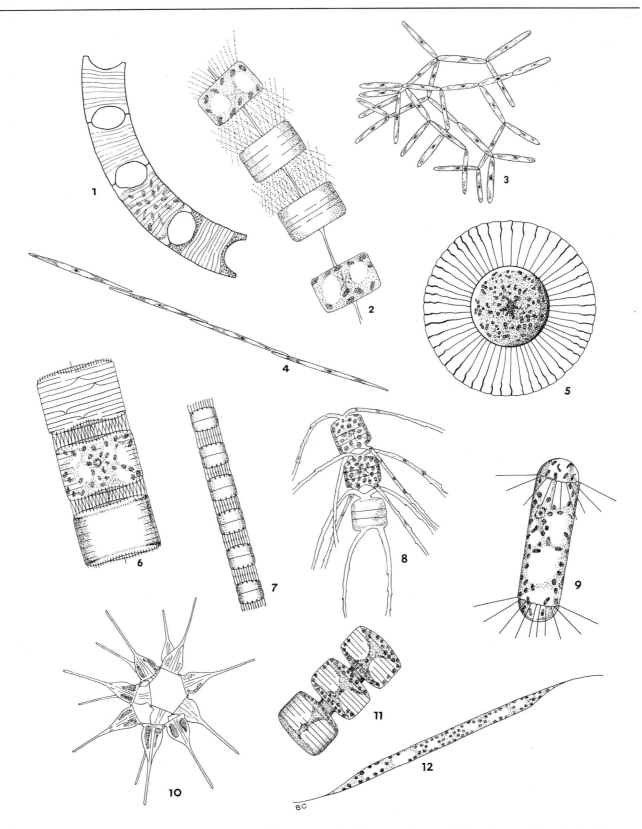

122. Diatoms (all very much enlarged) 1. *Eucampia zoodiacus*; 2. *Thalassiosira gravida*; 3. *Nitzschia frigida*; 4. *Nitzschia seriata*; 5. *Planktoniella sol*; 6. *Schröderella delicatula*; 7. *Skeletonema costatum*; 8. *Chaetoceros convolutum*; 9. *Corethron criophilum*; 10. *Asterionella japonica*; 11. *Porosira glacialis*; 12. *Rhizosolenia hebetata* forma *semispina*. (5 from Schütt 1892; all others after Hendey 1964)

Numerous algal classes represented by thousands of species comprise the phytoplankton. Diatoms, dino-flagellates, coccolithophorids, silicoflagellates, blue-green algae and a host of micro-flagellates from several algal classes have evolved mechanisms for a planktonic existence, some commonly shared, others unique, and some with attributes of unknown significance. New species are still being described. The salient character-istics of some of the more important forms represented in the phytoplankton, their ecology, and biogeochemical role in the ocean will be sketched. But there are other plants as well – bacteria, fungi, and yeasts – treatment of which lies beyond the scope of this account. Informa-tion on their myriad roles in transacting numerous biochemical processes in the sea, their diversity, and their ecology can be found in ZoBell's *Marine Microbiology* and Wood's *Marine Microbial Ecology*.

THE MEMBERS OF THE PHYTOPLANKTON
Diatoms:

The diatoms (figure 122), which possess a geological record going back to the Triassic (about 180 million years), occupy an interesting place in the plant kingdom. They manufacture a fairly thick (up to about 0·1 micron), hydrated, amorphous silica ($SiO_2.nH_2O$) cell wall (*frustule*) similar to opal. The use of silicon is remarkable since sea-water is undersaturated with respect to this element. Not only is their mechanism of silicon uptake and mineralization therefore intrigu-ing, but also the means by which diatoms prevent dissolution of their silicon frustule. Our knowledge of the latter is particularly scanty, although the presence of an outer organic membrane and/or a 'skin' of heavy metal ions has been suggested as protective mechanisms.

The evolutionary decision to elaborate siliceous cell walls required diatoms to solve the vital problems of the illumination of chloroplasts to permit photosyn-thesis, of nutrient and gas assimilation, of reproduction while encased in a rigid box, and of compensation for the heavy ballast of the silica cell wall in order to float.

The chloroplasts (cell organelles containing photo-synthetic pigments) are embedded in a thin cytoplasmic membrane adhering to the inner side of the silica frustule (figure 123). Since the opaline frustule is transparent, illumination of the chloroplasts is readily achieved. In fact, the diatoms might be considered as living in glass houses. Excessive light transmission, which commonly occurs at the surface in tropical ocean waters and during summer in colder waters, can bleach the chloroplast pigments and cause an inhibition of photosynthesis. Thus, a phytoplankton population might be simultaneously light-limited both near the surface because of high light intensities, and at the base of the euphotic zone because of low light levels.

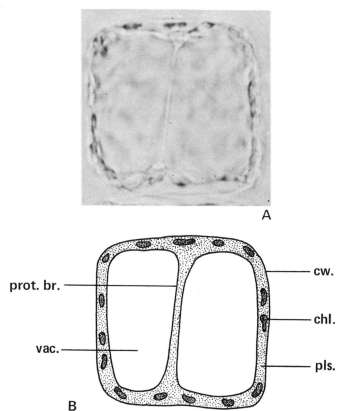

123. A. Photomicrograph of the major cytological features of the diatom *Lauderia annulata* Cleve, diameter approximately 30µ. B. Schematic presentation, girdle view: cw = silicon cell wall; pls = cytoplasmic layer; chl = chloroplast; prot br = protoplasmic bridge; vac = vacuole (from Smayda 1965)

Peculiar chloroplast movements in response to light intensity often occur, the chloroplasts clumping under high light intensities and spreading out under lower intensities. The former has been suggested as a mechanism to prevent light damage, and the latter to increase the capture of light for photosynthesis. The chlorophyll content of phytoplankton samples is often extracted with suitable solvents and then measured to approximate phytoplankton abundance and the poten-tial photosynthesis of this population.

Numerous perforations of variable size and arrange-ment present in the frustule, illustrated by the electron micrographs in figure 124, permit the diffusion of gases and nutrients into the cell. In addition to assimi-lation, secretion of dissolved organic substances ('ectocrines') occurs through these perforations in partial consequence of light conditions and age. These include organic acids which influence the availability of carbon dioxide needed by the cell for photosynthesis; polyphenolic compounds which may inhibit the growth

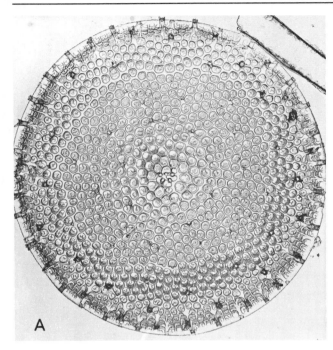

B

124. Electron micrographs of some diatoms
A. *Thalassiosira antarctica* Comber, valvar view, 2400x, a
centric diatom. B. *Fragilariopsis cylindrus* (Grun.) Krieger,
valvar view, 4100x, a pennate diatom

collapse or breakage. Architects might well derive inspiration from these 'glass houses' of the sea.

The cell walls of most plant and animal cells are soft and pliable, permitting volume changes and fairly simple fission of the cells to occur during reproduction. But the rigidity of the silicon frustule inherently precludes such volume expansion during growth or pinching-off of the diatom cell into two during cell division. The diatoms have resolved this problem by the magnificent and simple expedient of building their silicon cell wall with two halves, in which a slightly narrower, lower half (*hypotheca*) fits into the upper half (*epitheca*) pill-box style. During growth, the two halves pull apart telescopically with an asexual, vegetative cell division being consummated by the formation of a new bottom (inner) half for *each* of the parent cell's halves. Cell division is complete when the two daughter cells separate. This mode of cell division, which may occur up to three or four times per day during suitable growth conditions, poses the additional complication of a continuous reduction in cell diameter. With each cell division the slightly narrower lower half (*hypotheca*) of the parent frustule becomes the top half (*epitheca*) of one of the two daughter cells with a new, even narrower lower half being formed. The following scheme shows the fairly rapid reduction in average cell diameter accompanying diatom cell division. The number of cells after each division, starting with one cell, and their size frequency distribution (A = largest cells, F = smallest cells) reveal both the nature of their geometric growth (2^n) pattern and the rapid establishment of various cell sizes according to the binomial theorem.

Cell size	A	B	C	D	E	F	Total cell number
Parent cell	I						I
After:							
1*st* Division	I	I					2
2*nd* Division	I	2	I				4
3*rd* Division	I	3	3	I			8
4*th* Division	I	4	6	4	I		16
5*th* Division	I	5	10	10	5	I	32

Left unchecked, this progressive reduction in cell size would eventually prevent further division and even possibly lead to extinction of the species. The diatoms, however, have evolved an *auxospore* stage in their life cycle in which the vegetative cells become sexually differentiated into male and female cells. This occurs at a specific cell size, which varies from species to species, although both light intensity and duration,

of accompanying species, or even exclude them, or may protect the producer against bacterial attack. Other liberated substances may be growth-promoting substances, or may influence water quality by sequestering inorganic nutrients such as iron and other trace metals. The significance of this general phytoplankton attribute is still far from understood.

The type, size, pattern and number of frustule perforations are used by systematists as the basis for distinguishing between species. The use of microstructure for taxonomic purposes characterizes phytoplankton systematics in general, as will be further shown when discussing the dinoflagellates and coccolithophorids. This requires extensive use of electron microscopy, which has already revealed the astonishing architectural facility of the diatoms in building geodesic domes, ribbing patterns to minimize stresses, and general construction features needed to permit the elaboration of silicon into a porous, rigid matrix housing the diatom cellular contents without risk of

and certain trace metals, are known to trigger auxo-sporulation. Details of this matter are still being elucidated, but it is known that in certain species the male gametes swim to the female cells, where they enter through a small opening caused by a pulling apart of the epi- and hypotheca, and after migrating inwards, fuse with the eggs. Following fusion, the protoplasmic contents of the fertilized cell are extruded through the sperm entry pore, enlarge to the adult size normal to the species, silicify and then commence the vegetative cellular division described earlier. Thus, the pre-requisite of a sexual reproductive capability for habitation of the open ocean, as revealed by the *Sargassum* problem discussed earlier, is clearly met by the diatoms. One can only marvel at their unique solution to the photosynthetic, nutritional and repro-ductive restrictions imposed by the use of silicon as a cell wall material.

Solving the photosynthesis, nutrient diffusion and cellular division problems made it possible for diatoms to exist as attached (benthic) plants in the sea; but did not solve the problem of suspension in the euphotic zone. In fact, the diatoms are recognized as having two major sub-groups: the *pennate* forms (Pennales), which are bilaterally symmetrical, and the *centric* forms (Centrales) which are radially symmetrical. The pennate forms are invariably sessile, existing in coastal waters on the thalli of sea grasses and seaweeds, and predominate on or in the sea floor sediments, and outcrops. Many undergo jerky, propulsive movements of up to 20 microns per second, attributable to cyto-plasmic streaming in a structure *(raphe)* exposed to the environment. Some pennate diatoms undertake vertical migrations in the upper two or three centimetres of the sea floor, and these frequently cause diurnal discolora-tion of tidal flats. Within the open sea there is little pro-vision for attachment of these forms, although they have been found to grow attached to the flanks of some whales ('sulfur sides') and on *Sargassum* as a member of the 'Aufwuchs' ('growing on') community.

For diatoms to achieve a pelagic existence, it was required that the high density of silicon, which comprises from about 20–75 per cent of the dry weight, and the density of cytoplasm, which is greater than that of sea-water, be compensated in some way. Other-wise, rapid sinking from the euphotic zone would occur. The centric forms in particular have become success-fully adapted to flotation, although the mechanisms are not totally understood. Diatoms characteristically link individual cells of the same species into chains by gelatinous threads, spines, setae, silica rods and other protuberances (figures 122, 134, 135). Some are even enclosed in gelatinous envelopes to form pseudo-colonies, and still others remain solitary. It has been considered axiomatic that such colony formation is an adaptation for flotation, even though colony formation also characterizes *sessile* forms, both centric and pennate types. However, recent experimental observations reveal that colony formation invariably increases the sinking rate rather than decreases it, i.e. smaller colonies have a lower sinking rate than larger colonies. This suggests that colony formation is not an adapta-tion for suspension. In fact, the ecological significance of chain formation (figure 122) is unknown. It has also been speculated that it is an anti-predation device; increases the probability of successful auxosporulation by allowing production of male and female cells within the same chain; permits rotation of the chains (*rôtisserie*-like) which minimizes light damage of chloroplasts or equalizes illumination; or increases the contact area for convection by micro-turbulent currents.

Another view of the flotation mechanism is that selective discrimination against heavy ions (such as sulphate) in favour of uptake of lighter ions (such as ammonium) occurs. These ions, which would be concentrated in the large vacuole present in many species (figure 123), allegedly compensate for the excess ballast of silicon and cytoplasm and permit diatoms to float. Others have held that diatoms float because they produce and store fats, although this mechanism can now be demonstrated to be ineffective. Turbulence can be expected to play a role. Notwithstanding what-ever mechanisms may be operative to permit flotation, they eventually fail. This is readily apparent from the extensive deposits of diatomaceous ooze in certain areas of the open ocean, as well as in the diatomaceous earth deposits on land. Some believe that this deposition provided the organic precursors for the genesis of petroleum. If the sinking from the euphotic zone ordinarily occurs after successful growth, how does the population maintain itself? Many coastal species survive inimical environmental conditions by forming resting spores which settle on to the sea bed and germi-nate when environmental conditions permit. However, spore formation by open ocean phytoplankters would lead to rapid sinking below the euphotic zone with little chance of being eventually stirred back upwards by currents. It has been suggested that these species 'over-winter' within the euphotic zone by forming 'Schwebesporen' (floating spores) which germinate when environmental conditions permit. However, whether indigenous open ocean diatom (and other) populations are maintained in this way, or whether new populations are recruited through current movements is unknown.

Dinoflagellates:
The dinoflagellates (figures 125, 135), whose geological

record goes back to the Cretaceous (about 70 million years), possess two whip-like flagella permitting them to swim at speeds from about 0·05 to three centimetres per minute. They are capable of extensive diurnal vertical migrations in response to light, some species exhibiting a day rise to surface waters and others a night rise. Notwithstanding their motility, both their size and vulnerability to displacement by current movements make them *bona fide* and important members of the phytoplankton community.

Two main sub-groups are recognized. The armoured forms (Peridiniales) possess a cellulose cell wall subdivided into transparent, perforated platelets (2 in figure 125) whose arrangement and shape are important taxonomic characters. The unarmoured group (Gymnodiniales) comprises those forms which lack visible platelets (8 in figure 125) and are enclosed by a thin, structureless pellicle (cell wall); striations and excrescences are sometimes pronounced. The classical dinoflagellate cell (3 in figure 125) has one flagellum which lies in a groove *(girdle)* encircling it equatorially and acts as a rotator, the second flagellum is directed backwards and propels. In other species (10 in figure 125) the two flagella arise anteriorly and pull the cell through the water.

The dinoflagellates reproduce by binary fission in which the cells divide transversely without the concomitant reduction in cell size characteristic of the diatoms; sexual reproduction is poorly understood.

The dinoflagellates are an extremely heterogeneous group morphologically and physiologically, and include forms which bridge the plant and animal kingdoms. While many are capable of photosynthesis, a large number are capable of heterotrophic nutrition (i.e. utilize dissolved organic compounds as an energy source), either partially or totally. Some even consume particles *(holozoic)* which they ingest by opening a pore located at the site where the girdle meets the *sulcus* (the longitudinal groove in which the propelling flagellum is located). The form *Oxyrrhis marina,* in fact, preys on diatoms, while *Noctiluca,* famous for its bioluminescence, can even ingest fish eggs. Other representatives, members of the genera *Oodinium* and *Paradinium,* parasitize fishes, fish eggs and zooplankton (animal plankton) and often inflict serious tissue damage leading to high mortalities. The life cycle of most parasitic forms includes a free-swimming stage which has a readily recognizable dinoflagellate shape.

Still others occur as zooxanthellae in corals and sea anemones, for example, in which the dinoflagellate becomes sessile, but retains its photosynthetic capabilities and becomes involved in some sort of symbiotic association with the host animal. Reproduction can occur within the host, as well as in the free-swimming stage which sometimes results from a massive emigration of the zooxanthellae, as in corals suddenly exposed to a significant reduction in salinity accompanying hurricane rainfall. The ecological significance of these dinoflagellate (as zooxanthellae) – animal relationships is obscure. Although it is known that some zooxanthellae can use nutrients secreted by their hosts, that organic material secreted by these algae is available to the coral, and that the zooxanthellae assist significantly in the calcifying process, the fundamental role of the zooxanthellae in the coral economy remains unanswered.

The dinoflagellates are renowned for their bioluminescence in which the emission of light from special organelles is often detectable at night. Myriad, brilliant, short-lived bursts of light may dart across the sea surface, and the crests of rolling waves glow as the dinoflagellates activated into emission through water disturbances create an aquatic version of *aurora borealis.* Though open ocean displays may be brilliant, the most spectacular occur in Bahia Fosforescénte, Puerto Rico, and Falmouth Harbour, Jamaica, where prodigious numbers of dinoflagellates predominate, for reasons unknown. Fish swimming through the dinoflagellate swarms are readily apparent in the induced bioluminescent background. The biophysics of dinoflagellate bioluminescence is being actively investigated; its ecological significance is unknown.

Although phytoplankton are usually invisible to the unaided eye, periodically the explosive growth of some representatives to levels approaching population densities of 10^8 cells per litre causes a marked, readily apparent water discoloration. The most famous examples are the 'red tide' outbreaks of dinoflagellates which may discolour the water from a rusty-red to a tomato red. The development of these 'red-tide' outbreaks may be accompanied by the production of potent endo- or exo-toxins, dependent on the species, leading to the massive mortality of fishes and other animals. For example, 'red tide' blooms of *Gymnodinium brevis* in the Gulf of Mexico frequently cause such massive mortality, although it is not always easy to distinguish between mortality induced by a toxin or resulting from asphyxiation due to clogging of the gills by the dinoflagellates. *During the 1946–7 outbreaks more than 10^8 kilograms of fish were killed, with dead fishes washed on to the beaches of the west coast of Florida at a rate of 150 kilograms per linear metre!* The 'red tides' caused by certain species of the genus *Gonyaulax* (14, 15 in figure 125) are particularly insidious. A strong neurotoxin is produced which accumulates in certain shellfish ('paralytic shellfish poison') and which often leads to the death of humans consuming them. Among the many recorded deaths, 150 people died from it in 1799 near Sitka, Alaska.

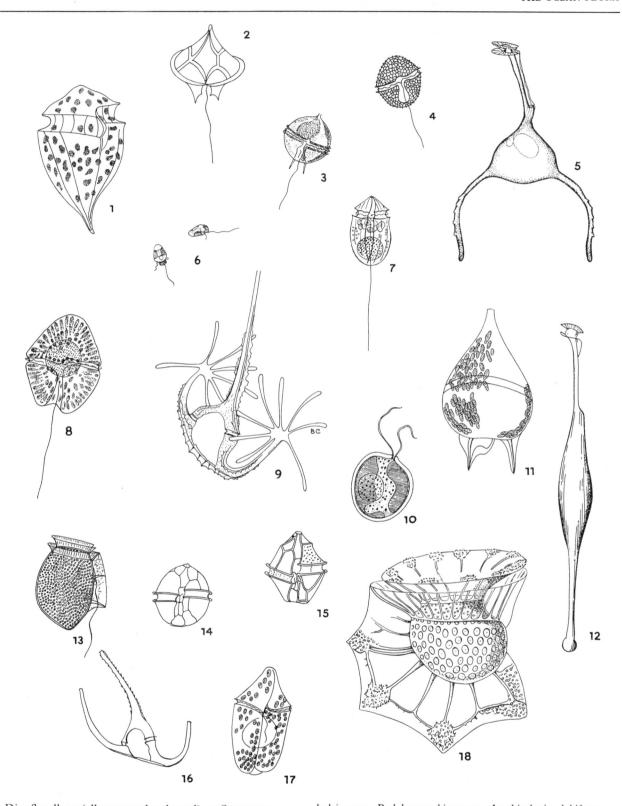

125. Dinoflagellates (all very much enlarged): 1. *Oxytoxum tessalutum*; 2. *Peridinium divergens*; 3. *Peridinium globulus*: 4. *Protoceratium reticulatum*; 5. *Triposolenia bicornis*; 6. *Massartia rotundata*; 7. *Amphidinium phaeocysticola*; 8. *Gymnodinium splendens*; 9. *Ceratium ranipes*; 10. *Exuviaella baltica*; 11. *Podolampas bipes*; 12. *Amphisolenia globifera*; 13. *Dinophysis acerta*; 14. *Gonyaulax tamarensis*; 15. *Gonyaulax polyedra*; 16. *Ceratium longipes*; 17. *Gymnodinium conicum*; 18. *Ornithocercus steinii* (1, 11 after Schütt 1895; 5, 12, 18 from Gran 1912; 9 from Schiller 1937; all others from Lebour 1925)

While the diatoms appear to be especially favoured by colder, nutrient rich water, the dinoflagellates are represented by a diversity of forms in all water types. They seem to be the most varied components of the phytoplankton in ecological character. Collectively, however, they appear to prefer warmer waters. While their powers of motility presumably eliminate some of the flotation problems confronting the diatoms, there is evidence to suggest that this ability permits them to search out more suitable layers within the euphotic zone, in addition to their phototaxic (or light orientated) movements. The warmer water-masses tend to be somewhat nutrient deficient, and perhaps more effective utilization of sparse nutrient concentrations accompanies swimming. Implicit in such a view is the suggestion that flagellates, including coccolithophorids, would be better adapted to tropical, nutrient-poor waters than diatoms. The general occurrence of these groups tends to support this deduction, although detailed analysis of this is required.

Coccolithophorids:

The geological record of the coccolithophorids (figure 126) extends back to the Cambrian (500 million years), and therefore they appear to pre-date the diatoms and dinoflagellates. They are extremely important micro-palaeontological components of the fossil record. The coccolith assemblages of deep sea cores reveal that a major faunal and floral change probably occurred within the Lower Pleistocene (about one million years ago) in which a change from colder to warmer conditions is suggested. Although the approximately 250 recent species are considerably fewer in number than the planktonic diatom and dinoflagellate species, the coccolithophorids are an extremely important group in tropical waters. For example, they appear to be the dominant component of the Mediterranean phytoplankton both numerically and in protoplasm weight (biomass).

The coccolithophorids are biflagellated organisms capable of the light-induced phototaxic movements exhibited by the dinoflagellates. The most unique aspect of the coccolithophorids (figure 126) is their manufacture of minute (1–35 microns in diameter) calcareous plates (coccoliths) which cover the external surface (figure 127). The number and type vary with species, although increasing evidence indicates that even within the life history of the same species a remarkable variability in coccolith cover may occur. Cells may be produced with or without coccoliths; those possessing coccoliths may or may not possess flagella; and the coccolith type and nutritional mode of the motile phase may differ appreciably from those in the non-motile, but still planktonic, phase of the same

species. These are but a few examples of the enormous plasticity of the coccolithophorid life history (of which our knowledge is extremely limited), but are adequate testimony to the amazing degree to which these forms have adapted to a planktonic existence.

The basis of coccolithophorid systematics lies in the crystalline structure of the calcite coccoliths (although the calcium carbonate may be present as aragonite or vaterite under certain conditions), a fact which requires the use of electron microscopy for effective investigation. It is still unresolved whether the carbon dioxide uptake during photosynthesis is at all connected with the uptake of inorganic carbon during coccolith formation. Notwithstanding the presence of coccoliths, sufficient light penetrates to the chloroplasts to permit photosynthesis in those forms where it occurs. The predominance of coccolithophorids in the highly illuminated tropical oceans has been partially attributed to an alleged light screening role of the coccolith layer preventing light injury of the chloroplasts and impairment of photosynthesis. Experimental work on the widely distributed *Coccolithus huxleyi* does not support this, and the overall reasons for the success of the coccolithophorids to thrive in the warm, nutrient impoverished, tropical waters are unknown.

The coccolithophorids reproduce by binary fission, in common with all components of the phytoplankton, and there is evidence that sexual reproduction can also occur. This simple mode of cell division permits explosive growth under suitable conditions, including the perplexing periodic and unpredictable development of 'chalky water' (figure 79) in the Norwegian Sea accompanying the enormous concentration (up to 10^8 cells per litre) of *Coccolithus huxleyi*. This milky green discoloration, attributable to the coccolith cover, is often accompanied by good fishing prospects, in contrast to the fish mortality accompanying the 'red-tide' blooms of dinoflagellates.

In summary, essential details of the morphology, life history, physiology and ecology of the coccolithophorids are still unknown. The coccolithophorids also conform to an apparent paradox of the phytoplankters. Notwithstanding the necessity for assuring flotation for successful participation in the phytoplankton, the coccolithophorids and diatoms mineralize calcium and silicon, respectively, with an attendant considerable increase in cell density that would not occur if 'naked' cell walls were formed. To a lesser extent, the dinoflagellates exhibit a similar trend by their formation of cellulose platelets. Even though the dinoflagellates and coccolithophorids can swim, such movements are seemingly encumbered by the calcium and cellulose ballast present, and impairment of motility quickly leads to sinking. That is, such cell

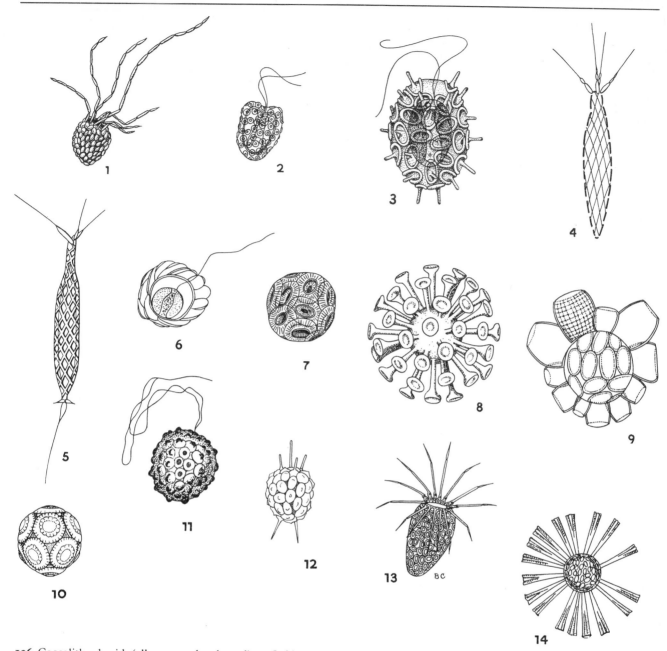

126. Coccolithophorids (all very much enlarged): 1. *Ophiaster formosus*; 2. *Hymenomonas roseola*; 3. *Syracosphaera subsalsa*; 4. *Acanthosolenia mediterranea*; 5. *Calciosolenia sinuosa*; 6. *Deutschlandia anthos*; 7. *Coccolithus pelagicus*; 8. *Discosphaera thomsoni*; 9. *Scyphosphaera apsteini*; 10. *Tergestiella adriatica*; 11. *Acanthoica acanthos*; 12. *Acanthoica quatrospina*; 13. *Michaelsarsia splendens*; 14. *Thorosphaera elegans* (6, 8, 14 from Schiller 1930; all others after Deflandre 1952)

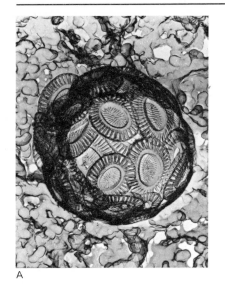

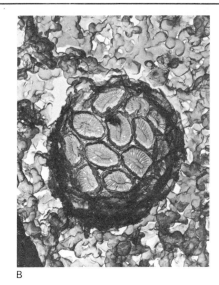

127. Electron micrographs of some coccolithophorids and their coccoliths : A. *Coccolithus huxleyi* (Lohmann) Kamptner, intact cell, 5330x. B. *Syracosphaera tuberculata* Kamptner, 4165x. C. *Discosphaera tubifera* (Murray and Blackman) Ostenfeld, 3375x. D. *Rhabdosphaera stylifera* Lohmann, 3188x. E. Coccolith from *Coccolithus huxleyi*, 10000x. F. Coccolith from *Rhabdosphaera stylifera*, 9000x

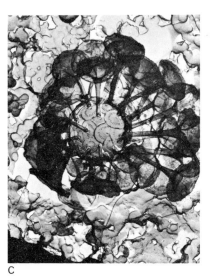

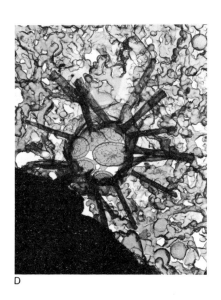

C

D

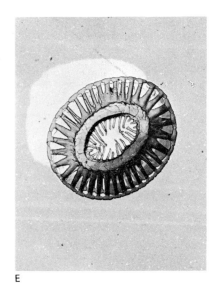

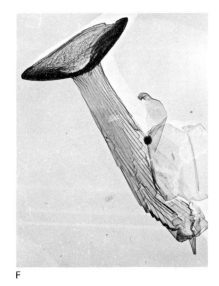

129. *opposite above* The little scarlet and white prawn *Parapandalus* (5 cm) has a light organ in the thorax and carries its lavender-coloured eggs on its swimming legs until they hatch

128. *opposite below* the splendid prawn *Sergestes* (10 cm) is almost transparent in life apart from its few splashes of colour

E

F

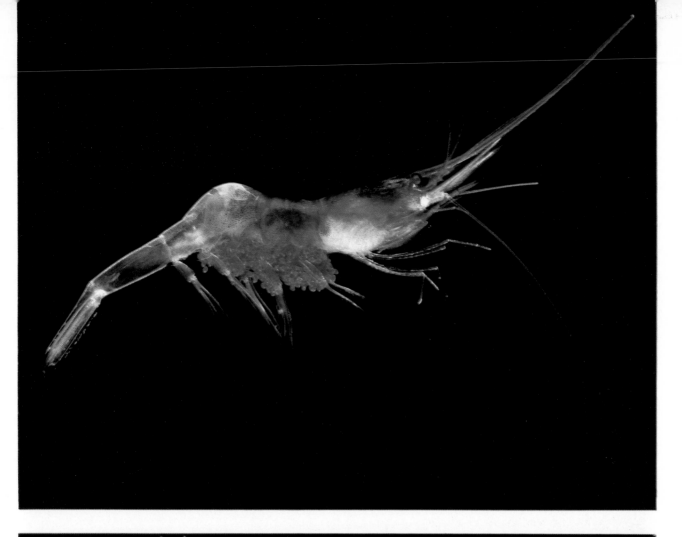

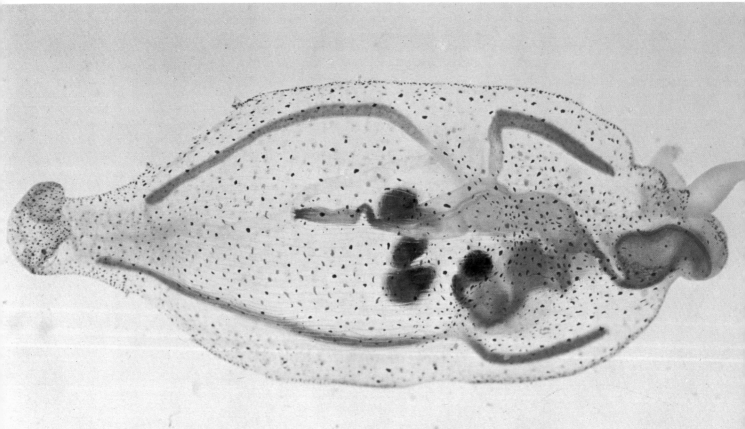

wall construction would appear to compound their suspension problem. But these phytoplankters are quite successful, for within the euphotic zone in tropical waters even a vertical zonation in certain species occurs. Near the base of the euphotic zone a 'shade flora' exists consisting of such giant, solitary diatoms as *Planktoniella sol*, characterized by a wing-like parachute (5 in figure 122), dinoflagellates belonging to the genus *Ceratium* (9, 16 in figure 125) and coccolithophorids such as *Deutschlandia anthos* (6 in figure 126) and *Scyphosphaera apsteini* (9 in figure 126) heavily laden with coccoliths.

Diverse forms:
Other components of the phytoplankton, of greater or lesser importance than the foregoing groups, also occur under certain conditions and in certain regions of the open seas. Our very limited knowledge of them precludes more than a cursory sketch of their biology and ecology.

1. *Silicoflagellates*: These bi-flagellated, solitary forms possess a siliceous exoskeleton of various patterns, and range in size from 10–150 microns (1 in figure 132). Their fossil record suggests that they were more important in past geological eras.

2. *Blue-green Algae*: Ørsted's revelation of the role of phytoplankton in the sea, described earlier, stemmed from his determination that the blue-green alga, *Trichodesmium* (2 in figure 132), which caused the water discoloration noted in his shipboard aquaria, was ubiquitous in tropical waters. In fact, the Red Sea derives its name from the slight water discoloration attributable to the reddish pigmentation (notwithstanding its blue-green algal affiliation) of the common *Trichodesmium erythraeum*. This microscopic, filamentous species, and other members of its genus, collect into loose bundles, which aggregate even further and often collect into readily visible windrows at the sea surface (figure 101). At times, enormous and extensive 'blooms' occur – in the South Pacific Ocean a 'patch' covering 20,000 square miles was observed. The behaviour of *Trichodesmium* is especially important in at least two related contexts. It often appears when the water mass is especially warm and nutrient impoverished, conditions taxing even to the versatile dinoflagellates and coccolithophorids, and this replacement of species would appear to be important to the food chain. The reasons for *Trichodesmium's* development may include a capability to use elemental nitrogen (N_2) in place of inorganic or organic nitrogen molecules. If nitrogen is indeed fixed, then this might help to reconcile the present inability of chemical oceanographers to balance the nitrogen budget in the ocean.

Small, spherical bodies called 'yellow-green' cells, thought to be blue-green algae, also occur. The vertical distribution of these cells extends deep into the *aphotic* (without light) zone, which suggests that they are capable of heterotrophic growth. Beyond this, we know very little about these plants, or the extent to which heterotrophic algal growth occurs in deep waters.

Aside from the primary predominance of blue-green algae in nutrient impoverished tropical waters, a number of forms appear to prefer brackish waters. Still others, however (just as some dinoflagellates exist as zooxanthellae), have established an internal relationship with other members of the phytoplankton community. Thus *Richelia intercellularis* lives internally within certain (usually tropical) species of the diatom genus *Rhizosolenia* (12 in figure 122). Small organelles, *phaeosomes*, present in the lists (wing-like structures) of the elegant dinoflagellate *Ornithocercus* (18 in figure 125), have recently been shown to be blue-green algal cells in some instances. Some blue-green algae are also *endolithic* (within stone), burrowing into coral reefs while apparently still capable of photosynthesis. The general details of these supposedly symbiotic relationships remain unknown and controversial, as does the matter of the overall contribution of the endolithic blue-green and zooxanthellae to the coral reef economy.

3. *Micro-flagellates*: This catch-all grouping includes members of several algal classes, all characterized by the presence of one or more flagella (figure 132). But beyond that, they collectively cover the gamut of known nutritional modes, some are transients as part of a complex life cycle, and many are so tiny (1–2 micron) as to be virtually undetectable. These characteristics have severely interfered with the systematic collection, assessment and investigation of these forms, with the unfortunate consequence that their importance, biology and classificatory relationships remain largely unknown.

METHODS OF PHYTOPLANKTON INVESTIGATION
Qualitative and quantitative investigations of phytoplankton behaviour have perforce required the development of suitable methodology for the collection, census taking, monitoring of physiological events, and even cultivation of these forms. The minuteness, diversity of nutritional mode, taxonomic heterogeneity, and general ecological behaviour of the phytoplankton have posed formidable problems in this regard. The initial, extensive use of conical nets towed through the water to collect samples is now universally recognized as permitting the escape of many of the smaller forms

130. *opposite above* The wasp-like stomatopod or mantis shrimp *Lysiosquilla* is common in the Indian Ocean, and can give a painful nip with its clasp-like pair of front legs

131. *below Phyllirrhoe* (2 cm) is a leaf-like mollusc whose surface is dotted with light organs and which is often found in close association with a small medusa; it has been suggested that the young and tiny *Phyllirrhoe* is parasitic on the medusa, but finally outgrows it so that the medusa becomes the parasite

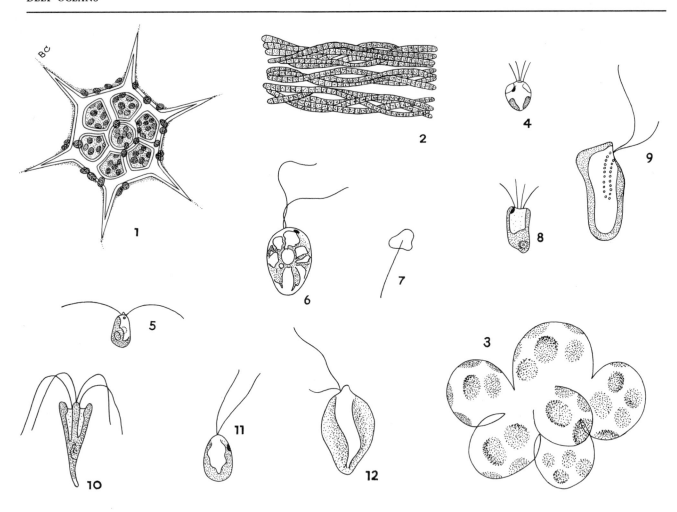

132. Diverse phytoplankton forms (all very much enlarged):
1. Silicoflagellate, *Distephanus speculum* (from Marshall 1934).
2. Blue-green alga, *Trichodesmium* (modified from Desikachary 1959). 3. Haptophycean, *Phaeocystis pouchetti* (from Lemmermann 1908). 4–12. Various microflagellates (after Conrad and Kufferath 1954).

through the net mesh. This selective capture led to a serious controversy early this century as to the role of phytoplankton in animal nutrition; a matter which will be considered later. Phytoplankton samples are now collected in closed samplers, preferably non-metallic, of various volumes, which are sent open to pre-selected depths, then closed, and the entrapped water sample retrieved. Sampling is complicated by the patchy vertical (and horizontal) distribution of phytoplankton in consequence of water movements, sinking, or their own phototaxic movements. A population census is made using suitable microscopic equipment, including counting chambers if a population as well as species census is to be made. It is necessary to preserve the sample to prevent the population from spoiling or changing when immediate analysis is impractical. This

causes the destruction of certain forms, such as unarmoured dinoflagellates and micro-flagellates, with a subsequent bias in evaluation. Nonetheless, this general approach can provide an adequate insight into the seasonal and regional variations in species composition and abundance, particularly when a proper sampling programme is followed. The disadvantage of this approach is the difficulty in converting numbers of phytoplankton cells into tissue weight, the amount of carbon available for predators, etc. This has led to the development of *gravimetric* techniques in which the sample is filtered on to suitable membranes, dried and weighed after various other pre-treatments. While this provides a weight estimate, such as carbon, contamination by non-living material can be significant. This treatment also destroys the material. More often, *proximate* analyses are made in which the amount of chlorophyll, following suitable extraction, or carbon, nitrogen or protein, etc., is measured. There are disadvantages shared with the gravimetric procedures, although it is often the better approach of the two. Clearly, these methods provide different information, and it is usually necessary to employ several different approaches in a particular study. It is also usually necessary to monitor various physical, chemical and biological parameters in field studies such as light, temperature, salinity, nutrients and zooplankton (animal plankton). Contemporary efforts have concentrated on developing gear and analytical techniques to permit continuous sample collection and simultaneous monitoring of many environmental variables.

Physiological events are monitored with natural populations by enclosing them in suitable containers and conducting experiments either *in situ*, on shipboard, or ashore. Measurements of primary production (rate of fixation of carbon in photosynthesis), treatment of which is beyond the scope of this article, are based on such an approach. It is becoming increasingly evident that a satisfactory knowledge of phytoplankton behaviour requires field studies to reveal what species are found in a given area, their abundance, and seasonal behaviour, in combination with experimental studies on natural and cultured species, which potentially can reveal why a species is found in a given area and why it is abundant or unimportant. Such studies can also provide valuable information on species' life histories, and their suitability as a food source. Thus, contemporary phytoplankton investigators use a variety of methodological approaches and qualitative and quantitative techniques in tackling problems dealing with natural and cultivated populations. The following sections will sketch some characteristics of natural phytoplankton communities, as well as several contemporary developments.

PHYTOPLANKTON REGIONS AND SEASONS IN THE SEA
Phytoplankton have successfully colonized the surface waters of the sea throughout its length and breadth. Even under the ice fields of Polar waters, green micro-flagellates respond to faint light intensities. And certain diatoms thrive on the undersides of ice, developing brown patches which absorb heat to a greater extent than surrounding clearer areas and which therefore lead to faster melting. Solar radiation then reaches the sea surface directly through these leads in the ice, and many of the attached diatoms then assume a planktonic existence. How other forms, apparently incapable of such growth, or those completing their growth pelagically, survive until the next freeze is unknown. Several typical coastal species have gained a

foothold within the Norwegian Sea, presumably because they are able to 'overwinter' frozen within the ice pack, which substitutes for the bottom of coastal waters. How this is achieved is also unknown.

A general review of phytoplankton distribution reveals that the diatoms are especially prevalent in colder regions, particularly Polar waters, while the dinoflagellates, coccolithophorids and blue-green algae are more prevalent in tropical waters. Within these general distributional patterns, further sub-groupings attributable to more circumscribed environmental conditions especially favourable to one of the phytoplanktonic groups are evident. Thus, within warmer water-masses, coccolithophorids and certain microflagellates are especially prevalent in the Mediterranean Sea; dinoflagellates are especially abundant in certain tropical coastal embayments contiguous with the open sea; and blue-green algae predominate where surface temperatures reach 25°C and above, to list some examples. The ecological basis for these general distribution patterns is virtually unknown.

Notwithstanding their apparent preference for certain environmental conditions, all groups are represented in every biogeographical area. For example, dinoflagellates and coccolithophorids also occur in Polar waters, though in diminished importance numerically and by species. Their general environmental preferences, however, are frequently reflected in their seasonal occurrence with such environments. Thus, diatoms grow well in tropical waters during the colder, more nutrient-rich periods, whereas coccolithophorids and dinoflagellates become more important during the summer months in temperate and colder regions. This phytoplankton group succession, however, is overshadowed by a more spectacular species succession during an annual cycle.

Clear geographical range boundaries generally do not occur, in part because occurrence and distribution reflect the growth-supporting potential of the watermass as well as its movements. It is now clearly understood that no single one of the three factors, temperature, light, or nutrients, is usually the ultimate determinant of whether a species will survive or not. Rather, it is the combination of these interacting factors which is more important. For example, experimental studies have shown that the apparent exclusion of a species from a given area because of high temperatures can sometimes be overcome if nutrient levels increase. The potential significance of this is that man may eventually be able to control, at least in coastal water, the nature of the phytoplankton community, selecting for more desirable species and even preventing 'red-tide' developments through proper environmental manipulation. The antithesis of this is that the increasing pollution of these waters, coupled with our still mediocre knowledge of the phytoplankton, may select for highly undesirable species. The practical significance of this factor interaction is that the distributional boundaries of species are somewhat variable from year to year, rather than fixed. This is further accentuated by the influence of current movements and systems on distribution. In fact, the association of certain species, such as the diatom *Planktoniella sol* (5 in figure 122), with specific currents has led to their use as indicator species of the presence of that current in a particular region. The effectiveness of currents in influencing species distribution is revealed by the sudden, unpredictable invasion of the diatom *Asterionella japonica* (10 in figure 122) periodically into eastern Norwegian coastal waters. This introduction apparently requires crossing of the Norwegian Sea by cells perhaps derived from northern England coastal waters. There is overwhelming evidence now that the various currents and eddy systems have their own particular assemblage of phytoplankton species. This illustrates an additional, more patchy, type of distribution characteristic of the phytoplankton.

Within a one-litre sample, the number of phytoplankton *species* present may vary from less than 10 to more than 250. Populations are essentially mono-specific during 'red tide' and 'chalky water' outbreaks, and most diverse during poor growth conditions. There appears to be a greater species diversity in tropical than in cooler waters.

Within any given area, seasonal oscillations in phytoplankton abundance and species composition occur in response to complex, seasonally changing interactions of light, temperature, vertical turbulence, plant nutrients (nitrogen, phosphorus, silicon, iron, manganese, molybdenum, vitamins, among others), and herbivore abundance. Phytoplankton cell numbers may vary seasonally from less than 10^3 to more than 10^8 per litre. Figure 133 depicts a very general type of annual phytoplankton cycle and some accompanying environmental trends. Two phytoplankton peaks are evident: a major pulse during spring, and a modest autumn bloom. Polar oceanic waters have a single phytoplankton maximum usually during mid-summer, whereas tropical oceanic waters may exhibit a single or double pulse at various times of the year depending on local conditions. Seasonal variations in abundance and environmental conditions are less pronounced in tropical than in Polar waters.

The causes of phytoplankton pulses have received considerable attention. Light and nutrients are especially important in this regard. Temperate and Polar waters usually have sufficient nutrients during winter and spring, but minimal growth occurs initially because of wind-induced mixing of the water-mass. The algal cells are then kept in fairly continuous vertical motion, moving into and out of the shallow euphotic zone, with the consequence that their residence time within the illuminated zone, where they photosynthesize, is barely sufficient to compensate for respiration and other losses while in the deeper, darkened layers. Growth is thus slow. The progressive vernal increase in light intensity and day length not only deepens the euphotic zone, but heats up the surface waters which dampens turbulence. (In Polar waters the reduction in surface salinity accompanying spring melting also helps to stabilize the water-mass.) The combined effect is that the phytoplankton now spend a greater period of time in the euphotic zone, which permits more prolonged nutrient uptake and photosynthesis resulting in rapid growth and the development of a phytoplankton peak (figure 133).

In tropical oceanic waters lack of nutrients prevents development of the main pulse. The warm, stratified waters and deep euphotic zone support little growth initially because previous growth has utilized most of the available nutrients. The severity of this nutrient limitation is demonstrated by the occurrence of brief, modest phytoplankton pulses following rainfall in certain areas of the Adriatic Sea. This growth is attributable, in part, to the feeble fertilization of the surface waters by the nutrients present in rain water. Nutrient limitation is usually overcome in tropical waters, however, by wind-induced mixing of the deeper, more fertile waters with that in the euphotic zone. The rapid consumption of newly available nutrients leads to a phytoplankton peak, the magnitude and duration of which is partially dependent on the wind behaviour which stirs nutrients upwards.

While the stimuli triggering phytoplankton blooms are reasonably understood, the factors involved in terminating these pulses are less clearly comprehended. Following the attainment of maximum populations, the apparent growth begins to wane leading to collapse of the bloom. Nutrients and grazing by zooplankton have been implicated as factors here. Enriching samples collected at the phytoplankton maximum and during its decline with various nutrients invariably stimulates growth. Nutrient concentrations often decrease to very low levels with increased growth (figure 133). This experimental and circumstantial evidence has led to the general conclusion that inadequate nutrient levels are usually involved in, if not responsible for, the termination of a phytoplankton bloom. Phosphate, nitrate, silicon and trace metals, such as iron, have been implicated.

The momentary abundance of phytoplankton is a balance between their rate of growth and their rate of

133. Schematic presentation of a type of annual phytoplankton cycle and accompanying general environmental trends

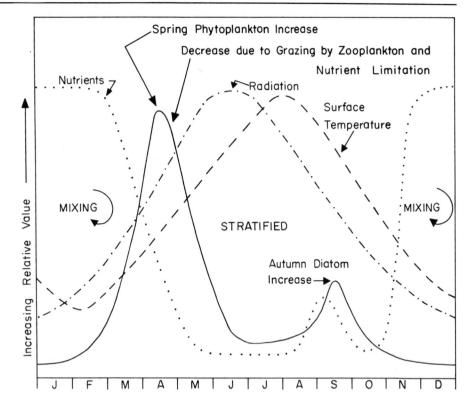

loss due to natural mortality, sinking from the euphotic zone, and grazing. Grazing is especially important since phytoplankton growth stimulates reproduction of the herbivorous zooplankton. It is imperfectly known, however, to what extent zooplankton predation is responsible for the termination of the main phytoplankton bloom, or controls the pre- and post-bloom population levels. At the turn of the century, there was a theory that phytoplankton abundance was too low to support the zooplankton populations, i.e. predation was a strong determinant of phytoplankton abundance. The zooplankton were therefore also required, this theory held, to assimilate dissolved organic substances through their body wall in order to survive. This erroneous conclusion was based on serious underestimates of phytoplankton abundance resulting from the use of net-sampling, the unsatisfactory technique mentioned earlier. Since simultaneous phytoplankton and zooplankton pulses occur, or the latter begins during active phytoplankton growth and reaches a subsequent maximum, circumstantial evidence suggests that zooplankton grazing might indeed be an important factor in termination of a phytoplankton bloom. In Polar and temperate waters the zooplankton pulse follows that of the phytoplankton, i.e. an *unbalanced* relationship. This population imbalance is partially attributable to the apparent requirement for a threshold phytoplankton population to stimulate zooplankton

egg production, and the subsequent effects of low temperatures on the hatching and larval development rates. Coincidental with the marked increase in herbivorous zooplankton numbers, phytoplankton levels usually begin to decrease (as do nutrient concentrations) – presumptive evidence for at least some involvement of the zooplankton in contributing to the phytoplankton decrease. Unbalanced cycles are characterized by the phytoplankton biomass exceeding that of the zooplankton during the main bloom, and by great seasonal fluctuations in phytoplankton abundance.

Tropical oceanic waters exhibit *balanced* cycles in which there is essentially a simultaneous development of phytoplankton and zooplankton pulses (high temperature levels permit rapid zooplankton development to maturity). Also, phytoplankton biomass is usually less than that of the zooplankton, and the seasonal fluctuations in phytoplankton abundance relatively small. This cycle has been viewed as one in which phytoplankton levels are especially controlled by grazing.

Colder areas exhibit a secondary, autumn phytoplankton pulse (figure 133). A relaxation of grazing intensity, coupled with the restoration of nutrient concentrations in the euphotic zone to higher levels through increased wind-induced mixing, has been cited as being an important factor in the development of this pulse. Depending on the region, grazing, nutrients and light have all been suggested as also being responsible for termination of this bloom. However, less is known about the dynamics of this pulse, where it occurs, than the major, annual bloom.

Thus, while there is considerable information about annual phytoplankton cycles, the underlying causes are still obscure in many respects. Most important, however, there are biological seasons in the sea in which populations wax and wane, rather than keep a steady abundance. This seasonal variability is mirrored by equally impressive regional variations in the sea's fertility. The sea thus has its oases and deserts, its winters and springs. This throb of cycles, converting sunlight into plant and fish, is so finely intermeshed that small fluctuations or variations one year can affect some aspect of this food chain several years later and hundreds of miles away. On a smaller, but no less important, scale, the uptake of insecticides, such as DDT, by phytoplankton in the Baltic Sea has endangered the fish hawk population at the upper end of the food chain.

The seasonal oscillations in abundance are accompanied by a remarkable species succession. This is a phenomenon in which a dominant species or community is replaced by another, which in turn is replaced, and so on. In some regions, species of *Thalassiosira* (2 in figure 122) dominate initially, to be replaced by *Chaetoceros* (8 in figure 122), these by *Rhizosolenia* (12 in figure 122), and this genus by dinoflagellates belonging to *Ceratium* (16 in figure 125). Not only does a species succession occur within the diatoms, but a succession from this group to the dinoflagellates and/or coccolithophorids, wherein there is also a species succession. The successional pattern and rate of species change differ regionally although major similarities in species exist. In general, the successional cycle is considered on an annual basis, although it is likely that a long-term succession may be operative in certain areas attendant with gradual *eutrophication* (nutrient enrichment), or any change influencing 'water quality' including increased temperature. The main point is that species succession is an important characteristic of phytoplankton populations.

Succession is more than a floristic phenomenon; significant biological consequences accompany these species changes as well. Not all algal cells, for example, are of equal food value to the zooplankton. Some animals depend on micro-flagellates for their successful development, and still others on diatoms, as vividly demonstrated for some barnacle species. Thus, zooplankton and benthic (bottom dwelling) animals whose pelagic larval stages eat phytoplankton must synchronize their reproduction to coincide with the occurrence of certain, palatable species. On the other hand, the appearance of a phytoplankton species may be disadvantageous to some component of the food chain. Thus, the occurrence of *Phaeocystis pouchetii* (3 in figure 132) is usually accompanied by the exclusion of herring from the area. *Phaeocystis* is a colonial form (but with periodic release of flagellated swarmers) enclosed in a gelatinous sac which imparts a strong odour to the water described as 'Dutchman's 'baccy juice'. This species, then, whose enormous population densities can slime nets, imparts substances to the water undesirable to the herring. The development of 'red tides' is a commonplace successional stage in upwelling areas – an event with potentially severe general consequences, as discussed earlier. Thus phytoplankton species succession also represents a succession of food availability, type and palatability, and water quality modifiers.

Within the phytoplankton themselves, the succession of forms is also accompanied by difference in rates of production and nutrient uptake, cell size, growth rate and life cycle, etc. While the floristic aspects of succession are generally well-documented, the physiological and general ecological attributes of these various successional stages are essentially unknown.

The mechanisms responsible for succession are unknown, even though natural and experimental observations now clearly indicate that a single factor is not basically responsible for the phenomenon. Rather, as for species distribution, a complex set of interacting chemical, physical and biological factors mediates succession. Some factors, such as light and temperature, are beyond control of the species, which are thus somewhat vulnerable to the seasonal and annual vagaries of such factors. Others, such as nutrient levels, are more directly influenced by phytoplankton, while selective grazing by zooplankton may be partially controllable through the abundance and colony size of the phytoplankton, although the evidence is scanty. An attractive hypothesis of the succession mechanism states that a form of 'biochemical warfare' is carried out by certain phytoplankton. Certain species secrete organic substances toxic to certain competing species, which prevent their occurrence; other compounds may function as growth promoters. Such views have been bolstered by admittedly astonishing examples of such algal antibiosis. For example, *Phaeocystis* (3 in figure 132), whose presence was reported above to disfavour predation by herring, has been implicated in the apparent absence of bacteria in the intestine of penguins. *Phaeocystis* is consumed by 'krill' (shrimplike crustaceans) in Antarctic waters which, in turn, are eaten by penguins. The gelatinous sac enclosing the

Phaeocystis colony contains acrylic acid (which becomes activated at the low pH high acidity) in the crustacean and penguin intestines. Acrylic acid prevents the development of the bacterial flora, because of its bactericidal nature – an activity which can withstand passage through the crustacean gut *en route* to consumption by the penguin! However, even such potent substances as acrylic acid are invariably operative only in conjunction with the temperature, light and nutrient levels in influencing phytoplankton succession, rather than being the primary mechanism.

PHYTOPLANKTON AND THE FUTURE

Prognostications are always risky, even when reasoned projections of the most probable course of events make such clairvoyance more precise. Unexpected methodological and scientific breakthroughs, or utilitarian needs, can set new demands, timetables and efforts spanning decades of reoriented research. Yet several trends suggest areas where phytoplankton studies will figure prominently. Man's increasing interest in harvesting the sea requires a knowledge of not only its ecology, but how to use this information to manipulate it to his benefit. Aquaculture or aquatic husbandry, in other words, is becoming relevant.

The seasonal variations in phytoplankton abundance discussed earlier mean in practice that seasonal variations in food availability occur. Nutrient limitation is somehow involved in establishing this cycle. Below the euphotic zone, significant nutrient supplies exist which can support phytoplankton growth, if they are made available. The previously discussed role of water-mass mixing in phytoplankton cycles not only reveals the potential contribution of these nutrients to the food chain, but also the mechanism through which they can fertilize the euphotic zone. Suppose then, that it were possible to upwell artificially these deep-lying nutrient reserves into the euphotic zone where continuous phytoplankton growth could occur to support a full-fledged food chain with a harvestable fish as the terminal product. This would introduce the requirement for confinement of these fish to minimize escape. A coral atoll would appear to be a natural site permitting such an effort. The clear, deep internal lagoon with abundant nutrient reserves at depth is surrounded by a barricade of coral islets and reefs easily modified, or extant, to minimize fish losses. What mixing rate should be used to maintain a phytoplankton abundance and species composition most efficient in permitting growth of the desired fish? The answer to this question is one goal of future phytoplankton studies. The answer isn't apparent. For example, near the Great Barrier Reef (Isle of Shoals) of Australia, virtually continuous mixing of the water-mass occurs with the euphotic zone extending directly to the bottom at 40 m. The phytoplankton populations do not exhibit a seasonal pulse and are rather low in abundance. This contrasts with conditions in the English Channel, off Plymouth, where pronounced seasonal phytoplankton variations occur – the cycle is *unbalanced* – with population levels considerably in excess of those at the Isle of Shoals. Yet, the zooplankton biomass is *greater* at the latter than at the presumably richer English Channel station.

The foregoing discussion considered not only phytoplankton abundance, but species composition as well. It is essential to understand species succession in order to permit future manipulation of the environment (primarily coastal now), to select for desirable species, and to prevent the establishment of less desirable forms. For example, what thermal pollution levels would elicit a favourable flora? This requires a knowledge of temperature requirements, and the interaction of this parameter with light and nutrients. The seemingly inevitable increase in this and other types of pollution requires rigorous control of water quality standards based on a knowledge of phytoplankton requirements and man's multi-purpose use of the marine environment.

Knowledge of the food value of phytoplankton will also become increasingly necessary as aquacultural, including hybridization, programmes attempt to rear marine animals for consumption. The inevitable colonization of the sea will bring with it an evaluation of the prospects of artificially illuminating the depths to provide phytoplankton gardens for sustaining desirable animal life and for oxygenation of man's submarine abode. The bioluminescence of dinoflagellates will also possibly be exploited, to illuminate the depths or underwater gardens.

But the most confident prognostication is ultimately the most valuable to man. That is, that there will be a growing awareness that the phytoplankton represent the most abundant, productive and widely distributed form of plant life on earth. Also that the successful colonization and mode of existence of the phytoplankton have been achieved by processes, adaptations and ecological adjustments and compromises which are worthy of study in themselves, to increase our understanding not only of these diminutive micro-algae, but of our total biosphere and its inhabitants.

The Harvest – Primary Production *Charles S. Yentsch*

Primary production (defined as the rate at which organic matter is formed from inorganic substances) on this planet comes about through the process called photosynthesis. During this process, carbon dioxide and water are chemically converted to organic carbon compounds with the energy derived from sunlight. A by-product of photosynthesis is oxygen evolution. Two principal sites of primary production are terrestrial and aquatic; photosynthesis is carried on by the plants that inhabit these two environments. People who customarily walk along the seashore quickly notice that every tide washes in large seaweeds, while other similar algae grow on rocks or nearby seawalls. These are large plants, easily noticed, and can be important producers in estuaries and embayments. Hence, it is somewhat of a surprise to find that microscopic algae (phytoplankton) floating freely but invisibly in the sea-water are the principal primary producers and the beginning of the food chain in the world's oceans. Principal members of the phytoplankton are diatoms and dinoflagellates. Both groups of organisms (figures 134, 135) have chlorophyll and, of course, are capable of photosynthesis. The main distinction between the two groups is the composition of their cell walls. In diatoms, the material is housed with a silica cell wall, whereas in the dinoflagellates the wall is made from organic cellulose material.

Scientists began to study these microscopic unicellular organisms as an outgrowth of their continuing interest in plant ecology of the earth and its oceans. Plankton nets were pulled through the water collecting organisms which were then identified for the study of evolutionary relationships. Even to the most casual observer it became apparent that there were very large-scale variations in the distribution of numbers of phytoplankton. In many areas the number of organisms appeared to follow certain seasonal trends (i.e. high in spring; low in summer). Oceanographic scientists began to study the ocean in an attempt to understand the causes of these variations.

In the ocean – a medium that is in constant motion because of a dynamic current system – it soon became apparent, in any one area, that a variation in the abundance of phytoplankton does not result from biological changes alone; one must account for the transport of phytoplankton into and out of the area. For experimental work on phytoplankton growth, oceanographers try to overcome these difficulties by containing the environment inside bottles, large plastic spheres or circular tanks. This allows the study of biological change without the influence of ocean current transport.

Much of our present knowledge of factors influencing primary productivity comes from indirect sources such as laboratory experiments with cultured phytoplankton, or by semi-laboratory experiments carried out in the oceans. In the latter, there are great experimental difficulties. This is because the concentrations of plants in the ocean are low compared with laboratory cultures, and techniques must be specific and highly sensitive. If one were to choose the factor most relevant to primary production it would be *light – its intensity and quality*. Let us examine some of the characteristics of light in the sea.

THE NATURE AND DISTRIBUTION OF LIGHT IN THE EUPHOTIC ZONE

In the oceans only the upper 100 m or less is illuminated with sufficient light intensity for photosynthesis. This is called the euphotic zone. This radiation varies as a function of the time of day, the season and latitude. The depth of the euphotic zone is dependent upon the amount of solar radiation, light reflected from the sky, reflection from the sea surface, and reduction of light as it passes through the water. The fate of light rays entering the sea surface is to be scattered and absorbed.

The amount of radiation entering the sea surface depends upon the altitude of the sun. On a clear day when the sun is at a high altitude approximately 85 per cent of the radiation comes directly from the sun and 15 per cent is reflection from the sky. When the sun angle is low, such as in late afternoon, the proportion of sky light becomes greater. The altitude of the sun also influences the amount of radiation reflected from the sea surface. One notices this when standing on the deck of a boat at sunset; the glare from the water is much greater. As much as 40 per cent of the incoming light may be reflected from the surface whereas at a higher sun angle as little as three per cent may be reflected. Sea surface conditions are important in this respect, and much more of the light is reflected when waves are formed or when there is ice cover.

If one takes a light detecting device (photometer) and measures the amount of light penetrating the sea surface, and then turns the device over so that it measures the light coming back out of the sea, one finds that five to seven per cent of the incoming light is lost from the sea. In other words the great majority of the light that enters the sea stays there, and only a small amount escapes. This tells us that the ultimate fate of light rays entering the sea is absorption. Probably rays are scattered initially, then absorbed, either by the water itself or particles, or by coloured substances such as phytoplankton.

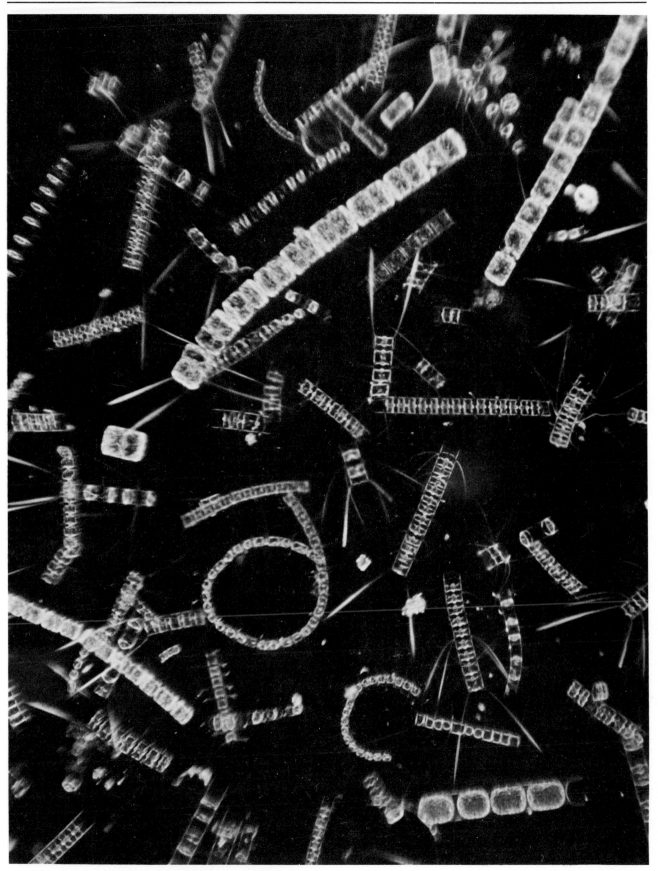

134. Living phytoplankton (× 175 approximately); chains of
cells of species of the diatoms *Chaetoceros* (with spines),
Thalassiosira condensata and *Lauderia borealis*

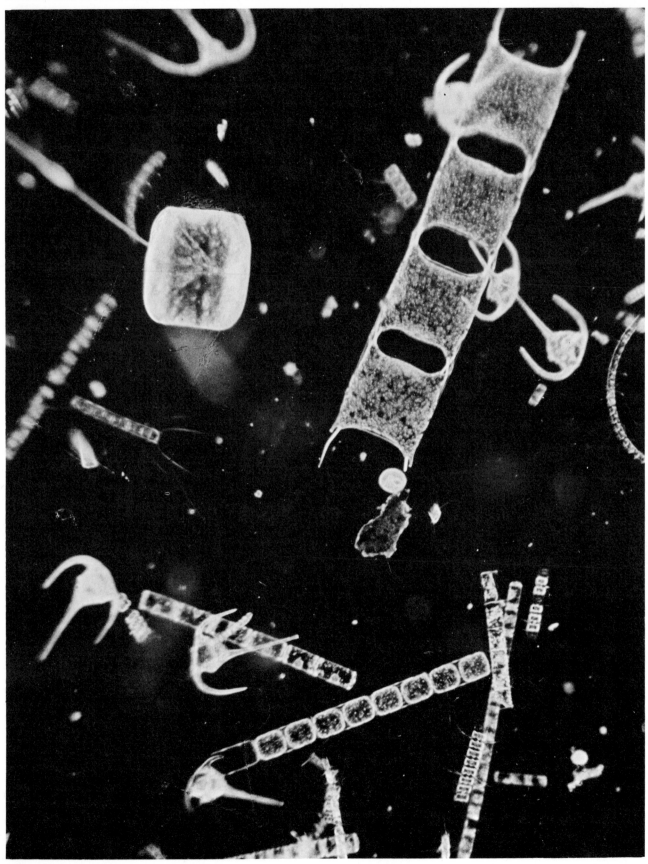

135. Living phytoplankton (× 220 approximately); chains of cells of the diatoms *Biddulphia sinensis, Rhizosolenia faeroense, Stephanopyxis borreri,* and *Chaetoceros* spp., the single pill-box like *Coscinodiscus conicinus,* and the anchor-like dinoflagellates *Ceratium* spp.

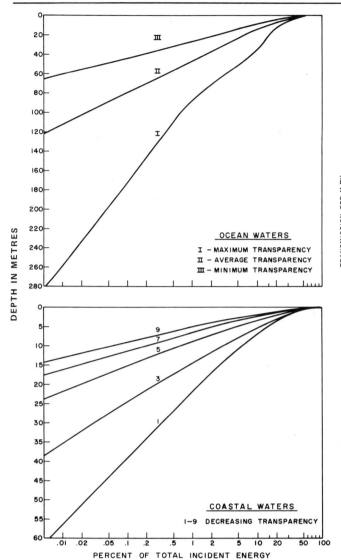

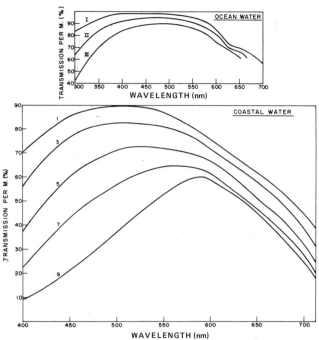

137. The transmission of wavelengths of light over a path of 1 m in waters of different transparency (from Jerlov 1951). Note the broad equal transmission of wavelengths in clear ocean water and how the wavelength of maximum transmission is shifted with decreasing light transmission. Part of the green colour of coastal waters is due to their higher productivity and hence, greater numbers of phytoplankton (numbers as in figure 136)

136. The penetration of light in ocean and coastal waters of different transparency (from Jerlov 1951). These measurements are made by lowering a waterproof photocell connected to a meter on deck by a wire. In full sunlight the depth of the euphotic zone occurs at about the one per cent light level. Note that in the clear oceans this is about 100 m, whereas in turbid coastal waters it would be not quite 10 m

Because of their proximity to land, coastal waters are generally less transparent than open ocean waters (see figure 136). Water itself greatly modifies the wavelengths which penetrate it. The red and dark blue ultraviolet wavelengths are quickly absorbed within the first few metres (figure 137). Clear water is extremely transparent to the blue-green region of the spectrum (see also Chapter 8).

The phytoplankton cell's ability to absorb submarine light depends upon the photosynthetic pigments located in special cell organelles, or chloroplasts of the algae. Some of the light absorption characteristics of phytoplankton and other photosynthetic organisms are shown in figure 102. Note that the reason green plants are visibly green is because chlorophyll absorbs red and blue light, and all photosynthetic organisms absorb the red and blue wavelengths. Photosynthetic bacteria, however, have the possibility of absorbing at the longer infrared wavelengths. In the blue-green region, absorption is principally by the pigments, chlorophylls and carotenoids. Absorption in the red and near infrared wavelengths is by chlorophyll alone. Red pigments of red and blue-green algae, called phycobilins, absorb the middle region of the visible spectrum.

Since diatoms and dinoflagellates are dominant organisms of the phytoplankton and their pigment composition is similar, it is important to consider their

138. The ability of phytoplankton to absorb light for photosynthesis at different depths throughout the euphotic zone. The pigments that absorb light are chlorophylls and carotenoids typical of those found in diatoms and dinoflagellates. Note that the red band of chlorophyll is not absorbing below 10 m, this is because all wavelengths in this region are absorbed primarily by water (after Yentsch 1962)

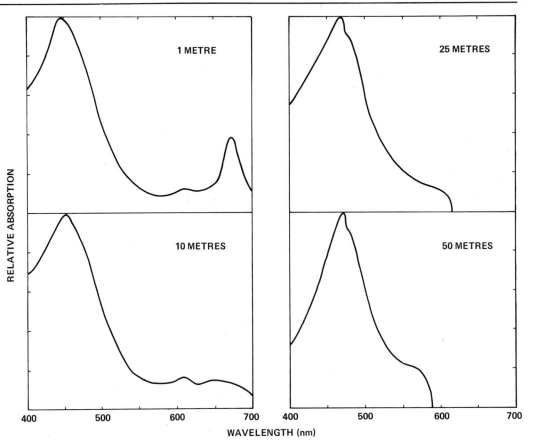

ability to absorb submarine light. Combining the absorption by algae with the wavelengths in the submarine light field shows that the red peak of chlorophyll cannot be active in photosynthesis much below 10 m (figure 138). This is because these wavelengths of red light have already been removed by water absorption. Blue absorption by chlorophyll is active to 50 m, and below 50 m it would appear that the carotenoid pigments are the principal absorbers of light. The utilization of light in photosynthesis depends, of course, upon its intensity; there must be sufficient intensity to supply the energy necessary for photosynthesis.

THE RELATIONSHIP OF LIGHT TO PHOTOSYNTHESIS
When light intensity is increased or decreased, the rate of photosynthesis (carbon dioxide fixation or oxygen evolution) responds according to the curve shown in figure 139. This is the so-called photosynthesis light curve and consists of two specific regions. In the first region, occurring at lower intensities, the photosynthetic rate increases in a linear measure with increasing light intensity. At a value of 10 – 15 per cent of full sunlight, the second region is entered, where further increase of light intensity increases photosynthesis only slightly; this is called 'saturation'. In some cases at very high light intensities, inhibition of the photosynthesis rate may occur.

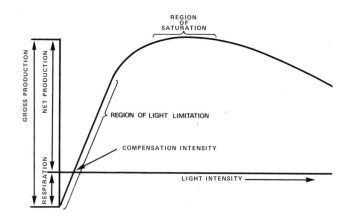

139. Relationship of the rate of photosynthesis to light intensity (heavy curve)

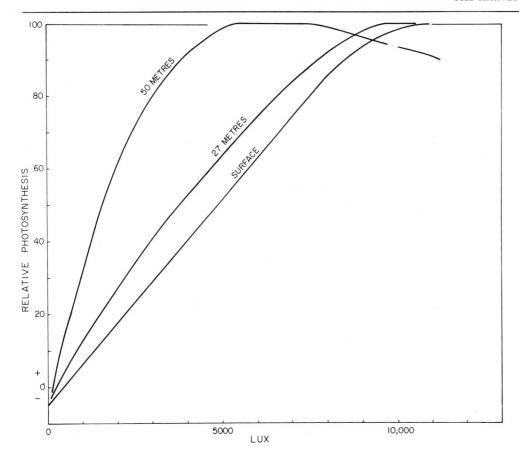

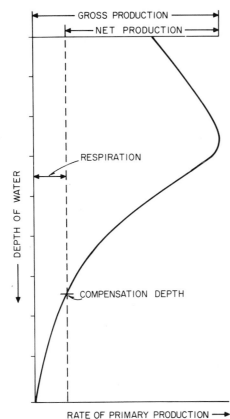

141. The rate of photosynthesis (heavy line) as a function of depth

The point where the photosynthesis curve intercepts the axis of the light intensity is called the compensation intensity. The amount of photosynthesis occurring below a certain light intensity (the compensation intensity) is the amount used for respiration. All photosynthesis above this intensity is called net photosynthesis. This means that this photosynthesis is in excess of that needed for respiration by the algae. The total of these two is termed gross photosynthesis. These parameters when considered with the exponential decrease of light in the water column are shown in figure 141.

At times, in the euphotic zone, environmental conditions arise which affect the maximum rate of photosynthesis. The effect of lowering the temperature is to decrease the maximum rate. The maximum photosynthetic rate is also depressed when there is a lack of carbon dioxide or nutrients. Moreover, if phytoplankton are at stationary levels in the euphotic zone, those at the surface are exposed to high light intensities, whereas the phytoplankton at depth are exposed to lower intensities. The result is that the plankton algae near the surface take on a photosynthesis light response similar to so-called 'sun plants' (figure 140). These plants have a high rate of photosynthesis at extremely high light intensities. On the other hand,

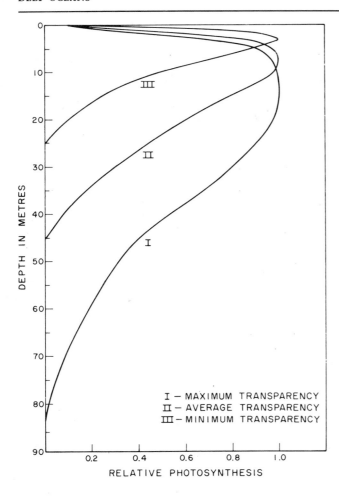

142. Integral or total photosynthesis in water columns of different clarity

phytoplankton near the base of the euphotic zone develop so-called 'shade characteristics' where maximum photosynthesis occurs at lower light intensities.

Despite these variations in the shape of the photosynthesis/light curve, accurate predictions of the vertical profile of photosynthesis in the euphotic zone can be made by using a general shape for the photosynthesis/light curve (similar in all phytoplankton) and taking into account the light intensity in the water column. These two factors are frequently used by aquatic ecologists in predicting integral or total photosynthesis in the euphotic zone. Examples of these predictions are shown in figure 142 and emphasize the importance of water clarity on the total amount of photosynthesis in a water column.

ENVIRONMENTAL FACTORS INFLUENCING PHOTOSYNTHESIS

Biomass estimates of phytoplankton are generally made by measuring chlorophyll in algae (biomass is the amount of living material expressed as weight).

The principal advantage in measuring chlorophyll is that as the major absorbing pigment it plays the essential role in the photosynthetic process. There have been many attempts to correlate the amount of photosynthesis with chlorophyll content. These values are variable because the rate of photosynthesis may change, quite independently of chlorophyll content. For example, temperature decreases the photosynthetic rate of algae in the Arctic and Antarctic, which photosynthesize less per unit chlorophyll than phytoplankton of temperate or tropical waters. Chlorophyll measurements may be hampered by the presence of chlorophyll debris in the particulate matter, which poses another difficult practical problem. Debris is a breakdown product that is generated with the death of the cell. These pigments are not easily distinguished from true chlorophyll, hence the total concentration of photosynthetic chlorophyll at times is in error.

Despite these problems, measurements of chlorophyll and photosynthesis show that not all areas of the ocean are equally productive, and the reasons are as follows: in the photosynthetic process, phytoplankton remove from water essential nutrients which, of course, make up particulate organic plant material. If the reader is a gardener he will know that amongst the most essential nutrients are compounds of nitrogen and phosphorus. In the incorporation of these compounds, the particulate plant matter becomes more dense than the surrounding sea-water and it begins to sink slowly. During sinking, the phytoplanktonic organic matter may be subjected to decomposition either by enzymatic processes after death or by destruction from grazing marine herbivores. In either event there is a net downward flux of nutrients, from the euphotic zone. Therefore, the essential nutrients at the surface, i.e. nitrogen and phosphorus, are transported to the deeper ocean waters. A number of oceanographers have assessed the rates at which this organic material is mineralized by oxidation. The depth at which all plant matter is oxidized back into mineral form is estimated to occur at 200 m. The nitrogen and phosphorus ending up at a depth greater than 100 m (i.e. below the euphotic zone) would be lost were it not for certain mechanisms. These nutrients return into the euphotic zone by the vertical mixing of the water column brought about by wind and wave action, by processes associated with ocean currents, and by 'upwelling' of deep waters.

It has been suggested that at times the oceans are 'out of balance', in the sense that many nutrients are transported downward to the bottom waters, whereas all of the light is retained in the surface waters. During the winter months the surface waters at mid-latitudes may become mixed to a considerable depth, and it is in this process that nutrients are brought back into the

143. Some of the principal factors controlling
primary production on,
and at the edge of, a
continental shelf in tem-
perate latitudes. Vertical
enclosed arrows indicate
intense vertical mixing.
Horizontal enclosed
arrows mean little verti-
cal mixing. N means
nutrients. The stipple
represents phytoplank-
ton growth. To the right
is a schematic represen-
tation of temperature
vs. depth (EZ = euphotic
zone)

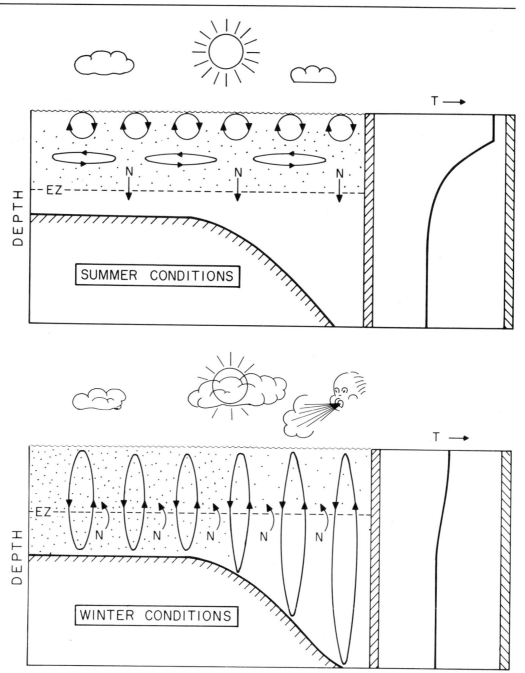

euphotic zone. Figure 143 will aid the reader in understanding this process. The mixing depth cannot be too
great or the growth of the phytoplankton population
will become limited by light. Generally speaking,
vertical mixing is considered detrimental when it
exceeds the depth of the euphotic zone (EZ). In some
areas of the oceans, vertical mixing can easily go to
depths as great as 200 – 300 m. 'Bloom' conditions in
the mid-latitudes coincide with the formation of what
is known as a 'seasonal thermocline', which is a region
in the water column where the rate of temperature
change with depth is greatest. It is this rate of tempera-

ture decrease that causes an increase in stability of the
water column (with the denser water at the bottom), and
the stability tends to slow down vertical mixing. During
summer months, with increasing solar radiation, surface waters become heated and hence less dense;
under these conditions the water-mass is considered
stable. In most oceans, during this period, photosynthesis and standing crops of plants are quite low,
largely because, as mentioned above, nutrients are
lost to deeper waters and not returned to the euphotic
zone. Under these conditions one must conclude that
the phytoplankton have to maintain their nutrient

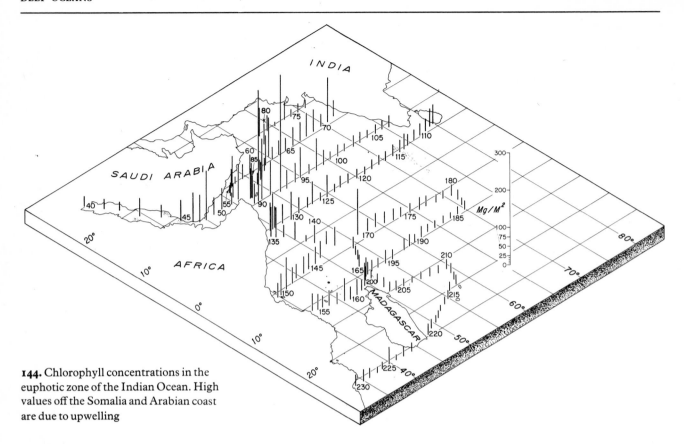

144. Chlorophyll concentrations in the euphotic zone of the Indian Ocean. High values off the Somalia and Arabian coast are due to upwelling

requirement by biochemical recycling as opposed to the physical processes of mixing water.

The cycle of vertical mixing starts when thermal stability is destroyed in the autumn. This is brought about by the decreasing solar radiation allowing the surface water to cool and then the wind mixing the water. As the thermoclinal layer deepens, the nutrient-rich waters at the base of the thermocline are mixed with the upper portion of the euphotic zone. In mid-latitudes, in winter months when solar radiation is low and mixing is intense, blooms still occur along the shallow coastal waters of some areas because, in these cases, depth of mixing is arrested by the bottom of the sea (figure 143). In deeper waters, however, mixing depths become much greater than the depth of the euphotic zone, and population growth is limited by the lack of light.

In the tropical oceans, variations in the intensity of solar energy are not as great as temperature variations at mid-latitudes, hence little or no seasonal variation in vertical mixing occurs. Production of phytoplankton in these areas is generally low for the same reasons as production is low during summer months at mid-latitudes.

The productive areas of tropical oceans are generally found in areas of large ocean currents and/or sites where water is 'upwelled'. In areas of large currents, certain conditions are set up near the high velocity edge of the current which aid in the upward transport of deep, nutrient-rich water. In certain areas of the oceans, upwelling of deeper waters appears to be a persistent feature, and these areas are very productive (i.e. Somalia, Arabian coast, West African coast and west coast of California) (figure 144). Upwelling is accomplished by the wind action transporting surface waters from near the coast back out to sea (figure 148). This displacement allows deep, nutrient-rich water to come to the surface.

So far in this discussion it has been emphasized that the factors augmenting primary production are light, nutrients and their interaction. Essential nutrients implicated are nitrogen and phosphorus; their occurrence and abundance is of importance.

Studies of the distribution of nitrogen and phosphorus compounds indicate that in the euphotic zone the removal by phytoplankton of the two elements runs parallel. Ratios of these compounds in sea-water are comparable to those in the elementary composition of live cultures of marine phytoplankton. Extremely low ratios of nitrogen to phosphorus indicate that phytoplankton have removed practically all of the inorganic nitrogen leaving only a residue of inorganic phosphorus.

Studies of cultures of marine phytoplankton show

145. *opposite* The little squid *Pyroteuthis* (5 cm) has multi-coloured jewel-like light organs round its eyes, at the tip of its tentacles and in its body

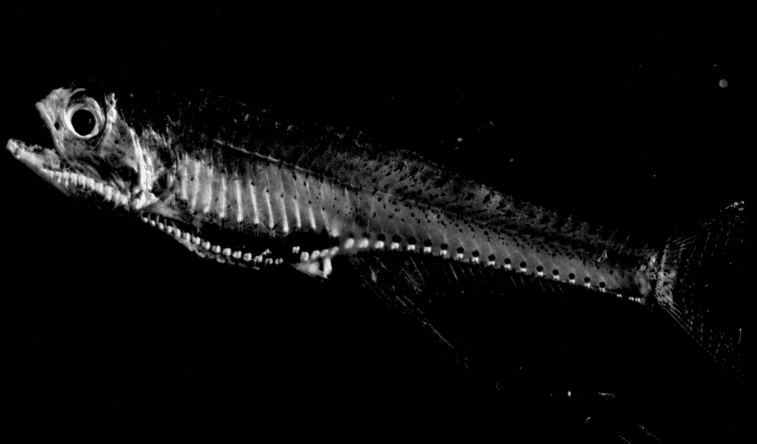

that when either nitrogen or phosphorus is exhausted from the culture media, the cells continue to grow and may divide for some time on their internal supplies, but 'deficient' cells develop in which the internal concentration of nitrogen and phosphorus becomes low enough to stop the cells dividing.

When all aspects of the phosphorus cycle in the sea are known, namely the concentration of particulate and dissolved organic and inorganic, it is possible to evaluate the phosphorus utilization of phytoplankton. Assuming a constant composition for phytoplankton, it becomes possible to estimate the phosphorus requirements for particular photosynthetic needs which occur throughout the year. When this computation is made it becomes apparent that phosphorus must be rapidly recycled in some fashion within the population to maintain the photosynthetic activity observed. The actual nature of this recycling or regeneration is poorly understood.

Nitrogen is available to the plants in the ocean in the form of inorganic ammonia, nitrate and nitrite. Whereas inorganic phosphorus is a sizeable fraction of the total phosphorus in the ocean, inorganic nitrogen is a minor factor in the distribution of the total nitrogen in sea-water. Of the inorganic fractions, nitrate is the most abundant and, as stated above, shows marked seasonal trends due to the photosynthetic activities of the phytoplankton. Recycling of the nitrogenous material also is poorly understood. It has been observed that marine animals (zooplankton) excrete nitrogen, largely ammonia, urea or uric acid (which are organic forms of nitrogenous material).

Extremely low concentration of nitrogen in the surface waters of tropical oceans has led to speculation as to the degree of biological nitrogen deficiency occurring there. In these cases, blue-green algae, such as *Trichodesmium*, are abundant largely because of their capacity to fix atmospheric nitrogen.

Other chemical compounds of importance to phytoplankton growth are silicate, iron, manganese and certain vitamins. The silicate requirement for diatoms, of course, is primarily to build diatom cell walls. A number of workers have shown that diatoms grown in water low in silicate develop a very thin cell wall. The availability of iron and manganese is sometimes very low in the open ocean, and their importance in the metabolism of certain organisms has led oceanographers to conclude that they may be responsible for limiting conditions for growth. Indirect evidence for growth limitation caused by the absence or small quantities of these metals can be shown by augmentation in growth after their addition to water already enriched with nitrogen and phosphorus. Other scientists have pointed out that most marine phytoplankton

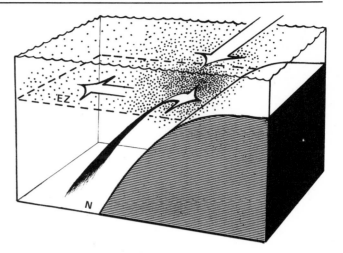

148. Schematic representation of upwelling along a coast. Wind blowing parallel to the coast creates a surface water transport 90° to the wind direction due to the rotation of the earth (Chapter 3). Deeper nutrient-rich water rises upward to replace that moved seaward. Stipple indicates high concentration of phytoplankton (EZ: euphotic zone; N: nutrients)

require vitamins, and principally vitamin B_{12}. The lack of direct measurements of the concentration of this vitamin has hampered our knowledge of its importance. Bio-assay techniques which can be used on a semi-quantitative basis have indicated that the concentration of this vitamin is indeed quite low.

SHORT-TERM FLUCTUATIONS IN THE NUMBERS OF PHYTOPLANKTON

The numbers of phytoplankton occurring at any given time are the result of the rate at which they are growing and the rate at which they are being removed. Phytoplankton are rapidly removed both by animals feeding on them and by sinking out of the euphotic zone. Because of the necessity of light for photosynthesis, the growth of phytoplankton is dependent upon adequate flotation within the euphotic zone. Buoyancy or the lack of it can be an active mechanism in changing the size of the population. Although some phytoplankton have mechanisms for swimming, a great majority of the species observed in marine populations are non-motile. Such cells, being slightly more dense than sea-water, tend to sink. Taxonomists and others have pointed out that the bizarre shape of diatoms – spirals and assemblies of long chains – are mechanisms to prevent or retard sinking. Since the amount of friction that a cell may exert on the surrounding medium is dependent upon the surface area, it is of utmost advantage for the cell to reduce its size if the surface area is to be large relative to the volume, and hence the density.

146. *opposite above* The female paper nautilus *Argonauta* (5 cm) carries its eggs in a shell secreted specially for this purpose. Both sexes are sometimes to be found riding on the backs of medusae such as *Pelagia* (figure 154)

147. *below* The gonostomatid fish *Bonapartia* (8 cm) has its ventral surface almost completely covered with rows of light organs

It has been argued that in very nutrient-starved conditions, sinking is an active process for phytoplankton, enabling them to enrich their metabolism. Consider a plant cell motionless with respect to the water. Under these conditions there is the possibility of the cell soaking up the nutrients in its immediate vicinity and thus building up a shadow of low nutrient water around the cell. The shadow effect would be overcome by sinking, since the medium surrounding the plant is being constantly changed.

Effects of phytoplankton sinking can be observed in the vertical distributions of chlorophyll. Where nutrients are low in the surface water and the density of the water is also low, maximum concentrations of chlorophyll are found in the lower limit of the euphotic zone. It has also been noted that these maxima occur where there is an increase in nutrients. Sinking experiments conducted by various workers have indicated that the buoyancy of the phytoplankton is somehow directly related to the physiological conditions of the cell. Hence a healthy cell, which is one enriched with nutrients, is more buoyant than a cell that is nutrient-starved. In contrast, when the stability of the water column is low and vertical mixing is active, maximum growth and concentration of chlorophyll is near the surface.

MODERN METHODS FOR ESTIMATING PRIMARY PRODUCTION

Estimates of the primary fixation of carbon are generally made by measuring the fixation of a radio-active tracer by the phytoplankton algae throughout the euphotic zone. Radio-active carbon 14 is added to water samples which are taken at specific depths throughout the euphotic zone. These samples are taken generally at the surface and at the depth to which 50, 25, 10 and 1 per cent of the surface light penetrates.

Water samples containing the phytoplankton population must be collected in water sampling bottles that are non-toxic, since small amounts of metals and trace impurities have been found to be toxic to the organisms. Samples of the population are placed in transparent glass bottles, and carbon 14 is then added to the bottles which are suspended within the euphotic zone on a wire or a rope for a period of some four to six hours in the middle of the day. The samples are recovered and the contents of the transparent bottles are filtered through a membrane filter, and the radio-activity in the particulate matter on the filter is counted and expressed as the quantity of carbon fixed during the period of exposure. Quite often the vertical profile of carbon fixation at each depth is integrated and the quantity of carbon fixed under a square metre of ocean surface is given. Carbon fixation is also

TABLE 9
Gross and net organic production of various natural and cultivated systems in grams dry weight produced per square metre per day. (From Ryther 1959; for data sources see original publication)

	Gross	Net
Theoretical potential		
Average radiation (200–400 g cal/cm²/day)	23–32	8–19
Maximum radiation (750 g cal/cm²/day)	38	27
Mass outdoor *Chlorella* culture		
Mean		12·4
Maximum		28·0
Land (maxima for entire growing seasons)		
Sugar		18·4
Rice		9·1
Wheat		4·6
Spartina marsh		9·0
Pine forest (best growing years)		6·0
Tall prairie		3·0
Short prairie		0·5
Desert		0·2
Marine (maxima for single days)		
Coral reef	24	(9·6)
Turtle grass flat	20·5	(11·3)
Polluted estuary	11·0	(8·0)
Grand Banks (April)	10·8	(6·5)
Walvis Bay	7·6	
Continental Shelf (May)	6·1	(3·7)
Sargasso Sea (April)	4·0	(2·8)
Marine (annual average)		
Long Island Sound	2·1	0·9
Continental Shelf	0·74	(0·40)
Sargasso Sea	0·74	0·35

frequently expressed as grams of dry algae.

Because of the high cost of ship operation, some oceanographers have found it profitable to utilize 'on deck' incubators. These are water boxes fitted with neutral density filters simulating the light intensities at different depths and flooded with sea-water. Other workers make estimates of carbon fixation using measurements of chlorophyll, light and water transparency.

COMPARATIVE VALUES FOR PRODUCTIVITY

In coastal waters, production values of carbon have been observed to exceed 3 grams per square metre per day (3 g/m²/day). Offshore, the carbon values range between 0·2 and 1 g/m²/day. In the open ocean, the values range near 0·3 g and probably do not exceed 0·8 g per day; however, not enough seasonal information has been obtained to establish this value clearly. The average annual production of carbon in the open ocean ranges between 25 and 100 g/m² of sea surface.

When primary production is considered in the

overall role of food chain efficiency in the sea, it becomes imperative to know what the efficiency of energy transfer is from one step in the food chain to another. This is one of the more poorly known aspects of the food chain dynamics in the ocean. It has been thought that in the open ocean a majority of the phytoplankton are consumed by zooplankton, including the very important small crustacean herbivores known as copepods. Most estimates of the food requirements for invertebrates living on the bottom of the ocean indicates that only a small fraction, one to ten per cent of the surface production, can ever reach the bottom and serve as a food source for these animals. It is a cardinal rule in discussing food chain dynamics to assign high efficiencies of food energy transfer whenever steps in the food chain are few. Characteristically, the largest vertebrate in the world, the blue or sulfur bottom whale, feeds mostly on a crustacean known as an euphausiid, which in turn feeds on diatoms or small copepods. This places the whale about three or four steps down the food chain sequence. Some of the larger fishes like tuna may be many more steps down. The decreasing efficiency with increasing number of steps in the food chain is largely due to the fact that there is only a small transfer of energy between each step, and the lion's share of the energy goes into maintenance of the organisms. Accurate estimates of the energy transfer from phytoplankton to other organisms have been badly hampered by a whole host of experimental problems. There is increasing evidence that one cannot generalize basically on this 'food chain length of step' thesis when studying energy transfer. The development of the organisms' feeding efficiency during its various growth stages must each be considered.

In Table 9 the productivity of some marine communities has been compared to that of land communities. Note that the maximum productivity of these communities is of the same order in practically all cases. However, it should be noted also that in the ocean planktonic communities, when the maximum is reached, it will only occur for an extremely short interval. Cultivated and cultured systems maintain a maximum for a much longer period. Also interesting to note is that coral reefs are highly productive. Yet they are found in ocean areas that are very poor in nutrients. Why? Maintenance of this high productivity must be due to the fact that new water is continually passing over the reef, and even though this water may be low in nutrients, the supply is virtually unlimited and continual.

On the whole one can say that the oceans, as compared to fertile regions of the earth, are virtually deserts. These data should serve as a warning to some who argue that production from the oceans will stem the famine situations arising from the increasing world population. *The oceans are only productive because of their size.* The total production of ocean is only two to three times that of land, whereas ocean covers 71 per cent of the planet. The real potential of food from the ocean is in exploitation of certain organisms that have a high potential yield of protein food, such as many bivalve molluscs. This manipulation of the ocean is now being called aquaculture and will eventually require the technical and scientific capability that has been applied to terrestrial agriculture.

6 Animal Miscellany: A Survey of Oceanic Families

Malcolm R. Clarke and Peter J. Herring

Animals (and plants) come in a bewildering variety of shapes, sizes, colours and abilities. The systematist has the problems of providing each sort of animal with a name and of arranging this biological miscellany into some sort of rational order or classification. He takes as his basic unit the **species,** which, for general purposes, can be defined as a collection of animals whose constituent members can interbreed. Species are grouped with other closely similar species in a **genus.** Common names of animals differ locally as well as nationally, and each species is therefore given an internationally recognised two-part name. The first part is the generic name, the second the specific name; in print both are italicized and the generic name is given a capital letter: thus man is *Homo sapiens*. The names are usually derived from Latin or Greek (the system was introduced by the Swedish naturalist Linnaeus in the eighteenth century, when Latin was the universal scientific language) but knowledge of their derivation is not important and the layman need not be discouraged by their use. Since many (indeed most) deep-sea animals have no common names, scientific names must needs be used.

For convenience similar genera are grouped into **families** which take their names from one of the component genera. Families are grouped into **orders,** orders into **classes** and classes into **phyla;** animals which are similar are usually more closely related than those which are dissimilar, so that the phyla may be considered as the large branches and the species as the twigs on the animal family tree.

In general discussion, groups are often mentioned indiscriminately without reference to their rank (it matters little that molluscs are a phylum while mammals are a class) and the endings of group names, which often give an indication of the ranks, are anglicized (molluscs instead of Mollusca, myctophids instead of Myctophidae).

This chapter discusses deep-sea animals within the general systematic framework of the animal kingdom while Chapters 7 and 8 examine the same kingdom from quite different viewpoints, that of the ecologist and that of the physiologist.

PROTOZOANS (PHYLUM PROTOZOA) (Figures 149, 150) The plant and animal kingdoms meet in this group of organisms without cells; those with chlorophyll may equally well be considered as plants. These acellular animals are not usually visible to the unassisted eye but a few giant Foraminifera may exceed four centimetres in diameter.

Protozoa are far more complicated and diverse in their structure than their acellular condition might suggest. Basically the protozoan consists of a jelly or cytoplasm enclosed in an envelope or pellicle and containing one or more nuclei, which are the reproductive and metabolic centres, and several kinds of other inclusions. Food may be enveloped by the cytoplasm anywhere round the margin, or a special 'mouth' region of varying permanence and complexity may be present. In some ciliates the 'mouth' can be widely distended to envelop food. Food vacuoles are often numerous and may follow a distinct path round the cytoplasm before emptying waste to the outside. A pulsing contractile vacuole eliminates water absorbed by many marine forms but this pulses much more slowly than in freshwater forms in which uptake of water by osmosis is more rapid.

The major groups of protozoans are characterized by their means of movement. In the **ciliates** the pellicle bears very many tiny beating hairs (cilia) which propel the animal about, whereas the **flagellates** utilise only one or a very few long whip-like hairs (flagella) instead. A third group (the **rhizopods**) can stream their cytoplasm as protrusions of the body known as pseudopodia (as in amoebae) either for movement or to envelop food. In one group of rhizopods (the Foraminifera) food is trapped in great webs of streaming cytoplasm, whereas in two others (the Radiolaria and Heliozoa) streaming is orientated along straighter pseudopods whose seemingly stiff threads can collapse in an instant.

Many forms have skeletons, and these may be organic as are the inner capsules of Radiolaria and the wine-glass shells of the ciliate family Tintinnidae, or inorganic such as the calcareous shells of Foraminifera, the silicate shells of Radiolaria, the strontium sulphate

164

149. Variation in skeletal structure in Radiolaria (very much enlarged): 1. *Cortinetta tripodiscus*; 2. *Hexancistra quadricuspis*; 3. *Paratympanum octostylum*; 4. *Cannosphaera antarctica*; 5. *Cinclopyramis infundibulum* (after Haeckel 1887)

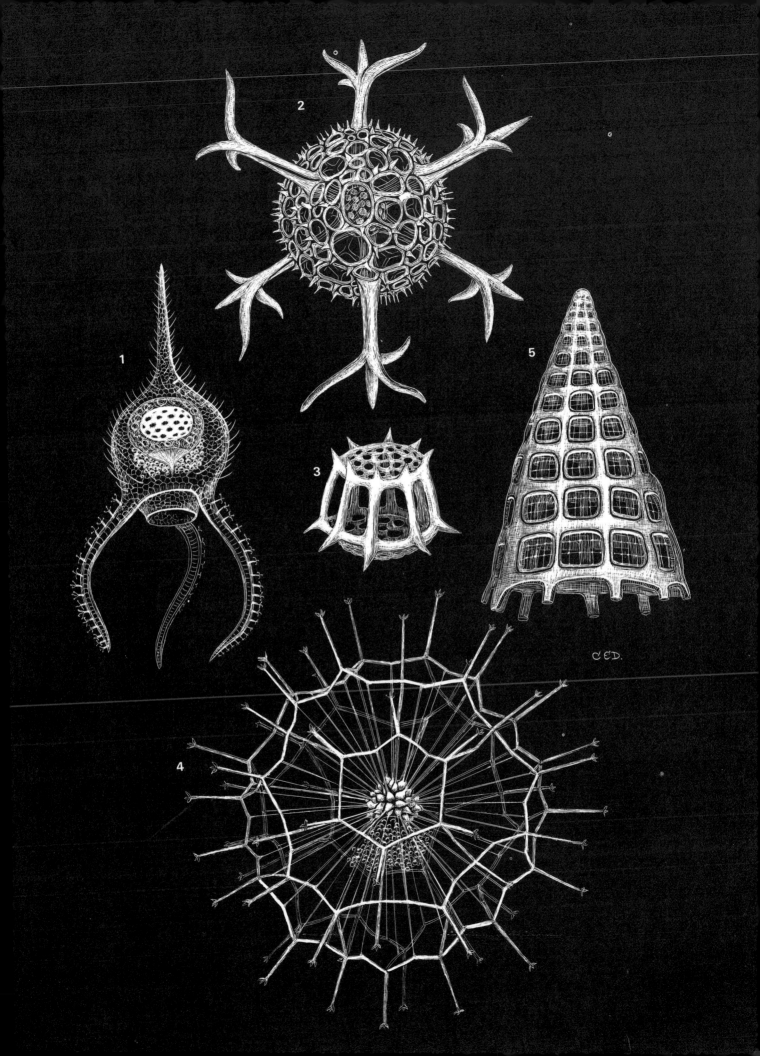

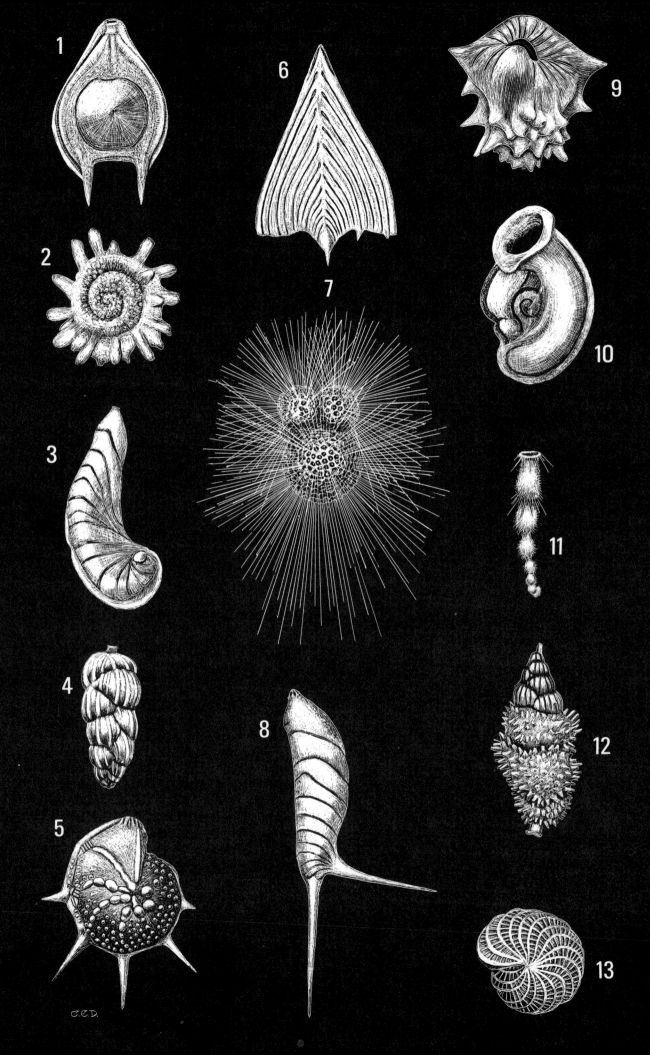

shells of Acantharia and the barium sulphate shells of the deep-living Xenophysophora (sometimes regarded as Foraminifera). Foraminifera shells are thicker the deeper their owners live and though they may be all manner of shapes they are always perforated. The outer cytoplasm in Radiolaria is often frothy and may contain minute symbiotic algae (plants). In dino-flagellates luminescent bodies (scintillons) may be present in the cytoplasm.

The most important Protozoa in the sea are the predominantly bottom-living Foraminifera and the planktonic Radiolaria. Individual planktonic fora-miniferan species (which may be important constituents of the plankton) exhibit great structural variability according to the water they are in. The *Globigerina* species have formed great deposits, the *Globigerina* oozes, over a third of the deep ocean floor where they are found at depths of 2,500–4,500 m with as many as fifty thousand shells in one gram of ooze. In deeper water these shells dissolve, and variation of solubility sometimes influences the species composition in sediments. In other regions radiolarian shells form oozes and these animals are sometimes important in primary production. The ciliate *Mesodinium* and the dino-flagellate *Noctiluca* 'bloom' at times, and in some sediments (not in the deep sea), living ciliates are known to represent over 90 per cent of the animals present and 0.4 per cent of the biomass; their impor-tance in the deep sea is probably only awaiting recognition.

SPONGES (PHYLUM PORIFERA) (Figure 151)
Sponges are more complex than Protozoa in that they are animals composed of many cells of more than one type, arranged roughly in layers. However, it is a very loose aggregation of individual cells, acting more or less independently and not forming discrete tissues. Indeed, so loose is the association that two species have been found with their cells intimately associated in a chimaeroid composite animal. There are no sense organs, nervous system or digestive system, and the adults are sessile. The body is built of a supporting meshwork surrounding a multitude of tiny pores, through which water is drawn by beating flagella on

150. Some examples of the variation in shell shape of species of Foraminifera (very much enlarged): 1. *Lagena seminiformis*; 2. *Calcarina defrancii*; 3. *Cristellaria compressa*; 4. *Uvigerina pygmaea*; 5. *Cristellaria echinata*; 6. *Frondicularia alata*; 7. *Globigerina bulloides*; 8. *Vaginulina spinigera*; 9. *Ehrenbergina hystrix*; 10. *Vertebralina auriculata*; 11. *Sagrina virgula*; 12. *Uvigerina aculeata*; 13. *Polystomella macella* (10, after Rhumbler 1909; others after Brady 1884)

special cells at the base of which a protoplasmic collar collects and engulfs particles of food drawn in on the current. The water passes into a central cavity and out through one or a few large exhalent openings. The meshwork is strengthened by skeletal spicules on the basis of which the sponges are classified. The mostly shallow water **calcareous sponges** have calcium car-bonate spicules, the deep-water **hexactinellids** have six-rayed silica spicules, and anchoring roots of enlarged spicules, and the **Demospongiae** either different-shaped silica spicules or the protein spongin or both. Spongin-supported sponges are common in warmer waters, while mineralised skeletons predomin-ate in cold waters. The spicules are secreted by specialised amoeboid cells and exhibit a multiplicity of definite geometric forms in different species; they may be an important component of the bottom sediments in some areas. The spicules make sponges unpalatable to most animals and they are therefore an excellent refuge for commensals. Indeed, over 16,000 shrimps have been found in a single large sponge! Additional protection is achieved by the secretion of potent antibiotics, or venomous toxins as in the aptly named fire sponge. Acid secretion, combined with mechanical action, makes many species very successful borers into calcareous reefs, where they often cause very considerable erosion. The growth form of some species is constant, but many grow irregularly like plants, the final form often depending on prevailing water movements. Their plant-like appearance is increased by the bright green colour produced in some by symbiotic algae living within their tissues, and others may appear yellow-green under water due to fluorescence colours induced by the blue light pene-trating the sea-water. A consequence of their loose cell association is a very high regenerative ability used by some in conjunction with budding as a reproductive method, but sexual reproduction with the formation of a little ciliated larva is more usual.

COELENTERATES (PHYLUM COELENTERATA)
The coelenterates are radially symmetrical animals built of tissue-forming cells in two distinct layers, or epithelia, separated by a non-cellular gelatinous layer, the mesogloea. There is a single blind-ending body cavity, and the animals usually bear specialized capture and defence organelles called cnidoblasts (page 246). A single species typically begins life as a fertilized egg developing into a ciliated larva, which settles down and becomes a sedentary polyp, reproducing asexually to form free-swimming medusae bearing the sexual organs. The group thus exhibits 'polymorphism', or multiple forms of a single species, and an alternation of sexual and asexual generations. Polymorphism may be

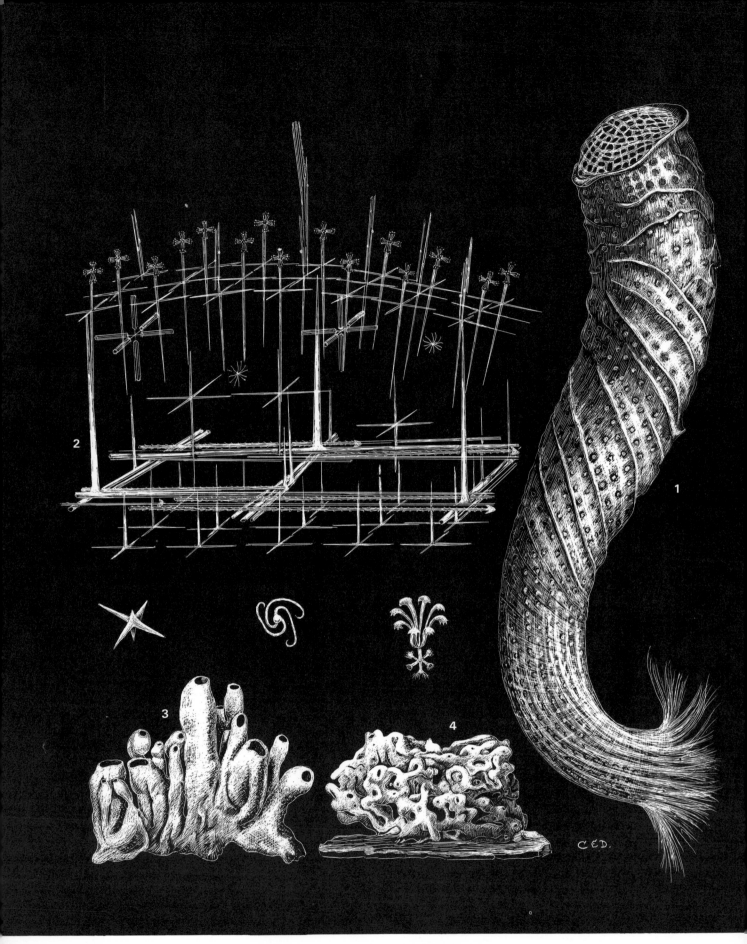

151. Deep-sea sponges: 1. the hexactinellid *Euplectella asper-gillum* (22 cm high) and 2. the arrangement of the siliceous skeletal spicules in the related species *E. nodosa*; 3. a colony of the horny sponge *Psammoclema vosmaeri* (7 cm high); 4. the calcareous sponge *Heteropegma nodus gordii* (3 cm high) (1 and 2 after Schulze 1887; 3 and 4 after Polejaeff 1884)

further complicated by the presence of several types of polyp in one species, particularly colonial species, or by the reduction or loss of either the polyp or medusa form. The characteristic cnidoblasts are quite independent units but may be stored in batteries on the tentacles as well as all over the animal's body. Their penetration and poisonous effect on man is also very variable, most being harmless, some merely painful and one or two occasionally lethal.

The group is divided into two, the sea-anemones, jellyfish and siphonophores in one group, and the comb jellies. The first group is further separated into the **Hydrozoa,** in which species have both polyp and medusa forms, the **Scyphozoa,** with the medusa the dominant form, and the **Anthozoa,** with only a polyp.

Hydrozoans often form small polyp colonies with characteristic growth forms, and the individuals of the colony may be supported by a chitinous sheath, or be naked. Polyps in the colony may be specialised as feeding, protective or reproductive individuals. Though most of the colonies live in shallow water attached to weed or stones, some may be found in the open ocean growing on certain fishes. The little separate sex medusae from such colonies may be found in the open sea, but the only hydrozoan medusae common in the deep ocean are the group which have reduced or lost the polyp form, the trachyline medusae. Some of these are very poisonous and many are luminescent, as are a number of polyp colonies. The luminescence in some forms is unusual in being derived from a remarkable protein which lights up simply on addition of calcium. Also included in the Hydrozoa are two small groups of coral, the millipores and stylasterines.

The siphonophores are very complex and beautiful, often luminescent, hydrozoans whose relationships have long been a matter for dispute. They are now believed to be floating colonies, with the medusoid individuals much modified to form swimming bells or floats and leaf-like bracts, and the polyp individuals forming the tentacular and digestive units. In its simpler form the colony consists of a long stem, with (physophoridan siphonophores) or without (calycophoran siphonophores) a gas float, with swimming bells (medusae) at the top, and with a series of similar groups of individuals at intervals along its length (figure 152). Many of the apparently more complicated forms can be explained by very great shortening of the main stem and telescoping together of the separate groups (figure 153). Those with gas floats are believed to move up and down by varying the amount of gas, and are often implicated in deep scattering layers, for their gas floats are excellent sound scatterers. Probably the best known member of this group is the very colourful but poisonous Portuguese man-of-war, *Physalia* (figure

152. A whole specimen of the calycophoran siphonophore *Praya galea* (total length 60 cm) showing the similar units at intervals along the contractile stem (inset: an enlargement of a single unit), all supported and propelled by the two large swimming bells at the top (after Haeckel 1888)

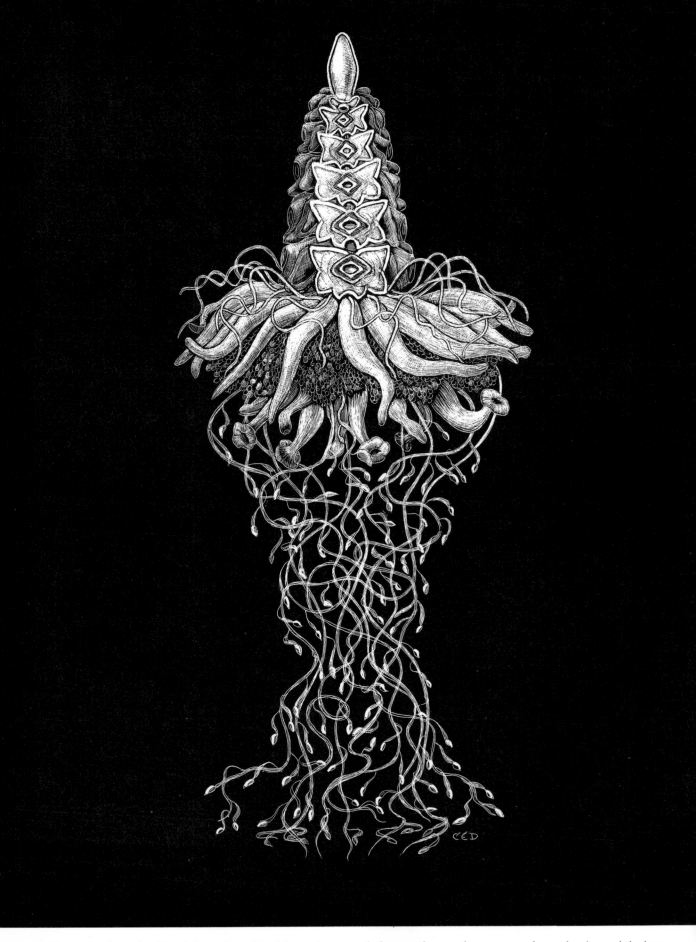

153. An entire physophoridan siphonophore *Discolabe quadrigata* (6.5 cm high excluding tentacles) with the gas float at the top, and below it a column of tightly packed swimming bells; below are the mouth, sensory and gonad units, and the long dangling tentacles (after Haeckel 1888)

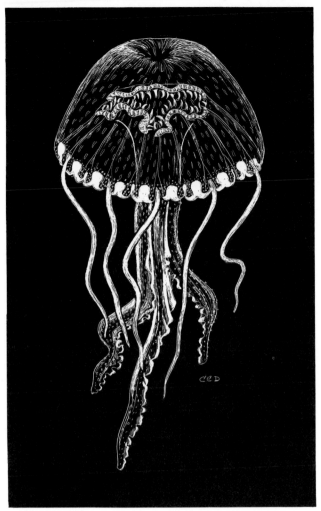

154. The common warm-water luminescent medusa *Pelagia noctiluca* (6 cm diameter), with gonads in the umbrella, and long frilly mouth lobes (after Mayer 1910)

155. A deep water stauromedusan *Tesserantha connectens* (6 mm diameter), caught by the *Challenger* expedition in the south east Pacific (after Haeckel 1882)

222), which floats at the surface. The other surface floating forms, bright blue in colour, have a thin chitinous gas-filled chambered disc float, above which a diagonal sail projects in *Velella* (figure 164) but not in *Porpita* (figure 163), and below which hang the individuals. They have been regarded as being derived from a single floating polyp, hanging upside down, and there are indeed floating hydrozoan larvae very similar in plan. *Velella* has a deep-water larval stage, a curious reversal of the surface larvae of bottom-living invertebrates.

The 200 species of Scyphozoa or jellyfish have reduced polyps or none at all, and have separate sexes. It is by great expansion of the gelatinous mesogloea that they attain their size; the common *Cyanea* and some deep-water species are very large. Trailing tentacles catch their prey as they swim along and it is transferred to the long mouth for digestion. The mouth is basically four-cornered but extension of these corners into long frilly lobes characterises one, usually coastal, group of medusae (figure 154), including the very poisonous *Dactylometra*. Loss of the tentacles and elaboration and fusion of the frilly lobes leads to the condition in *Rhizostoma* in which cilia waft small food particles into the network of mouths, or in *Cassiopeia* which does the same lying upside down on the bottom in shallow tropical waters. One group of small medusae have even become secondarily attached to seaweeds. Most deep-water medusae are both luminescent and beautiful deep maroon or purple in colour, and their jelly may be so firm as to be almost cartilaginous (figures 155, 117). They, and the surface species, are

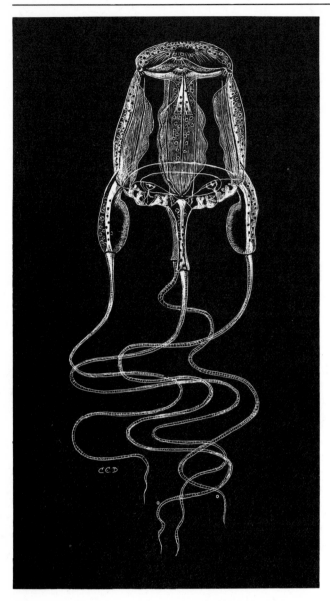

156. The common Mediterranean cubomedusan *Carybdea marsupialis* (4 cm diameter); this species probably lives in deep water when young but swims up to the surface when adult (after Mayer 1910)

forms as well (figure 157). One large subdivision includes those with an endoskeleton, usually of spicules. These may be of calcium carbonate and separate in the soft corals or fused in the organ-pipe coral. The remarkable blue coral, coloured by a bile pigment, is the only one of this subdivision to have a massive calcareous skeleton. The gorgonians, sea fans and sea feathers, have a skeleton of either calcareous spicules or the protein gorgonin or both. They are found down to 4,000 m and are often orientated according to the water currents. The pennatulids or sea pens have a single long central polyp with smaller polyps on each side of it and a skeleton of calcareous spicules. The other major subdivision includes the sea-anemones, true corals and black corals, with, respectively, no skeleton, a massive calcareous exoskeleton, and a horny skeleton. The anemones are usually shallow-living, some growing to 1 m in diameter, though there are deep-living (figure 157), floating, burrowing and swimming species. The true corals show great variation in growth form, and the warm-water reef-builders usually only emerge from their skeleton at night. It is believed that the presence of vast numbers of symbiotic dinoflagellates in the tissues of reef-builders is essential for the deposition of such massive amounts of calcareous material, and they are almost absent in other corals. Some corals, such as the fire corals, are poisonous, and all require animal food. The black corals are deeper-water species with a slender black branch-like skeleton, rather similar in appearance to the gorgonians.

The last group of coelenterates, the **comb jellies** or ctenophores, have no cnidoblasts, only lasso-like organelles for entangling their prey. They are voracious carnivorous hermaphrodites swimming by means of eight rows of fused cilia, whose movement produces the beautiful iridescence of most of the almost transparent members of this group. Some have contractile tentacles with which they catch their prey, but others engulf their prey with their very wide mouths. The tentaculate forms are usually globular as in *Pleurobrachia* (figure 212), but some have become laterally flattened, such as the ribbon-like Venus' girdle of warmer waters which also swims by whole body undulations, or dorso-ventrally flattened in several creeping species. Like so many other coelenterates comb-jellies often produce spectacular luminescent displays.

ARROW WORMS (PHYLUM CHAETOGNATHA)
These small, stiff animals are extremely important predators in the sea. They have proved very useful as 'indicators' of different water-masses in coastal areas. Particular species are found in water-masses with particular physical properties, and sometimes discrete

important plankton predators. Several shallow species are often found with certain animals (e.g. *Hyperia galba* figure 169, or *Argonauta* figure 146) riding upon them. The warm-water sea-wasps (Cubomedusae, figure 156) have achieved a bad reputation as a result of the venom of their cnidoblasts, which may prove fatal to man.

The exclusively polypoid Anthozoa include 6,100 species and are best known from the sea-anemones and reef-building corals, but include many deep-water

157. Deep-sea anthozoan coelenterates: 1. the pennatulid *Pennatula naresi* (23 cm); 2. the red gorgonian *Siphonogorgia pendula* (25 cm); 3. a solitary deep-living anemone *Antheomorpha elegans* (2 cm); 4. a group of individuals of the solitary coral *Caryophyllia profunda* (6 cm high) (1 after Kolliker 1880; 2 after Studer 1889; 3 after Hertwig 1882; 4 after Moseley 1881)

158. The planktonic tomopterid polychaete worms swim with their numerous oar-like appendages and are generally flattened and transparent

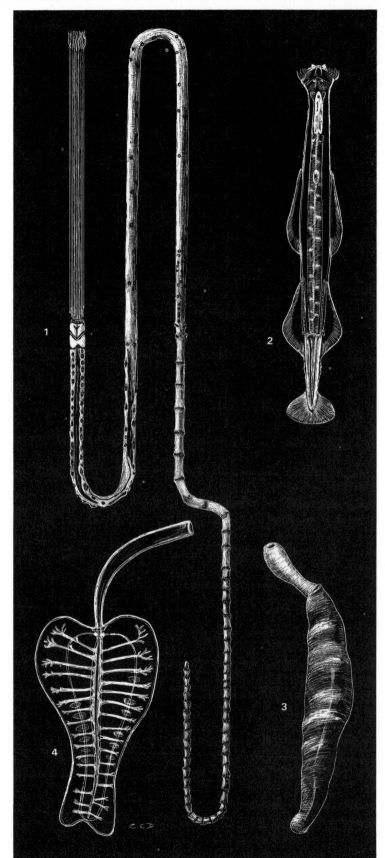

159. Three different groups of worm-like animals: 1. the beardworm or pogonophore *Lamellisabella zachsi* (17.5 cm) removed from its tube to show the crown of tentacles and three different body regions; 2. the warmer water chaetognath or arrow-worm *Sagitta planctonis* (3.7 cm); 3. *Bathynemertes hardyi* (11 cm), a red bathypelagic nemertine worm; and 4. the transparent nemertine *Pelagonemertes rollestoni* (4.5 cm) showing the branching digestive tract, which is coloured yellow and red in life (1 after Ivanov 1963; 2 after Dakin and Colefax 1940; 3 and 4 after Wheeler 1934)

water-masses may be identified from the species present, even though the more easily studied physical properties show little or no difference. There are about eight genera containing 65–70 species. All are planktonic except *Spadella*, which creeps about on the substrate, usually seaweeds. They are active, rapid swimmers having a characteristic darting movement. Their mouths have large hooks to grip their prey, which is swallowed whole and consists of crustaceans, *Tomopteris* (figure 158), young fish and other chaetognaths. Diatoms have been found in their stomachs but this may be fortuitous. Water vibrations elicit an escape reaction unless they are of the right frequency and amplitude, when the animal makes an accurate movement to catch the prey. Feeding is more active at night. The two eyes cannot produce images and probably only function in timing vertical migration which is sometimes diurnal, sometimes seasonal. In *Eukrohnia hamata* the depth range is about 0–400 m at 60° North and South but, like many other animals, is deeper, at 800–1,200 m, in the tropics. Chaetognaths have been caught as deep as 6,000 m. Genera and species which are thought to have existed for the longest period, and more advanced, mature specimens, live deeper than the youngest genera, species or specimens. The life span seems to vary from six weeks to over two years according to the species. Breeding occurs all the year round in both high and low latitudes, but is probably seasonally accelerated in temperate waters. While these animals display relatively little variation from the pale transparent 'standard' form (figure 159), some species have red or blue coloration.

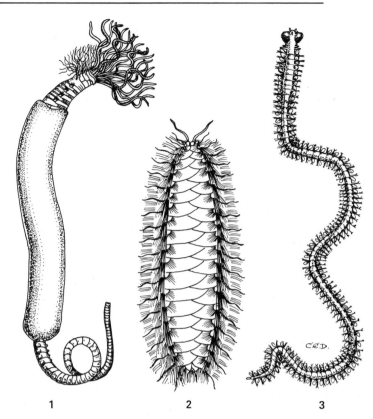

160. Three annelid worms from different habitats: 1. the Antarctic *Amphitrite kerguelensis* (15 cm) is partly encased in its tube of fine mud; 2. the scale-worm *Laetmonice producta* (10 cm) is found free on the bottom; 3. *Callizona Mobii* (11.5 cm) is a large-eyed pelagic species (1 and 2 after M'Intosh 1885; 3 after Apstein 1900)

ANNELID WORMS (PHYLUM ANNELIDA) (Figure 160)
The annelid, or segmented, worms include the leeches (with a few marine representatives), the earthworms and their relatives (all land or fresh water), and the bristle or polychaete worms. The latter are almost all marine, with swimming larvae and numerous segmental bristles. The 60 or so species of pelagic polychaetes have flattened oar-like projections from each segment with which they swim, and are usually transparent. They are voracious carnivores with large jaws and big eyes and often long tentacles as in *Tomopteris* (figure 158). One species is believed to suck the juices of chaetognaths. Most bristle-worms are bottom-living. Some, rather like the planktonic forms but with reduced swimming appendages, wriggle over or near the bottom feeding on small animals. Others have become more sedentary, living in tubes or burrows and feeding on small particles or mud, with loss or reduction of the jaws. Some project out of their tubes and by means of elaborate crowns or fans of ciliated tentacles sweep the mud or water for food.

Smaller particles are eaten, larger ones rejected and used to build up the outside of the tube, whose foundation is often a parchment-like secretion of its owner. Sometimes these tubes may be portable, but in other cases they are heavily calcified and cemented to surfaces in great stony masses, often presenting a considerable fouling problem. Species buried in the bottom have U-shaped tubes through which they pump the food-containing water, filtering it with elaborate mucus webs. Such a water circulation is also essential for their respiration, and some have developed feathery gills, and even haemoglobin, to assist in their poorly aerated environment. Protection afforded by the tube is not available to those running over the bottom in search of their prey, and some have developed strong armour-like overlapping scales all down the body (scale-worms, figure 160), or have taken to sharing accommodation such as another worm's tube, or a hermit crab's shell. Others develop stinging bristles or foul secretions, or in the case of the planktonic species, roll themselves up in times of danger and

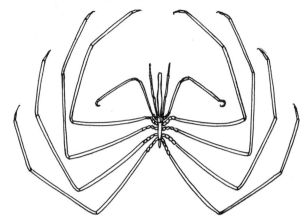

161. The large, bright-red, deep water pycnogonid *Collosendeis macerrima* may span 25 cm (after Bouvier 1917)

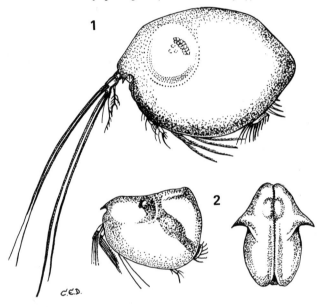

162. Two planktonic ostracods. 1. the large (6–7 mm) deep-water chocolate-brown *Macrocypridina castanea* and 2. *Fellia cornuta* (2 mm) seen from above and from the side, showing how the animal is almost entirely enclosed within the bivalved shell (after Müller 1894)

sink to greater depths. Many are luminous, though the function of the light is generally unknown. Reproduction of the bottom species often involves the development of special swimming appendages, just like those of the planktonic forms, and ascent of the sexually mature adults to the surface where they shed the eggs and sperm before sinking to the bottom again; a single female may have one million eggs. Alternatively the sexual segments may break off and swim to the surface by themselves. Huge swarms of such species often occur at the surface, linked to a lunar periodicity as in the Palolo worms of several oceans, which at such times are an important source of food not only for

fish but also for waiting fishermen. The individuals in these swarms are often luminescent, presumably to attract one another. Many adult worms can regenerate bitten-off portions, and some intentionally bud and fragment as a means of asexual reproduction. The eggs usually develop singly, but certain scale-worms carry their eggs around under their scales in a primitive form of parental care, and a few species are viviparous.

INSECTS, ARACHNIDS AND CRUSTACEANS (PHYLUM ARTHROPODA)

The arthropods are characterised by their hard exoskeleton and jointed limbs, and include the insects, arachnids, and crustaceans. In contrast to their huge success on land there are no permanently subsurface marine **insects,** and the only one whose life is entirely bound up with the sea is the bug *Halobates* which skates over the surface of the warmer oceans. The **arachnids** have several wholly marine species including some tiny sand-living tardigrades, the shallow-water tank-like horse-shoe crab, and many species of mites, some of which occur as deep as 5,000 m. A wholly marine subgroup of the arachnids are the sea-spiders (pycnogonids), whose gut extends into their legs and which are benthic animals (figure 161).

The **crustaceans** are the insects of the sea in their successful colonisation and spread in the marine environment, and comprise at least 70 per cent of almost any haul of marine animals. There are very many groups of different crustaceans – itself a tribute to their success – but only the more important ones will be considered here. The **branchiopods** (water fleas, fairy shrimps and brine shrimps) are a primitive group having a large number of body segments, most of which bear leaf-like respiratory and swimming legs. Most are primarily freshwater forms, the marine representatives being restricted to a few species of water-fleas, found in surface waters, occasionally in swarms. The **ostracods** are a more important group in the sea, and are almost entirely enclosed in a bivalved carapace or shell from which only the antennae and the extremities of the few limbs emerge. They have only two pairs of trunk limbs, less than any other crustacean, and are almost all very small. They are to be found on and in the bottom sediments as well as throughout the depths of the open ocean, the deeper pelagic species generally being red or brown in colour (figures 162 and 194) and with more sensory hairs than the shallower species. Though most planktonic forms are eyeless, some species have compound eyes; and the largest ostracod, almost globular, about 1 cm in diameter and appropriately called *Gigantocypris* (figure 194), has elaborated its eyes into two huge forward-facing headlamp-like reflectors, which perhaps assist in its

163. *opposite* The blue warm-water siphonophore *Porpita* floats at the surface by means of buoyancy chambers in the disc (5 cm diameter); it rhythmically sweeps the water by opening

and closing its net of tentacles which are armed with knobs of stinging cells

capture of the small fish, chaetognaths and copepods found in its stomach. Though a few species live in the nostrils and gills of fishes and other animals, the pelagic species are mostly carnivores or scavengers eating small animals, dead material, and perhaps fecal pellets. Bottom species are detritus feeders, the currents set up by their limbs wafting food particles into the shell, where they are bound together by the mucus from special glands and eaten. Despite many being apparently blind, there are a number of species which put out a luminous secretion from glands round the mouth, the chemistry of whose luminescence is the best-known of any such system. Though some are believed to be parthenogenetic it is probable that the female stores a packet of viable sperm for a long time after mating and can therefore fertilise subsequent batches of eggs in the absence of the male. A few species, including *Gigantocypris*, retain the eggs in the shell during development, but most lay them directly into the water. As in many other small animals, the thermocline acts as a barrier to the vertical movements of some species. The shells of benthic species are often thickened and elaborately sculptured, and these persistent empty shells are much used as geological markers in the search for oil deposits and the classification of sediments. Though ostracods may occur in large surface swarms in certain areas they are not usually a numerically very important component of the plankton.

The **copepods** on the other hand, and in particular the 1,200 or so species of the pelagic group known as the calanoid copepods, are the most important single group of marine animals in the economy of the sea, often making up 70–90 per cent of the biomass of a zooplankton sample and not infrequently occurring in vast swarms. These little animals (usually not more than a few millimetres in length) never have a carapace or compound eyes, only have six pairs of limbs behind the head and none on the abdomen. It is with these oar-like limbs that they swim, often assisted by flicks of the abdomen or of the long antennae. Most species filter the water currents, set up by their limbs, with bristly mouthparts, eating the filtered-out phytoplankton or detritus. They are thus the grazers of the ocean and the primary consumers of the phytoplankton harvest. Though most are small, their size increases to a maximum at 500–2,000 m depth, one species reaching 17 mm, but in still deeper water the size of the species decreases again. Many of the larger deep-water species as well as the shallower species are luminescent. Body form is very variable (figure 166); one large group, the tiny harpacticoids, are largely interstitial animals with tiny antennae and an almost wormlike body. The antennae are also short in many cyclopoids,

a group many of whose planktonic members have remarkable (though not compound) telescope-like eyes (figure 166), which are believed in *Copilia* to be 'scanning' eyes working rather like a television camera. Much-modified simple eyes are also found in some blue surface-living copepods (figure 179), some of which have two pairs of lenses on top of the head, others a huge eye with an accessory lens below the head. Eggs are either carried by the females in sacs attached to the abdomen or released separately into the sea, in which case they may be provided with flotation devices. In each case they hatch as nauplius larvae, and go through a number of moults before reaching adult form. Males are so far unknown in many species and may be rather short-lived or very much fewer in number than the females. It has also been shown that high hydrostatic pressure can reverse the sex of the nauplii of some species. Though most species are free-living, a number are much modified parasites attached to the outside of fish and whales (where some species may reach 30 cm in length!) or living inside sea-squirts. Others are commensal in or on echinoderms, polychaetes, molluscs and many other invertebrates. The numbers of a copepod species can fluctuate very greatly, particularly at the surface, though not so much in bottom-living species, and their fecal pellets are so numerous as to be an important food supply for deeper water plankton. Copepods supply the main food source of very many marine animals, particularly fish; sardines are wholly dependent upon the smaller species, and 60,000 copepods have been found in one herring stomach. In addition, they are often so numerous as to supply a large fraction of the food of sei whales in the Northern Hemisphere.

The fourth major subgroup of crustaceans, the **barnacles** (cirripedes) is characterised by the fixed life of the adults and consequent development of heavily calcified protective plates. The best-known barnacles are the stalked barnacles, important fouling organisms of ships, and the sessile barnacles common on the shore, though there are other deep-sea sessile species as well as several groups of parasitic barnacles. Though not immediately recognisable as crustaceans they may be fairly accurately regarded as crustaceans standing on their heads and kicking their food into their mouths. The legs emerge from the shell and sweep the water clear of food, be it detritus or phytoplankton in the smaller species or larger animals in the bigger species (some even eating coral), which is then dragged into the shell and eaten. They are the only crustaceans highly adapted to a sessile life, and species are to be found settled on a variety of bases, including driftwood, fuel oil, fish skin and jaws, dolphin teeth, whale jaws, sea-snakes, turtles and

164. *opposite above* The siphonophore *Velella* (3 cm) has a diagonal crest projecting from its oval float, and is preyed upon by the purple mollusc *Janthina*, which is itself buoyed up by a float made of bubbles of air

165. *below* The frilled silver and blue or black nudibranch *Glaucus* (3 cm) floats upside down at the surface of the sea, swallowing air to maintain its buoyancy

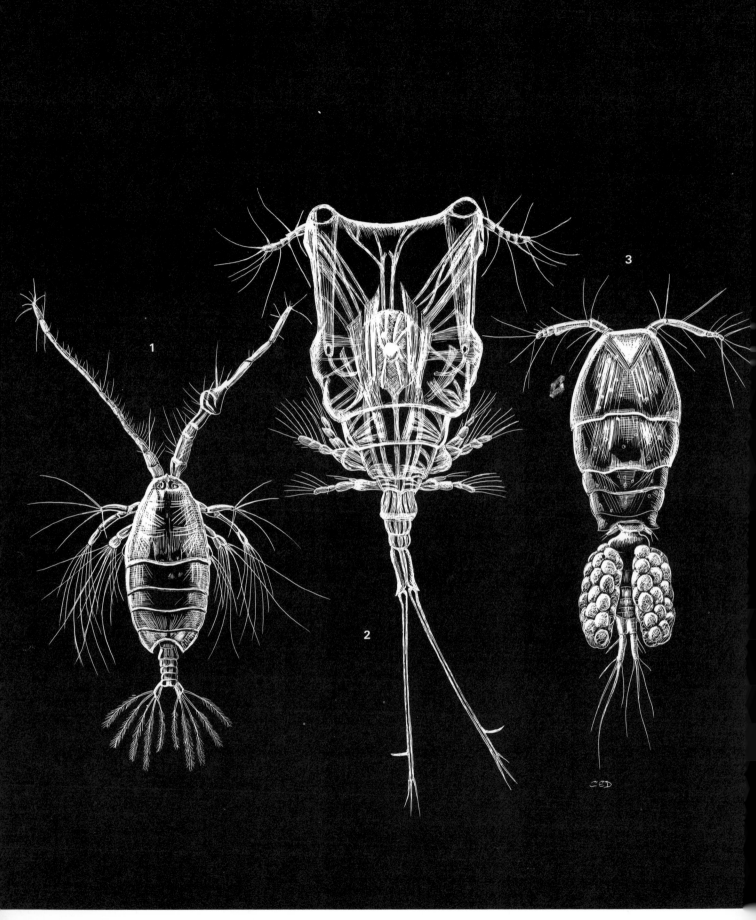

166. Planktonic copepods. 1. a male *Pontellina plumata* (2 mm), a near-surface blue-purple calanoid species with the long right antenna modified for grasping the female; 2. the cyclopoid *Copilia quadrata* (3 mm) has remarkable telescope-like 'scanning' eyes; 3. a female of the more typical cyclopoid species *Oncaea venusta* (1 mm) carries the eggs in a pair of sacs attached to the abdomen (after Giesbrecht 1892)

167. The oceanic barnacle *Lepas fascicularis* often settles as a larva on a lump of floating oil, but as it grows it forms a raft of bubbles to keep it at the surface

sea-cows. One species of stalked barnacle even secretes its own float hanging from which it drifts around the oceans (figure 167), and colonies of a related species are known to eat young flying fish; when one animal is satiated the remaining food is taken by another. Eggs hatch as free-swimming nauplius larvae which develop into the attachment stage or cypris larvae, these occasionally occurring in such numbers as to discolour the water. This larva chooses a suitable site, often attracted by a layer of adsorbed protein provided by the presence of other settled barnacles, settles by means of a cement gland in the head and grows very rapidly to the hermaphodite adult stage. Parasitic barnacles are found mostly on other crustaceans and are often reduced in the adult to a bag of eggs and a mass of roots in the host tissue, though retaining a nauplius larva. Apart from their economic importance as fouling organisms, barnacles are eaten in many countries (large species may reach half-a-pound in weight and the size of a tea-cup) and some are grown in Japan for harvesting and conversion to fertiliser.

All other crustaceans are included in the sub-class

168. The rare deep water crustacean *Nebaliopsis* (2 cm)

Malacostraca, with compound eyes (often stalked) and typically a carapace which covers the thorax and is often hardened to form a shell. The small group including the rare deep-water *Nebaliopsis* (figure 168) are probably the least modified except that their inflated carapace appears almost bivalved. One species bores into the carcasses of sharks caught on long-lines off Japan. The **cumaceans** also have an inflated carapace covering the body and head, probably largely an adaptation to overcome the respiratory problems of their muddy sediment environment, though they may be found in the plankton at night. The **tanaids,** found to at least 6,000 m, are also relatively uncommon small bottom dwellers normally living in tubes or burrows.

Of much more importance are the **opossum shrimps** (mysids), in which the carapace covers but does not fuse with the thorax, and which often have obvious statocysts (balancing organs) in the tail-fan. Their name arises from their characteristic of hatching their eggs in well-developed brood pouches beneath the thorax, and they are common in estuarine and shallow coastal water as well as being found in the deep oceans at all depths. The smaller species often have no gills, and respire through their thin carapace, as in the surface living species *Siriella thompsoni* (figure 169), but

169. *opposite* Typical malacostracan Crustacea. 1. the surface-living opossum shrimp (mysid) *Siriella thompsoni* (10 mm); this is a male, the female has much smaller abdominal limbs; the statocyst (S) on the tail fan and the lack of gills is characteristic of many mysids. 2. the euphausiid *Euphausia superba* (5 cm), the krill of Antarctic waters; this large species shows the branched gills (G), typical of euphausiids, at the base of each leg. 3. *Systellaspis debilis* (5–6 cm), a bathypelagic shrimp which has light organs on the thorax and abdomen (circles and heavy black lines) and at the base of the abdominal limbs; pincers on two of the five pairs of walking legs characterise it as a decapod. 4. the amphipod *Hyperia galba* (2 cm), often found on the surface of medusae, seen from above (a) and the side (b); similar views of the common European isopod *Eurydice pulchra* (0.7 cm) (5a, 5b) show the different ways in which amphipods and isopods are flattened (1, 2, 4 and 5 after Sars 1885, 1895, 1899; 3 after Kemp 1906)

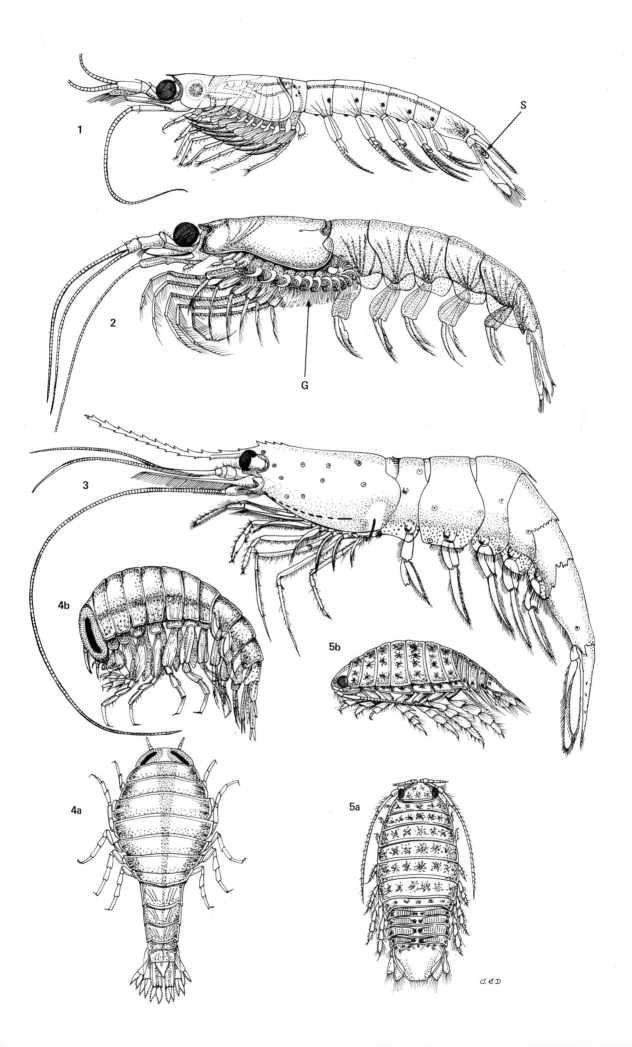

1

S

2

G

3

4b

5b

4a

5a

CED

170. A large (5 cm) deep water isopod, coloured pink in life

larger species such as the deep-water scarlet *Gnatho-phausia* have gills beneath the carapace. This mysid emits a luminous cloud when handled and has a very hard spiny carapace, whereas most other deep-sea mysids have long thin legs, soft bodies and very often divided eyes, two different areas on the eye looking in different directions. Most species are believed to be filter feeding, or to eat small particles, but little is known about the biology of the deep-sea forms.

The **isopods** have no carapace, and are normally dorso-ventrally flattened (figures 169 and 170). Apart from the first thoracic leg (modified as a mouthpart), all the legs are similar. Most species are to be found on the shore and in coastal waters and are omnivorous. The relatively few living permanently in the open ocean are mostly carnivorous scavengers, though there are many species found on the bottom at depths of up to 10,000 m, many of which are apparently restricted to very deep water. Deep-water species are

often blind and reddish, though a few are black, and are to be found much nearer the surface in Polar regions. Some open ocean surface species are associated with floating debris and are often found on lumps of floating oil, while certain coastal species are associated with chitons. A more permanent association has led to many becoming parasitic, especially upon fish and crustaceans, and though many attach to the outside of the host some have become completely internal parasites. Dwarf males are found in many of these parasitic species. From an economic point of view one isopod, the gribble, causes considerable damage by boring into wood pilings, and other species frequently destroy the bait in crab and lobster pots.

Amphipods appear similar to isopods except that they are generally laterally flattened and the thoracic legs are of more than one form (figure 169). One large, least modified group of amphipods, the gammarids, are found in both fresh water, the shore region

171. The large amphipod *Cystisoma*, reaching at least 12 cm in length, is almost completely transparent, and its huge inflated head is largely composed of eye. The white mass (orange in life) is a brood pouch in which the early larvae are kept

(often most obviously as sand-hoppers) and the shallow and deep sea. They are largely omnivorous though many shore forms probably utilise mainly vegetable food. Some amphipods have become modified for life as interstitial fauna, and others, the caprellids, have become partially walkers rather than swimmers, their abdomens becoming greatly reduced. Many caprellids climb hydroid colonies on which they may feed, others are commensals with starfish, and eat diatoms and small animals, and abyssal caprellids have been found. The hyperiid amphipods are the most important pelagic group of these animals and have many interesting characteristics. In many the eyes are enormously enlarged (figure 118), taking up almost the whole head in *Cystisoma* (figure 171), or divided into areas looking in different directions. Some have enormously elongated heads and eyes (figure 119), and in *Rhabdosoma* the whole body is greatly elongated and no thicker than a large needle. Some are luminescent,

and many hyperiids such as *Hyperia galba* (figure 169) are believed to be associated with, and feed upon, surface or deeper water medusae. The remarkable *Phronima* (figure 172) eats out a *Pyrosoma* or a siphonophore bell until only an empty barrel remains, and then lives within the barrel, carrying it about and rearing its young within it. Amphipods probably evolved in the cool temperate regions where they are still most abundant, but are of little economic significance apart from causing occasional damage by boring into fish caught on long lines.

Though there are only about 90 species of **euphausiids** they are an important group in the ocean. All are pelagic, have the carapace fused with the thorax, no statocysts, and have feathery gills on the thoracic legs (figure 169). All but one have prominent complex light organs, particularly on the ventral side of the body. Only the deep-water bright red *Bentheuphausia* lacks these organs. As in some opossum shrimps and

172. The amphipod *Phronima* (3 cm) lives within a barrel formed from the test of a salp or a swimming bell of a siphonophore

amphipods a number of euphausiids have divided eyes, with different areas looking in different directions, and this feature seems to be generally associated in these species with an elongated pair of legs, the carrying of the eggs in membranes by the females, and reduced or absent vertical migration. Though these species carry their eggs until they hatch, most euphausiids do not, and there is no special brood pouch as there is in the superficially similar opossum shrimps. The majority of euphausiids are able to take live prey but some can also filter-feed or eat detritus. Many species, in all latitudes, are found at the surface in immense swarms, and it is these swarms (krill) that are utilised by the baleen whales in the Southern Hemisphere and which may provide an economic source of protein (Chapter 9). *Euphausia superba*, the major food source of the baleen whale in the Antarctic, is also the largest surface species, though some deep-water species are larger, reaching 150 mm in length. Euphausiids are also an important component of the diet of many fish. They contain large amounts of vitamin A and are probably the source of the rich deposits of this vitamin in the livers of many fish. Apart from their potential as a food source (whales have been estimated to consume 150 million tons of euphausiids in a 3–4 month summer feeding period), they may also be important in the accumulation and transfer throughout the oceans of fallout products such as the radioactive isotope of zinc ^{65}Zn.

The **decapods** are very important among the larger animals in the sea. They have a carapace fused with the thoracic segments and covering the gills, and five pairs of walking legs some of which are pincer-like (figure 169). The shrimps and prawns divide into two groups, one laying their eggs into the sea where they hatch as nauplius larvae, the others carrying their eggs on the abdomen and hatching in a more advanced stage. The crawling forms, the lobsters, crabs, crawfish and squat-lobsters all carry their eggs until hatching in a similar way. Decapod larvae often bear no resemblance whatsoever to the adult and have frequently been classified as different animals until the relationship was known. The slow change in structure to the adult occurs through a long series of moults. The deep-water shrimps and prawns are very similar, deep red or orange, often with highly complex light organs beneath the cuticle, or even formed from modified liver tubules, and sometimes with reduced eyes and soft carapaces (figures 128, 129 and 195). Surprisingly, decapods are not found deeper than 6000 m. Benthic species are usually larger than pelagic species, often have enormously elongate antennae and are long-limbed (as are the spider crabs). Like all other decapods they are probably basically carnivorous. Pelagic forms usually feed on smaller prawns, euphausiids and fish. Crabs and lobsters may have a form of social organisation and dominance in their communities, and crabs in particular may be associated with molluscs, anemones or fish. Several species of crab have become free-swimming by the development of oar-like hinder legs, and these swimming crabs, which are occasionally met with in huge swarms, are very voracious. Many decapods have rather bizarre habits, such as the cleaning shrimps which clean the mouths and bodies of fish, and the sargassum shrimps, disguised as pieces of weed. There are burrowers, noise-makers, and even some crabs that use stones for fighting each other. The dense populations of many types of decapod make it easy to catch them in quantity, and members of every group are exploited commercially. The dense sporadic aggregations of others may perhaps be exploitable in the future (Chapter 9).

The last group of malacostracans found in the sea are the **mantis shrimps** (stomatopods figure 130), so-named because of their curious mantis-like claw that folds upon itself like a clasp-knife. These animals, though occasionally abundant in local swarms, are usually burrowing animals in coastal waters, where their larvae are frequent members of the zooplankton.

MOLLUSCS (PHYLUM MOLLUSCA)
Second only to the insects in number of described species (between 65,000 and 100,000), the molluscs

comprise an important part of terrestrial, fresh-water and marine communities. Over 31,000 species are marine but the great majority of these are from the continental shelf or shore. The phylum consists of the classes Monoplacophora, Aplacophora, Polyplacophora, Scaphopoda, Gastropoda, Bivalvia and Cephalopoda; as these are all found to some extent in deep water and display great diversity, each group is treated separately.

In 1952, the *Galathea* expedition collected from the bottom at 3,570 m ten specimens of a 'living fossil' which was subsequently named *Neopilina galatheae*. Since then, several other specimens representing a further three species of *Neopilina* have been collected. This genus belongs to a group of molluscs (the **Monoplacophora**), which were previously thought to have become extinct 350 million years ago. These small, mud-dwelling, limpet-like animals which are less than four centimetres long caused considerable biological excitement because they have distinct body segmentation, a condition not found in any other molluscs. This gives strong support to speculation that molluscs developed from segmented animals such as the annelids.

There is a group of worm-like molluscs (the **Aplacophora**) which live exclusively on the bottom and extend down to a depth of at least 4,000 m. Members of one of the two families are always buried head downwards with their posterior gills projecting from the surface. Species of the other family are nearly always found coiled round corals and hydroids upon which they feed. They are all less than 30 cm long (usually less than 5 cm) and are not important numerically.

The **chitons** (Polyplacophora), with a series of armour-like plates on their backs, are well known on the sea shore but are rare in deep water. Members of one family extend to at least 4,200 m on the bottom and some species are not found shallower than 1,000 m. With increase in water depth they decrease in size and become paler.

The **elephant tusk shells** (Scaphopoda) resemble hollow, round teeth. They have been found at depths of nearly 7,000 m on the bottom. The shallow-living forms select Foraminifera from the mud in which they lie half buried and drag them back to the mouth with special tentacle-like organs. The group is not important in deep water.

The **snails** (gastropods) are important at the surface, in midwater and on the deep-sea floor. Of these the prosobranchs (with gills at the front) are mainly bottom-living down to the greatest depths but they include some remarkable species adapted for life away from the bottom. *Janthina* is snail-like and secretes a fluid

which it wraps round bubbles of air to make a float (figure 164), and one species overcomes the danger of loss of eggs from the surface by being viviparous. A special gland secretes a purple fluid which is thought to anaesthetise its siphonophore prey, *Velella* and *Porpita*; its diet also includes the insect *Halobates,* copepods, and even its own kind. Heteropods are completely transparent prosobranchs which swim upside down by means of a fin. *Atlanta* has a keeled shell (figure 181), *Carinaria* (figure 180) a very reduced shell, while *Pterotrachea* has lost the shell completely (figure 173). Sexes are separate, and large suckers on the fins aid copulation. Heteropods grow to about 25 cm in length and are quite common in midwater net hauls.

Opisthobranch snails (with gills at the back) include several groups specially adapted for midwater life. The many species of thecosome pteropods have special wing-like structures for swimming and delicate transparent shells. The largest, Venus' slipper (figure 173), is completely transparent and may be five or six cm long, but most species are less than one cm. Gymnosome pteropods have no shells. They often have hooks and tentacle-like processes with suckers and are active predators of the shelled pteropods (figure 173). They are not common and although about 35 species are known, *Hydromyles globulosa* greatly outnumbers all other species in the tropical Indo-Pacific, and *Clione limacina* is by far the most numerous in the boreal Atlantic. Great swarms are sometimes seen, and *Clione* is said to have been important in the diet of the Greenland whale.

Several sea slugs (also opisthobranchs) are also adapted for a pelagic life but are not common. *Glaucus* (figure 165) floats by means of gas-filled pouches of the gut which lie inside curious elongate processes of the body. In the ends of these processes it stores for its own protection stinging cells (cnidoblasts), taken from its siphonophore food. It lies against the surface film of the water and is dark above and silvered below. *Phyllirrhoe* (figure 131), is flattened, swims by graceful undulations of the body and has a surface studded with tiny light organs. It attaches itself to siphonophores and is unusual in having symbiotic algae in its tissues.

The **bivalves** (lamellibranchs or pelecypods) occur on the bottom to depths exceeding 10,000 m and, in all, about 10,000 species are known. With increasing depth they become smaller and rarer. Some lamellibranchs are specially adapted for life at greater depths. In deeper living *Abra* species the gill becomes smaller and, besides being respiratory, acts as a conveyor belt for transporting incoming sediment from the siphon to large flaps bordering the mouth (palps). Deeper forms have larger palps and a longer hind gut to

facilitate absorption of food from the nutrient-poor sediments of deeper water.

The **cephalopods,** entirely marine, include *Nautilus*, cuttlefish, squids and octopuses and grow to large size (20 m maximum overall), are extremely numerous and are very important predators and food in the ocean. *Nautilus* is a 'living fossil', the three living species being the only survivors of over 2,500 species which flourished for several million years during the Palaeozoic era. Its main interest is the light it has thrown on fossil studies. *Spirula* the 'fag-end fish' rarely exceeds four cm in length, is the only living cephalopod with a coiled, internal shell (figure 174) and lives entirely in deep water; its nearest relatives, the cuttlefishes, are limited to the continental shelf. The deepsea squids are by far the most important group ecologically, being found in all seas in very large numbers; they form the food of many cetaceans, seals, fish and birds and are important predators of smaller animals. Squids generally have ten tentacles which bear hundreds of suckers armed with either cutting, toothed rings or sharp hooks. They usually swim extremely rapidly forwards or backwards by means of a jet of water pushed from a small tube under the head by contraction of the body muscles. This may be powerful enough to shoot the squids out of the surface in the few flying species. Some, however, have larger fins for swimming and several retain ammonium salts in their body which give them buoyancy. *Spirula*, *Nautilus* and the cuttlefish achieve buoyancy by sucking water out of the cavities within their shells. Light-producing organs are found in many species;

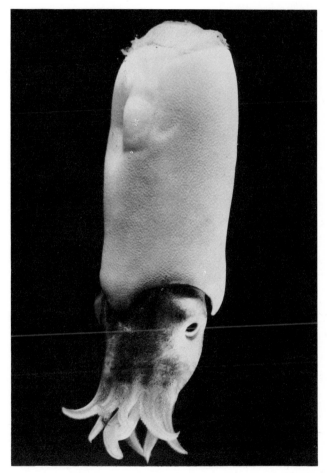

174. The small squid *Spirula* (5 cm) and an X-ray showing the spiral internal shell

173. *opposite* Planktonic molluscs: 1. the thecosomatous (shelled) pteropod *Creseis acicula* (1.5 cm) in its needle-like shell with the two wings projecting from the top; 2. *Cymbulia peroni* (Venus' slipper) (8 cm) is a thecosome that has replaced its hard shell with a knobbly gelatinous pseudo-shell which it carries like a hood, from ventral (2a) and from dorsal side (2b); 3. the gymnosomatous (naked) pteropod *Spongiobranchaea* *australis* (2 cm) has small wings and sucker-bearing arms; 4. *Carinaria mediterranea* (22 cm) is a heteropod with a small shell from which the gills protrude, while 5. *Pterotrachea coronata* (26 cm) is a heteropod with no shell (1 after Meisenheimer 1905; 3 after Pelseneer 1887; 2, 4 and 5 after Vayssière 1904, 1915)

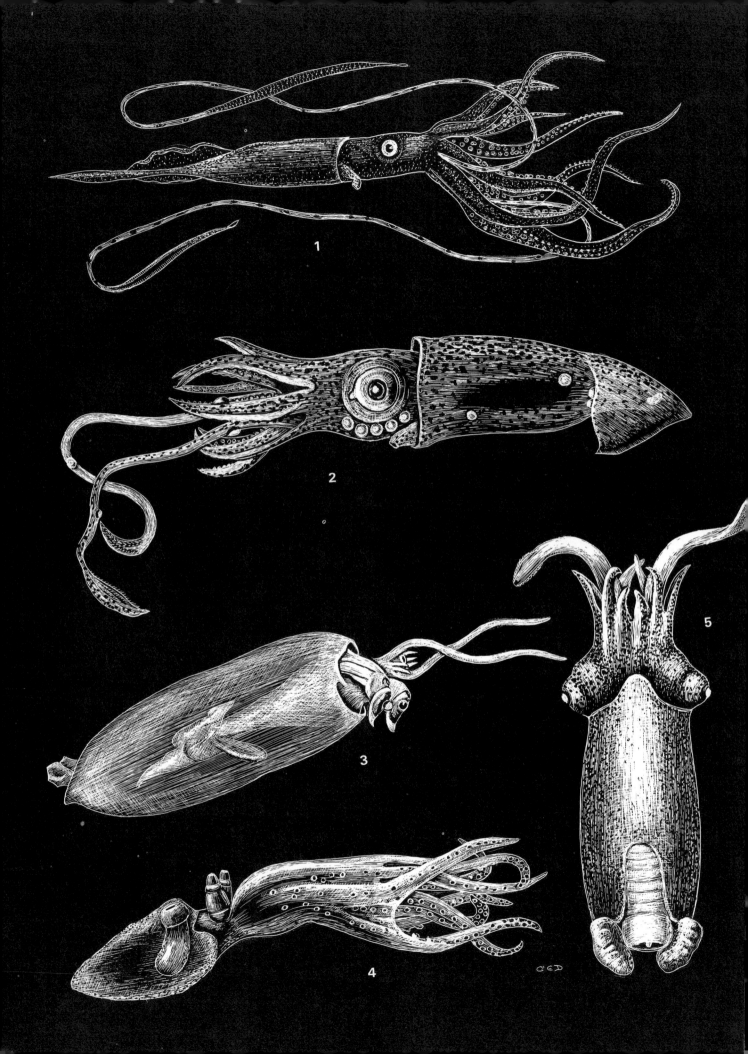

they may be spread over the ventral surface (figure 226), on the under side of the eyes and liver (figures 175, 145), on the tips of long tentacles as in *Chiroteuthis* (presumably to lure prey) (figure 175), or may even produce a luminous shower of sparks in the water in order to startle and temporarily blind attackers, as in *Heteroteuthis* (figure 210). From almost complete transparency (figure 176), they may be silvered, reddish (figure 226), blue, purple or black. Oceanic octopods have many similar adaptations, but also have a few individualistic tendencies. The near surface *Argonauta* female secretes a paper-like shell into which she lays her eggs (figure 146); folds of skin, secreting and stretching over the shell, also trap small organisms and sweep them to the mouth. Argonauts often ride on the backs of jelly fish and eat their ovaries. A near relative, *Tremoctopus*, holds coelenterate tentacles along its arms to use the stinging cells for its own purposes, and another, *Ocythoë*, is viviparous. *Amphitretus* (figure 175) and *Japetella* are almost transparent and the former has tubular eyes. Other octopods have broad powerful fins. The closely related black *Vampyroteuthis infernalis*, has four such fins as a larva but loses one pair during growth.

ECHINODERMS (PHYLUM ECHINODERMATA)

The echinoderms are entirely marine and are well known on the sea shore as well as being very important in the economy and ecology of the bottom of the deep sea. They show no true segmentation, are usually radially symmetrical, usually have rows of hydrostatically operated tube feet and have calcareous skeletons composed of ossicles, plates or spicules.

The **sea lilies** and **feather stars** (class Crinoidea) were very prolific in the past but at present there are only about 630 living species known. Some extinct species had stalks up to 21 m long, but now there are only 80 species with stalks, none of which exceeds half a metre in length (figure 177). The stalkless crinoids or feather stars are normally loosely attached by cirri, but can swim slowly by moving their arms up and down, and they may even run on the tips of the arms before swimming. The mouth, which lies on the disc from which the five or so arms arise, faces upward.

Feather stars extend down to about 1,500 m on rocky bottoms, while stalked crinoids live on muddy bottoms at 200 – 1,500 m. They are common in cold water and sometimes form dense aggregations; 10,000 specimens were caught in one haul from 340 m.

175. *opposite* Some unusual cephalopods: 1. *Chiroteuthis imperator* (body length 27 cm) is a gelatinous squid having ammonium salts in its tissues for buoyancy and light organs on the tips of its tentacles; 2. *Lycoteuthis diadema* (body length 3.5 cm) is a small squid with light organs on its eye and in its body (seen through the transparent body wall); 3. *Sandalops melancholicus* (body length 1.5 cm) is a transparent squid with its eyes on stalks; 4, *Amphitretus* (total length 6.5 cm) is a

gelatinous midwater octopod with telescopic eyes; 5. *Spirula* is a small (body length 5 cm) sepioid squid having a spiral shell and a light organ on the tip of its body between the fins (after Chun 1910)

176. *above* The cranchiid squid *Liocranchia* (8 cm) is almost completely transparent apart from its liver, which is silvered to render it invisible to predators

177. Deep-sea echinoderms: 1. the dusky-purple crinoid or sea-lily *Metacrinus costatus* (the lower end of the stalk is broken off, total length as shown is about 23 cm); 2. the starfish *Lophaster stellans* (5 cm across); 3. *Pourtalesia hispida* is a bilaterally symmetrical echinoid or sea-urchin with a very thin test and is 5–6 cm long; 4. *Scotoplanes globosa* (18 cm) is an elasipod holothurian or sea-cucumber with the body processes typical of these animals, and the feeding tentacles in a frill round the mouth (1 after Carpenter 1884; 2 after Sladen 1889; 3 after Agassiz 1881; 4 after Théel 1882)

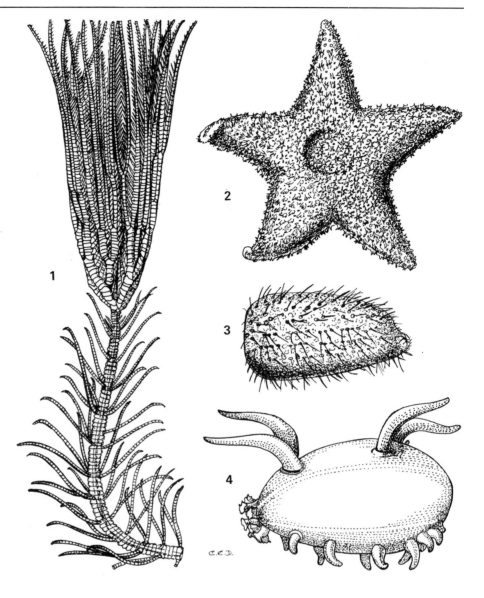

Sea stars or **starfish** (class Asteroidea) have five to fifty thick arms running from the central disc (figure 177). There are probably more than 1,200 species, varying from 1 – 2 cm to about 50 cm across and all live on the bottom. They move by rows of tube feet and eat mainly molluscs and other echinoderms but their diet may include crustaceans and even fish, and some species may have very catholic tastes. Prey may be swallowed whole, or digestive juices may be poured into a bivalve and the digested contents later sucked out. A coral-eating species known as the 'crown of thorns' is in process of destroying large parts of the Great Barrier Reef off Australia. Sexes are usually separate. Eggs and young are sometimes brooded under the disc between the arms or even in pouches of the stomach. Brooding species usually have less than 1000 eggs, while non-brooding species are known to lay up to 200 million. Deep waters harbour littoral species as well as some characteristically deep-sea species. Although the deep sea has a rich starfish fauna, many species are very limited in geographical range.

The **brittle stars** (class Ophiuroidea) have five slender, usually unbranched, arms with a central disc and a hard articulated skeleton. They are small and most of the 1,600 existing species have a very similar appearance (figure 201), although a few have branching arms (e.g. *Gorgonocephalus*, the basket star). Many littoral species extend into abyssal regions, while some exclusively abyssal species have a very wide distribution. The commonest and widely spread abyssal species, *Ophiomusium lymani*, lives at 700 – 4,000 m. Brittle stars sometimes occur in very large aggregations in deep water (figure 202). In shallow water they have

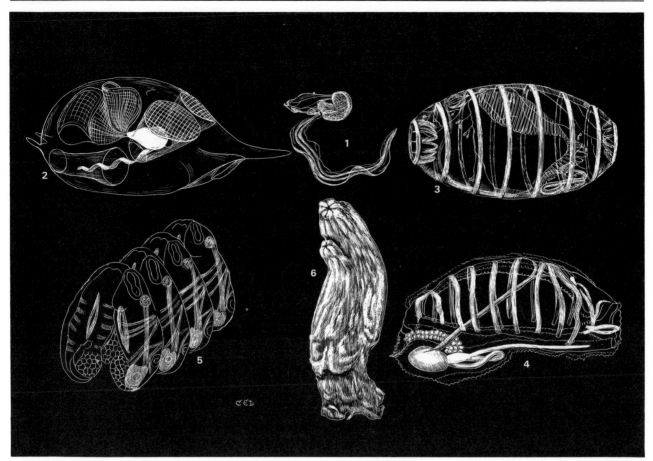

178. Protochordates: 1. *Folia aethiopica* (0·4 mm) a tadpole-like larvacean, and 2. the 'house' of a similar species showing the animal within it and the mucus filtering webs; 3. the asexual form of *Doliolum chuni* (7 mm) has symmetrical muscle bands, and valves at each end; 4. the solitary form of *Salpa gerlachei* (27 mm) with the large 'nucleus' or stomach at bottom left, and 5. a group of *Salpa* aggregate individuals just broken off the stolon; 6. *Ascidia challengeri* (17 cm high), a sea-squirt, here has the lips of the inhalent and exhalent siphons tightly pursed (1 after Lohmann 1896; 2 after Hardy 1956; 3 after Neumann 1913; 4 after Foxton 1961; 5 after Berrill 1961; 6 after Herdman 1882)

been seen to collect into big bunches in response to light. These relatively small animals (up to perhaps 15 cm across) are very agile and their nervous system is surprisingly well developed. They feed on polychaetes, small crustaceans, echinoderms, molluscs and detritus. They are mainly hermaphrodites, changing sex during their life, and a few species have tiny dwarf males which cling mouth-to-mouth to the females. Some species are viviparous and brood their eggs in special internal pouches.

The **sea urchins, heart urchins** and **sand dollars** (class Echinoidea) are globose, oval or discoid, have a closely knit calcareous armour, long protective spines and usually five 'ambulacral' grooves containing long protrusible tube feet (figures 177, 188). They occur in all seas and on every type of bottom down to 5,000 m, but apparently go no deeper. Regular, radially sym-metrical urchins favour hard bottoms, while irregular bisymmetrical ones favour sand. Many shallow forms extend into the deep sea, but exclusively abyssal species predominate at these depths. Species have wide distributions but none are cosmopolitan. Some urchins live from four to eight years and spawn after one year, and hybrids are not uncommon.

The 500 living species of **sea cucumbers** (class Holothuroidea) have elongate cylindrical bodies with a mouth surrounded by a circle of tentacles near one end and an anus near the other end. The body wall is leathery, slimy, and contains a skeleton of minute ossicles. Nearly all the deep-sea forms are in the order Elasipoda. Most of these live at 1000 to 5,000 m and some, such as *Onoirophanta mutabilis*, are cosmopolitan. They may live in great multitudes (figure 202) and up to ten species may be caught at one station. They are sluggish, sedentary animals but some species have been seen to make bounding movements or sinusoidal swimming movements. Locomotory podia lie underneath and elasipods often have upper papillae modified into sails, rims or tails (figure 177). One holothurian genus, *Pelagothuria,* is specially modified for a pelagic life. The few specimens, from the Pacific, Indian Ocean and Atlantic, have been grouped into four species but these may prove to be but one. They have been collected from the surface to a depth of 1000 m.

PROTOCHORDATES

The protochordates have several characteristics such as gill clefts and the presence of a skeletal stiffening rod (the notochord), which indisputably link them to the vertebrates. They include the phylum Hemichordata (which includes both mud-eating, burrowing **acorn worms** (figure 182) and tube living deep sea forms feeding by means of ciliated arms), and the primitive chordates (which with the vertebrates comprise the phylum Chordata) namely the **tunicates** and the **lancelets** (cephalochordates). The **tunicates,** so-named because their body is covered by a test of tunicin – a cellulose-protein complex similar to some found in vertebrates – are important in the marine economy and are divided into the sessile forms (sea-squirts or ascidians) the pelagic forms (salps, diliolids and pyrosomas) and the tiny plankton larvaceans. Sea-squirts are bag-like animals attached to the substrate either directly or by a stalk, and which use cilia to pass currents of water in through one siphon, through a filter of mucus-covered gill slits, and out from the exhalent siphon. Some species are colonial, and by asexual budding form a mat of individuals over rocks or weed. Sexual reproduction in ascidians involves the formation of a little tadpole-like larva which swims freely and finally attaches securely by means of a secretion from a cement gland. Deep-water species are common, either attached by long stalks or covered in projections of the test to prevent them sinking into the ooze. Species without such adaptations are always anchored to a solid object (figure 178). They are normally small and though most still consume plant detritus some are modified to eat small whole animals. Deep water species are also usually solitary. The blood of ascidians contains high concentrations of vanadium, and different species accumulate other metals as well, such as nickel, tantalum, titanium, chromium, manganese and perhaps zirconium, molybdenum and tungsten, associated with free sulphuric acid, whose function may be related to the synthesis of tunicin.

Of the pelagic forms of tunicate, the salps and doliolids are individually barrel-like while the pyrosomas are colonial. Salps are hermaphodites, eggs develop with the aid of a sort of placenta and these are liberated as 'solitary' asexual forms. These produce from their underside a reproductive cord or stolon, off which groups of new salps bud. These little groups remain linked in chains as they break away from the parent as the sexual 'aggregate' forms (figure 178), but eventually separate and repeat the whole process. They thus exhibit both polymorphism, since the solitary and aggregate individuals are different in structure, and alternation of sexual and asexual generations, just as do the coelenterates. These animals are basically very similar to the sea-squirts in organisation, except that the inhalent and exhalent openings are at opposite ends and swimming is assisted by the rhythmic contraction of muscle bands round the body, taking water in at one end, sieving it through the gill clefts and propelling the animal forward as it is pumped out at the other end. Salps often occur in huge swarms particularly in warmer waters; one swarm was so dense that a boat could only be launched into it with difficulty. Doliolids are very similar (figure 178), but with more symmetrical muscle bands, and an even more complex polymorphism and alternation of generations. The eggs do not develop in the body of the female, but through a tadpole-like larval stage similar to that of the sea-squirts; this loses its tail and becomes barrel-like before embarking on the very complicated asexual budding. Pyrosomas (figure 213) may be regarded as colonies of individuals embedded in the wall of a long gelatinous cylinder open at one end, with each individual having its inhalent siphon on the outside and its exhalent siphon opening into the cylinder. The cumulative outflow of all the individuals thus leaves the cylinder by its open end continuously jet-propelling the colony. The colony size is increased by budding, some 10 m by 1·3 m having been seen by skin-divers off Australia, and new colonies are formed by sexual reproduction, the eggs being fertilised *in situ*, developing and budding to form a tiny colony of about four individuals before being released from the parent colony. Pyrosomas also occur occasionally in swarms, but are most remarkable for their steady luminescence. The last class of tunicates, the larvaceans, become sexually mature while retaining the organisation of the tadpole-like larva of the doliolids. All are tiny, never more than a few mm long, and are remarkable for their feeding mechanism. Larvaceans are very common members of the plankton and provide an important component of the food of many larger animals, particularly fish larvae.

The remaining group of protochordates, the **lancelets,** occur during the day in the bottom sand and muds of shelf areas, and ascend into the plankton at night. They are small fish-like animals with segmented muscle bands but feed by filtration of a ciliary water current

179. *opposite* The blue copepod *Pontella fera* (0.4 cm) is typical of the animals living in the uppermost layers of the warmer areas of the oceans. The dark spot between the antennae of each animal is a very large and elaborate eye

182. The animal photographed in the act of making this
curious trail (at 5000 m) may be a type of acorn worm

through mucus-invested gill clefts. They sometimes
occur in seasonal aggregations related to spawning and
in the Far East are regarded as a gastronomic delicacy.

FISH

Over 25,000 species of living fish are known and
these account for about 42 per cent of all recent
vertebrate species. Only the more important or
interesting marine fish will be mentioned here. The
class **Agnatha** includes fish with no jaws. Of these the
hagfishes spend most of their time buried in mud
except for their snout but come out to worm their way
into dead or disabled fish and eat them from the inside
outwards (figure 245). They can be locally abundant to
depths exceeding 1000 m. The lampreys use their
sucking discs to attach to the outside of other fish.
They are rarely taken in very deep water. The class
Chondrichthes embraces cartilaginous fish with jaws,
namely the rabbit-fish (chimaerids) and the sharks,

rays and skates (elasmobranchs). Rabbit-fish have one
gill opening, breathe through their noses, are found
down to almost 3,000 m and live a rather inactive life
catching invertebrates on the bottom. Elasmobranchs
have five to seven gill openings and though nearly all
of the approximately 550 species are voracious
carnivores the largest shark of all, the whale shark
(figure 183), has a sieve-like gill apparatus which is
used to catch plankton and small fish. Most shark
species are found fairly near to land but some live near
the surface in the open ocean while others, black or
dark brown in colour and having very large fatty
livers, dwell entirely on the bottom to depths of 3,000 m
(figure 184). In certain limited areas such as Durban,
South Africa, sharks are dangerous to man. Gruesome
evidence of this has sometimes been removed from
shark stomachs, but how one assesses the discovery in
one shark's stomach of 47 buttons, three leather belts,
seven leggings and nine shoes, is open to conjecture!

180. *opposite above* The heteropod mollusc *Carinaria* (22 cm)
swims with its single muscular fin, and its gills project from
beneath its small shell. The tubular eyes probably assist in the
capture of its active animal prey

181. *below* The purple heteropod *Atlanta* (0.5 cm diameter)
lives in a flattened spiral shell with a wide keel, into which the
head with its long proboscis and large eyes can be retracted

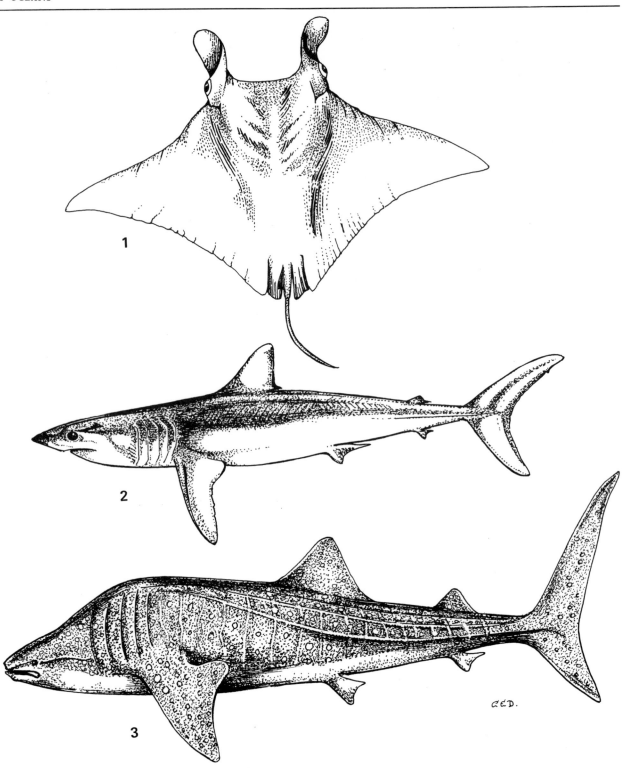

183. Three species of open ocean elasmobranch; 1. the Giant
Devil ray or Manta ray *Manta birostris* may reach over 7 m
across; 2. the ferocious mako shark *Isurus glaucus* (up to 3m
long) is a warm water species known to attack man; 3. *Rhinco-
don typus* the huge filter-feeding whale shark may reach over
15 m in length but is quite harmless (1 and 3 after Bigelow &
Schroeder 1948, 1953)

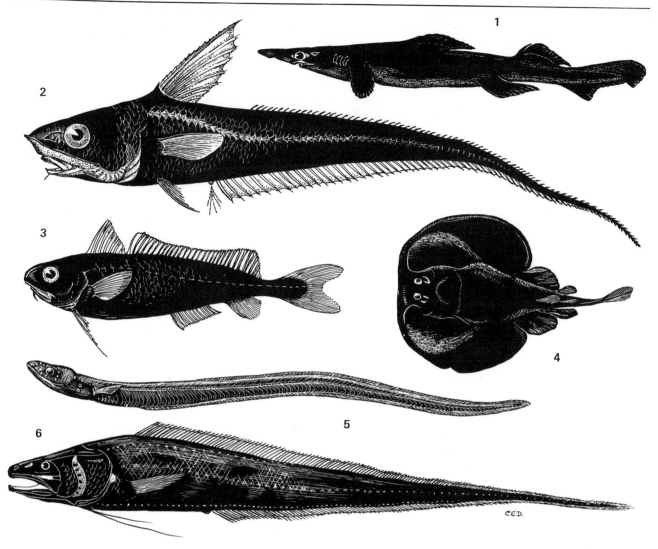

184. Fish typically caught on or near the bottom: 1. the shark *Deania calcea* (1 m); 2. *Macrurus guntheri* (36 cm) is a small macrurid, rat-tail or grenadier which, like *Deania*, has a ventral mouth; 3. *Mora mediterranea* (60 cm) is a deep water cod; 4. the electric ray *Torpedo nobiliana* reaches at least 85 cm in length; the electric organs lie in the broad crescentic areas on either side of the head; 5. a pale coloured deep water eel *Synaphobranchus pinnatus* (28 cm); 6. the small (45 cm) brotulid *Dicrolene filamentosa*. Related species of all the above fish, except for the electric ray, grow very much larger than these (1 after Garrick; 2 after Collett 1896; 3 after Goode & Bean 1895; 4 after Bigelow & Schroeder 1948; 5 after Roule 1919; 6 after Garman 1899)

Skates and rays have flattened bodies and breathe in through holes on the upper surface, the spiracles. The enormous manta or devil ray (figure 183) grows to 23 feet (over 7 m) across and almost 1,500 kilograms in weight, and the saw-fish (Pristidae) reach over 6 m in length. Most skates and rays lie and swim close to the bottom, but the eagle and cow-nosed rays swim in midwater some of the time, and the devil rays live near the surface and may sometimes leap out of the sea. These large fish sift plankton in a similar manner to whale sharks, while most skates and rays feed on bottom invertebrates. The majority of species live on the continental shelf but the electric rays are found down to at least 1000 m and the skates to 3,000 m. Electric rays (figure 184) vary in their discharge but 220 volts has been measured from one species, and men have been knocked over by a shock. The rays probably use the shocks to stun their food, and man has used them as a cure for disease of the spleen, chronic headaches and gout; the presence of one of these rays in a room used to be considered a sure stimulus to easy delivery for a pregnant woman. Some rays bear long poisonous stings which have been utilised by man to tip spears and whips.

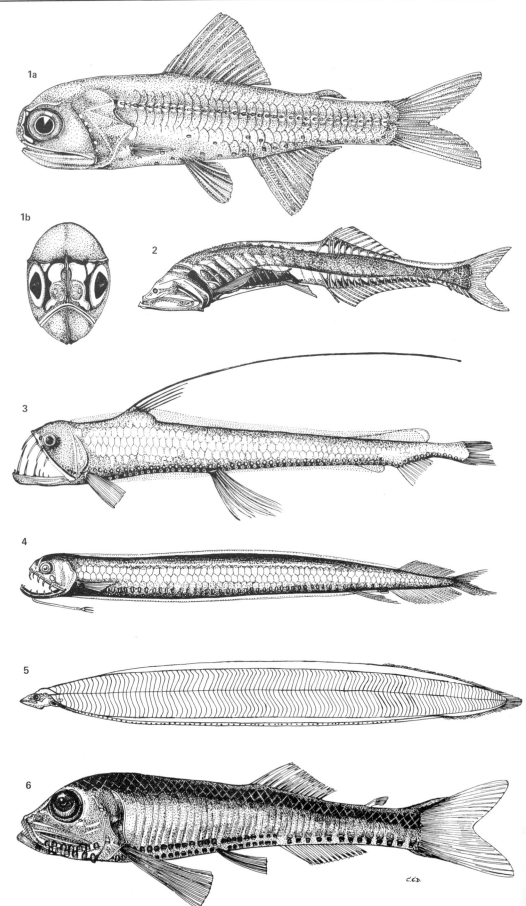

185. Some common meso-pelagic and bathypelagic fish: 1. (a) *Diaphus elucens* (5.5 cm) is a myctophid with large light organs between the eyes (b) as well as on the body; 2. *Cyclothone pseudopallida* (7 mm) is a typical member of this abundant genus; 3. *Chauliodus sloani* (30 cm) has light organs inside its mouth, and at the tip of its long dorsal fin which it is reputed to dangle in front of the head like a fishing line; 4. *Stomias affinis* (9 cm) has a luminous barbel; 5. many species of young eel resemble this conger leptocephalus larva, which in life is almost completely transparent; 6. *Vinciguerria attenuata*, related to *Cyclothone*, has several rows of ventral light organs (1 after Brauer 1906; 2 and 6 after Zugmeyer 1911; 4 after Nafpaktatis 1968; 5 after Roule 1919)

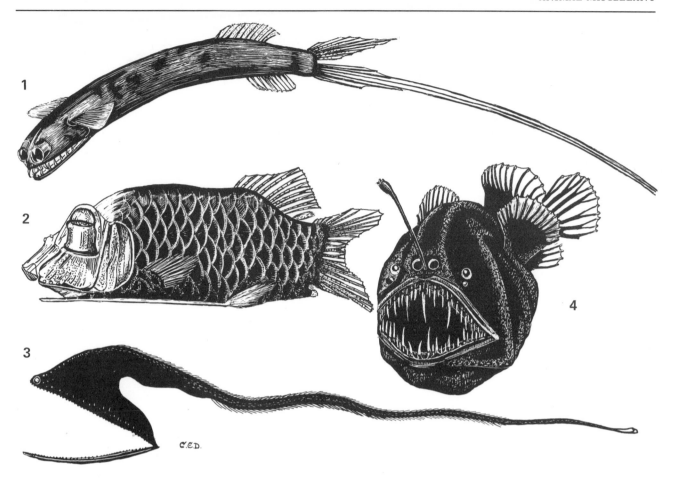

186. Four uncommon deep water fish: 1. *Gigantura chuni* (body up to 11 cm) has forward-looking tubular eyes; 2. *Opisthoproctus soleatus* (5–6 cm) can illuminate its flattened underside, and its tubular eyes look upwards; 3. the gulper eel *Eurypharynx pelecanoides* (60 cm) has a light organ on the end of its tail and a mouth and stomach capable of huge distension; 4. *Melanocoetus johnsoni* (6 cm) is an angler fish with smaller free-living males (after Brauer 1906)

The class **Osteichthes** comprises bony fish, and the sub-class Actinopterygii (16,000 species) contains all but the lungfishes (freshwater) and lobefins. The coelacanth is the only remaining representative of the lobe-fins which thrived 300 million years ago, and it has changed little in that period. So far only about 60 specimens have been caught. The order Clupeiformes includes all the commonest deep-sea midwater fish, most of which belong in the sub-order Stomiatoidea. The family Gonostomatidae includes many of the midwater fish (figure 185). *Vinciguerria attenuata* only reaches about 5.6 cm in length, is bathypelagic in water exceeding 1000 m, is an important food of tuna and albacores and is extremely numerous in the Pacific. *Cyclothone* must be the vertebrate genus with the most individuals living. These fish reach a maximum of eight cm length. They are bathypelagic in all oceans except the Arctic and are most common deeper than 300 m. Light coloured species are found around 500 m and dark coloured ones from 500 – 1,500 m. They have been seen near the bottom at 2,290 m but are also responsible for scattering layers in many regions. Two other closely related species are *Gonostoma* and *Bonapartia* (figure 147). The hatchet fishes (family Sternoptychidae) are flat from side to side, bright silver on the sides and have large light organs, often magenta coloured, on their underside (figure 227). *Chauliodus* is in a family on its own, attains a length of 30 cm and has a thick gelatinous coat during life (figure 185). It has very large teeth, a large, often brightly coloured, light organ behind the eye, and a barbel hanging from its jaw. Fishes of the family Stomiatidae, containing *Stomias* and *Macrostomias*, reach 40 cm in length and also have large teeth, jelly and barbels (figure 185). *Opisthoproctus* (figure 186), giving its name to its family and its sub-order, is remarkable in having upwardly directed tubular eyes with sides of silver and gold, and a light organ inside its anus. This organ shines along a columnar channel having a dorsal reflector so that, by moving the large scales on the flat sole-like underside of its body, it can shine light downwards between the scales.

The order Myctophiformes contains many odd forms such as *Omosudis* with an enormous head and mouth capable of eating fish and squids as long as itself; large bathypelagic species of the genus *Alepisaurus* (the lancet fishes) reach over 2 m in length and eat many species of fish and cephalopod which have never been directly caught by man; *Bathylychnops exilis*, the 'four-eyed' fish, has a double lens system on each of its two eyes. Besides these more bizarre types the order contains the lantern fishes (Myctophidae) (figures 185, 209), which are common on the surface at night and are important food fish in midwater. They have characteristically delicate scales, large eyes and a row of light organs underneath.

The order Saccopharyngiformes or gulper eels (figure 186), have enormous mouths, minute eyes, and are poor swimmers but manage to swallow very large prey. One theory is that the light organ on the tail lures prey which swims straight into the mouth.

The order Anguilliformes includes the snipe eels (family Nemichthyidae) (figure 187), which when seen from submersibles are often hanging vertically, head down in the water.

The order Beloniformes includes the half beaks, some of which are little more than 1 cm long when fully grown, and the flying fishes (figure 192).

The order Gadiformes includes the many species of rat tails (family Macruridae), abundant on the bottom (figures 184, 244).

The order Lampridiformes includes the moonfish or opahs (family Lamprididae) growing to over 2 m and weighing 300 kilograms, and the oar fishes (family Regalecidae) with their beautiful silver bodies and coral-red fins with long streamers. One of these fish, the 'King of the Herrings', may reach about 7 m and weigh 300 kilograms and has probably given rise to some sea serpent stories.

The order Lophiformes includes the angler fish (family Ceratiidae) with their enormous mouths, light-bedecked lures, minute tails and in some species tiny parasitic males which attach and actually grow on to the larger females (figure 186).

REPTILES

There are about 50 species of sea snakes (family Hydrophiidae) and they are only very rarely found in water deeper than 200 m. They are limited to the Indian and Pacific oceans, include some very poisonous species and have a laterally flattened tail for swimming.

Marine turtles breed and generally live in or near the

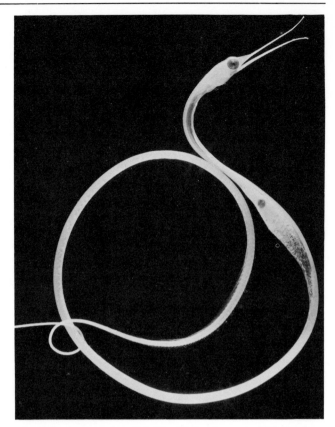

187. Snipe-eels, such as this *Nemichthys* (45 cm) which has just consumed one of its fellows, are common in midwater trawl catches

tropics. There are five species of which the green turtle *(Chelonia mydos)* is most well known. At least a proportion of females, which live on turtle grass along the Brazilian coast, probably make a 1,400 mile journey back to Ascension Island every two or three years to lay eggs. That this represents a considerable navigational feat is clear when one considers that the island is only five miles across and is 'upstream' from Brazil. The females walk up the shore at night and lay 300 – 700 eggs in several visits. Eggs are laid about a metre down and the young turtles have a hard dig to the surface when they hatch 57 days later. After great speculations on the means employed by the newly hatched turtles to go in a straight line to the sea, even when this is out of sight, it was found they are sensitive to the general brightness of the light above the sea and merely have to keep this balanced between the two eyes to take a direct route. Salt glands under the eyelids remove excess salt from the body as tears.

BIRDS

Many species of birds are seen over the oceans but they all go to land or ice to nest. Many have very wide ranges and some carry out formidable migrations each year.

Penguins (family Spheniscidae) cannot fly but are wide-ranging oceanic birds, very well adapted to long periods of swimming with their wings in the open ocean. Limited to the Southern hemisphere except for one species on the Equator in the Galapagos Islands, they are particularly abundant in the Antarctic. The largest are the Emperors, which reach over 1 m in height and breed at the edge of the ice cap during the Antarctic night. Penguins may locate their food by sonar, utilising not their own sound source (as do dolphins), but the noise produced by the collapse of cavitation bubbles formed as the penguins swim.

The order Procelliformes contains all the most oceanic of birds, the 'tubenoses'. The largest of these are the thirteen or so species of albatross (family Diomedeidae), which are renowned for their skill in gliding and their wing-span of up to 4.5 m and body lengths of as much as 1.4 m. Most albatrosses rarely go further north than 30°S but four species are found in the North Pacific. The petrels, shearwaters and prions (family Procellariidae) vary in size from the Giant petrel, almost as large as an albatross, to the tiny prions or whalebirds of the Antarctic. There are about 60 species, some of which are found throughout the world's oceans. Storm-petrels (family Hydrobatidae) are about the size of starlings and have a rather fluttering flight. There are about twenty species and they nest in burrows on islands during the breeding season.

Skuas (family Stercorariidae) comprise five species ranging far out to sea in the North Atlantic. Skuas chase other birds until they vomit and then deftly catch the food before it hits the sea.

The three species of phalarope (family Phalaropodidae) probably spend most of their time at sea.

The only other wide-ranging oceanic birds are a few of the tern species (family Sternidae) and one of the gulls (family Laridae), the kittiwake.

MAMMALS

Sea mammals include the whales, dolphins and porpoises (order Cetacea), which are reviewed on page 263, the manatees and dugongs (order Sirenia; sea cows), the sea otters (family Lutridae), which do not go into deep water, and the seals.

There are 32 'seal' species (sub-order Pinnipedia), which include the walrus with its large mollusc cracking tusks, five species of sea lion (with small external ears and hind feet which turn forwards on land) and a large variety of seals (with no external ears and hind feet joined to the tail by skin). They are all good swimmers and divers, often spend long periods

in the open sea and may migrate over long distances. The largest is the elephant seal of the Southern Ocean, which weighs up to three and a half tons.

OTHER GROUPS

In addition to the groups of marine animals outlined above there are a number of other phyla of lesser (or perhaps unappreciated) importance in the economy of the oceans, and it is convenient to treat them together.

The **sea-mats** or moss-animals (bryozoans) are generally shallow-water forms having a superficial resemblance to some of the hydroids. They grow either as plume-like or encrusting colonies, the individual animals living in tubes which are often strengthened by calcareous deposits. Occasionally they are found growing on animals such as sipunculid worms, and are very important fouling organisms. One species causes the allergy known to North Sea fishermen as 'Dogger Bank Itch'. They collect their food by means of ciliated tentacles, a method also employed by the small, worm-like tube-dwelling **phoronids** and the **lampshells** (brachiopods), which are quite unrelated to the bivalves though having a superficial resemblance. These animals may be attached to rocks or live in mud and are to be found at all depths; some grow no bigger than 1 mm and are part of the interstitial fauna of sand. One genus, *Lingula*, appears, from the fossil record, to have remained unchanged for some 500 million years – surely a record. Among the very tiny animals living in mud and detritus are the **kinorhynchs**, segmented spiny animals no more than 1 mm long feeding on diatoms or bottom material and often with zooxanthellae in their tissues, and the **gastrotrichs**, also feeding on detritus and having not only spines but cilia and adhesive tubes by which they may attach themselves to sand grains. Rather similar are the wheel animalcules or **rotifers**, ranging from 0·04 to 2 mm in length with a ciliary crown at the head end by which they swim. There are relatively few marine species, most of which are shallow water and bottom-dwelling detritus feeders, though one genus is found attached to the gills of *Nebalia*, and several others are pelagic.

The carnivorous ribbon worms (**nemertines**) (figure 159) are also usually found on the shore or in shallow water, though often having ciliated planktonic larva, but some have been modified for a pelagic life by becoming gelatinous or flattened, and the deeper living ones are often reddish; at least one species is also luminescent. One group of nemertines has become commensal in the mantle cavity of bivalves. The **roundworms** (nematodes) are important members of the populations of smaller animals found in the bottom sediments, since although these free-living species are usually only a few millimetres in length, four and a half million have been recorded per square metre of sea bed. There are no pelagic species but very many parasitic ones in both vertebrates and invertebrates, often very much larger than the free-living species; one in a whale measured 8.4 m in length. Entirely parasitic are the **acanthocephalan** worms, larval forms living in the bodies of invertebrates, the adults in vertebrates such as seals, turtles, fish and birds. Similar separation of the larval (intermediate) host and adult (primary) host is found in the parasitic **flatworms** (platyhelminthes), comprising the flukes and tapeworms; the former are internal or external parasites, the latter all internal parasites. These occur in a wide variety of marine animals, the tapeworms in particular often achieving considerable length; one in a whale's bile duct reached 21 m. There is also a considerable number of free-living flatworms occurring in the sea and while most are bottom living, though none are found in great depths, there are some which are truly pelagic and others that have pelagic larvae. The pelagic species are almost entirely limited to warm waters. The **threadworms** (Nematomorpha) have free-living adults, but the larvae of these animals are internal parasites of a variety of species of crustacea. Several curious groups of burrowing worm are found among the bottom fauna including the **sipunculids,** some of which have also been found buried in the head skin of fishes, the **priapulids,** which lie buried in the mud, are predacious

carnivores, and are probably related to the kinorhynchs, and the **echiuroids.** The latter group includes several abyssal species and one family in which the larvae normally develop into the large females but, if they come into contact with an adult female first, develop into tiny ciliated males, living on the proboscis of the females. Animals, which since their relatively recent discovery have attracted considerable attention, are the **beard-worms** (pogonophores) (figure 159). These long thread-like worms, found in depths of up to 10,000 m, usually live in chitinous tubes projecting out of the mud and collect their food by means of ciliated tentacles. They may be quite numerous on the bottom, and undoubtedly many remain to be discovered; a recently described species apparently lives in old rope and wood. Though simple in appearance, they are believed to be distantly related to the vertebrate stock.

Despite the great diversity of marine animals included in the groups discussed above there are doubtless many animals living in the sea that have yet to be discovered. Every cruise probably catches several new species of animals, though it is only the larger ones or the 'living fossils', such as the coelacanth and *Neopilina*, that catch the public imagination. The plankton net, the trawl and the dredge have almost certainly not yet intruded into the lives of many of the inhabitants of the deep ocean.

7 Animal Ecology

N.B. Marshall

LAND AND SEA

Though the sea covers over two-thirds of the earth, marine life has fewer species than land life. We know about a million kinds of animals, and three-quarters of these are insects. Insects are so successful that we may well think it curious that they are virtually unrepresented in the ocean. After all, there are sea-spiders (pycnogonids), some of which are deep-sea forms, while their distant relatives, the horse-shoe crabs and various mites, find a good living on the ocean floor. But insects are bound to the land by the very factors that have contributed most to their success; their powers of flight and their great dependence on plants, especially the flowering species, for food and cover.

The nearest counterparts of insects in the ocean are the crustaceans. Indeed, the copepods remind us of insects in their smallness, abundance and ubiquity. Like the insects, the crustaceans have, as it were, put their jointed external skeleton and fast-moving muscles to intricate and diverse uses, this time for moving and feeding in water, a very viscous and inert medium compared to air. Moreover, the limbs of crustaceans, more numerous than those of insects, have evolved to suit a diversity of habits. The outstanding feature of the ocean is its wealth of minute food organisms and organic particles. Though the plants of the plankton are microscopic forms, many crustaceans rhythmically sweep their finely bristled limbs to filter or grasp this pasturage. Foremost among the herbivores are many copepods and certain euphausiid shrimps. In turn, members of these two groups are the staple diet of the largest whale-bone whales.

If few crustaceans are at home on land, this is not altogether true of the snail (gastropod) group of molluscs. Even so, most of the 50 thousand species are marine, as are about four-fifths of the 11 thousand species of bivalves, which are, of course, an entirely aquatic group. Again, some of the marine snails and most of the bivalves are intricately designed to feed on minute food organisms and organic particles. But one purely marine group of molluscs, the cephalopods, are powerful predators. Indeed, the squids truly rival the fishes in dynamic form and submarine skills. They are the sharks of their group and, like the sharks, their hold on the ocean does not depend on their diversity. There are some 400 cephalopod and 250 shark species. Much the most diverse of the fishes are teleosts, comprising over 20 thousand species, about two thirds of which are marine. The backbone of fishes is based on a flexible rod, the notochord. Hence they are called chordates, which brings us to their very remote relatives, the protochordates and beard-worms, all of which are marine, and once again, largely dependent on minute forms of food. Remoter still, but having certain affinities to the protochordates, are the echinoderms, classic forms of marine life, moved by an extraordinary system of hydraulic machinery.

The present intention is not to survey every group of marine animals, although many groups, other than those mentioned above, are well represented in the deep ocean as described in Chapter 6. In the end, if we exclude the insects, the sea holds more diverse forms of animal life than the land. But life first arose in the sea, and this is reflected in the near-oceanic salt balance of animal body fluids. The sea is also more benign than the land in its heat balance. Water has a very high specific heat: it can absorb large quantities of heat with relatively little change in temperature. It also loses heat slowly. Thus, the climate in water is more equable than on land. Compared to much of the land, seasonal and daily changes in aquatic temperatures are much less severe. Is it surprising then, that the two most successful groups of land vertebrates, birds and mammals, are warm-blooded and able to regulate the temperatures of their bodies?

Water also buoys up its inhabitants. More precisely, the buoyant effect of water is about 800 times that of the atmosphere. Hence, in more than one way, the sea can support the huge bulk of whales, giant fishes and squids. In quite a different way, the buoyant properties of the sea have led to its exploitation by fragile forms, which need little in the shape of supporting tissues. We have only to think of jellyfishes, siphonophores and the pelagic tunicates (figures 153 and 178). Lastly, the

188. Bottom photograph taken at a depth of 2086 m showing a sea urchin, various animal tracks, a mound, a hole, and at top an arm of a brittle star

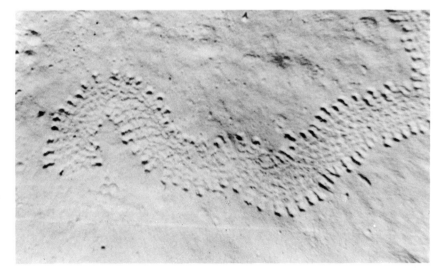

189. Many bottom photographs show very definite tracks on the soft ooze, few of which can be assigned with certainty to a particular animal: one such is this trail about 11 cm in width at a depth of 3442 m

slight negative buoyancy of the tissues of marine animals has meant that comparatively simple systems can be used to achieve neutral buoyancy: swimbladders in fishes, floats on certain siphonophores, and various chemical means of adapting the body fluids for flotation (see Chapter 8). One significant biological feature of being weightless in water is, of course, that animals can stay at a preferred level with the minimum expenditure of energy.

In certain respects, though, life on land is easier than in the sea. The atmosphere, as we know very well, is much more transparent than the hydrosphere. Even in the clearest waters of the ocean, sunlight reaches little beyond a depth of 1000 m; and since its mean depth is about 4,000 m, most of the ocean is pitch dark, apart from sporadic sparks of living light. But many deep-sea animals, largely through the evolution of light organs and sensitive eyes, have met this challenge very well. Life on land is also easier in that a given volume of air contains about 40 times as much oxygen as can dissolve, even under ideal conditions, in the same volume of water. Indeed, the concentration of life and the rate of circulation in the upper mid-waters of the deep ocean is such that the oxygen tension is greatly reduced within certain areas. But there is still life, and it may be abundant, in almost anoxic layers, which extend, for instance, over much of the eastern tropical Pacific and the Indian Ocean. Just how the animals of these layers manage to exist is still an enigma.

THE TRANSITION FROM COASTAL TO DEEP-SEA WATERS

Beyond the tide marks the sea floor continues to slope gently downward to the edge of the continental shelf, which lies at a mean depth of 180 m and forms the threshold of the deep ocean. Once this threshold is crossed, the ocean deepens rapidly as great submarine escarpments, the continental slopes, cant steeply down towards the more level reaches of the deep-sea floor (figure 2). More precisely, the slopes merge with more gently sloping continental rises at depths near 3,000 m. The outstanding change in passing from coastal to oceanic waters is not so much a transition from relatively restricted shelf waters to vast oceanic reaches, but rather to a massive expansion of underwater space. As we shall see, distinct strata of this space, from the very surface to levels near the deep-sea floor, are occupied by distinct communities of animals.

For the physical features of the deep ocean the reader is referred to Chapter 3. Certain of these features will need to be considered at later points in this chapter. Meanwhile, we pass to the ocean floor and its life. Here, perhaps, the most impressive biological change from coastal to oceanic reaches is the great reduction in the amount of cover. Seaweeds grow only in the well-lit fringes of the ocean, down to depths of about 100 m. Diverse invertebrates and fishes find shelter, support and a living in this vegetation. The only comparable living spaces in the open ocean are not on the bottom, but among the floating tangles of Sargassum weeds. In the clear shallows of the tropical ocean, the most intricate cover, occupied by diverse fishes and invertebrates, is formed by coral reefs and atolls. There are also deep-sea corals, but the cover they afford is very small compared to their tropical counterparts. Still, some deep-sea corals, such as those on the continental slopes in the North Atlantic, grow extensively. Here some branching corals form thickets down to a depth of about 2,000 m, thickets that shelter and support certain sponges, sea anemones, soft corals, bryozoans, annelids, barnacles, starfishes and brittlestars. But beyond the slopes, as hundreds of photographs (figures 188, 189) and various cine-films show very well, much of the deep-sea floor is like a vast monotonous plain, interrupted here and there by the mid-ocean ridges, sea mounts and islands. Even apart from the scarcity of certain kinds of food, which we shall consider later, the restricted nature of the benthic fauna living on the deep-sea oozes is understandable.

In one respect, though, the deep sea floor affords just as much cover to animal life as the bottom of coastal seas. Living just below the surface of the sediments is a fauna of small invertebrates, especially of polychaete worms, bivalves and nematode worms, and though this fauna becomes less diverse with depth, it may still comprise more than a hundred species at depths of 3,000 m or more. Here, then, there is a diversity of life in miniature about two miles below the productive surface waters of the ocean, where there is no sunlight, temperatures are close to zero, and the sea exerts pressures of several hundred atmospheres. Yet there is a kind of flora in the surface of the oozes, a flora of bacteria, some of which turn the intractable remains of plants and animals into living substance. In more ways than yet known, bacteria must be key organisms in the economy of life on the deep-sea floor.

ORDER IN THE OCEAN

All that makes the life of organisms is ordered in a hierarchy of intricate ways. The persistence of this life depends on forms of external order, both physical and biological, which have emerged after changes over many millions of years. During such evolution many more species have perished than now exist. The organization of the successful species is such that they continue to come to terms with the physical and biological order around them. In ecological aspects, adaptive types of organisms have evolved to form stable community systems. Such features of oceanic order will now be considered.

VERTICAL PATTERNS

Animal life in the deep ocean depends almost entirely for food on suspensions of surface-dwelling plants, diatoms and various kinds of flagellates (Protozoa). The biology of this planktonic microflora is considered in Chapter 5. Here we need only recall that for active photosynthesis the plants are limited to well-lit surface waters, reaching, at most, to a level of 100 m. As we shall see, in forming the base of food pyramids, the plants, whether directly or indirectly, sustain the ecological integrity of animal life. Vertical patterns, now to be considered, would soon vanish if phyto-plankton productivity failed.

THE EPIPELAGIC ZONE (Figure 8)

Many kinds of animals, from whales to protozoans, live within or close below the plant-producing surface layers. This is the epipelagic zone, which merges below with a twilight (mesopelagic) zone around a depth of about 200 m in the clearest tropical waters. The epipelagic zone contains the seasonal thermocline, so called because it forms only during the warmer part of the year in temperate waters. The thermocline is a transition region between a sun-warmed surface layer, mixed by the winds to an isothermal condition, and cooler underlying waters. Over the tropical and sub-tropical belts, the thermocline tends to persist through-out the year, and it lies at depths between 20 and 150 m. As shown in Chapter 5, plant productivity, and hence animal productivity, is closely tied to the stability of the thermocline. At times, epipelagic fishes such as albacore, big-eye tuna, and swordfish are fished near the thermocline, where they presumably find some of their food. The thermocline also seems to pose a barrier to the daily vertical migrations of various mesopelagic animals. In tropical and subtropical regions, which are the principal headquarters of deep-sea animals, upward migration through the thermocline can involve an animal in changes of temperature of around 10°C, from say, 15°C to 25°C. Even so, food in the productive surface layer is so fundamental to the economy of mid-water life that many animals have acquired the requisite physiological tolerance to breach this thermal barrier. Since the sharpest thermoclines may be no more than 10 m in thickness, the most active dwellers in oceanic space, nektonic animals such as squid, and fishes, could be through the transition region in a matter of seconds. Even small planktonic animals are surprisingly active. For instance, the copepod *Calanus*

finmarchicus, about three mm long, can swim upwards at about 30 m per hour. Here, in passing, we should realize that though the zooplankton drifts in the ocean, the constituent organisms are not entirely at the mercy of the waters, not even in the turbulent surface layer. Unlike the motes in a sunbeam, they are not randomly distributed; even the smallest forms are attracted or repelled by one another.

SURFACE DWELLERS

Insects, we recall, are virtually unrepresented in the ocean (page 205). More precisely, several kinds of bug (*Halobates*) live *on* the ocean, where, like their counter-parts in freshwaters, they use their long legs to stride over the surface. They are said to feed on small animals trapped by surface tension. Species of *Halobates*, which have a body length around five mm, are confined to warmer oceanic regions.

Pleustonic animals are held at the surface by some kind of float. The Portuguese man-of-war (*Physalia*) (figure 222), an odd kind of siphonophore, is a classic instance. Under its gas-filled float, which bears a crest-like sail, hang the digestive, tentacular and reproductive parts of the body. The set of the sail is such that *Physalia* tacks over the sea, trailing its mane of bright blue tentacles below, which at full stretch in a mature individual extend for 10 m or more. Prey organisms from fishes to zooplankton are paralysed by the tentacles, and these now contract, drawing the food to the flexible mouths of the digestive polyps. But various fishes actually live among the tentacles, either avoiding them or being in some way unharmed. Even supposing there is occasional sacrifice to a deadly refuge: in the long run may not such partial protection be advantageous to the life of these fishes?

Though its headquarters are on the warm ocean, *Physalia* sometimes drifts into temperate seas, which is true also of its relative the purple-sail (*Velella*). *Velella* and its two purely tropical allies, *Porpita* (figure 163) and *Porpema*, have a float with labyrinthine air pass-ages based on a chitinous framework, but the two latter genera have no sail. All these coelenterates are so cunningly contrived that no one would predict that certain of their distant relatives, the sea-anemones, could copy their way of life. However some anemones, known from Indo-Pacific tropical regions, use their foot to shape a bubble float. Thus upheld, they drift, fishing with their tentacles, like a *Physalia*.

Certain molluscs have also taken to a pleustonic life. Species of the snail *Janthina*, which have a handsome purple shell, make a raft of mucus-coated bubbles. They are often found with *Velella*, on which they prey (figure 164). *Fiona* and *Glaucus* (figure 165), two sea-slugs, also attack *Velella*.

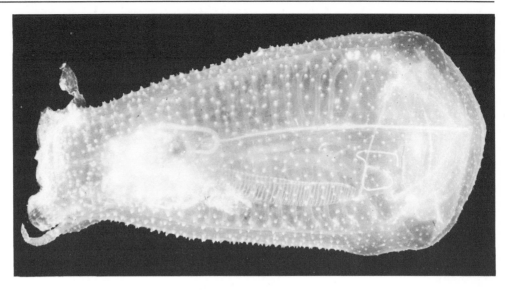

190. The very large salp *Thetis* (15 cm) has a thick pimpled test through which some of the gill slits and muscle bands can just be seen

Like these floating predators, the animals living just below the face of the warm ocean look as though they have been dyed in deep-sea blue. These near-surface forms, called the neuston, belong to diverse groups: medusae, arrow-worms, tunicates, crustaceans, molluscs and fishes (larval stages). Of the crustaceans, the copepods, particularly the pontellids, are the most vividly blue (figure 179). Salps, squid (e.g. *Onychia carribaea*), the argonaut and young fishes, are more delicately coloured. But whatever the density of colour we are bound to wonder: why are all these animals coloured in this way? The pigments are blue carotenoproteins which best absorb light in the orange-red part of the spectrum and the most beguiling idea is that blue forms will be camouflaged in some way. For instance, if blue prey and predator are at the same level, the former will tend to merge into a deep blue background, but if the predator is below, the prey will be silhouetted against the down-welling light. All the same, imperfect camouflage may well be effective in the context of natural selection and the long-term life of species. If this is so, why should powerful predators such as *Physalia* be protected as well as herbivorous forms (e.g. salps and larvaceans)? The leathery turtle certainly eats salps and jelly-fishes and most probably, like the hawksbill turtle, feeds on *Physalia* as well. If these turtles have colour vision – and the Caspian terrapin certainly has – their prey may enjoy some protection. What about a blue copepod and a blue arrow-worm? If the copepod is moving nearby, the arrow-worm may use its vibration detectors to find the copepod – certainly not its simple light receptors. But both may be missed by a flying-fish, which is likely to have a sense of colour. Clearly, there is much uncertainty, but close study of the neuston is only a few years old. When we know more

of food chains, pigment biochemistry and visual powers much may become clear.

Apart from this deep blue film of neuston on the warm ocean, the planktonic animals of the epipelagic zone are mostly transparent or translucent in optical texture. Even experienced observers take some minutes before they recognise all such forms in a sample of living plankton: thus we 'see' how transparent members of the zooplankton, especially when they are still, may escape the notice of sharp-eyed predators, such as squid and fishes. Is it surprising that virtually defenceless filter-feeding species, which depend largely on plant food are, at most, translucent forms? For instance, countless herbivorous copepods (mainly from one to four mm in length) are exposed to predators as they move slowly through the sea, screening micro-plants on fine fringes of bristles that emerge from limbs that twirl just behind the mouth. Some of the euphausiid shrimps, though not entirely transparent, are outstanding herbivores, particularly *Euphausia superba*, the whale-food of Antarctic waters. But euphausiids and even copepods have quickly-acting neuromuscular means to dart away from enemies, a form of escape not given to pelagic tunicates, suspension-feeders *par excellence*.

Apart from their digestive nucleus, which may be green or red, salps and doliolids are beautifully transparent, the first with cylindrical or prism-like bodies, the second barrel-shaped (figure 190). Minute plants that chance to enter the branchial region are caught on moving curtains of mucus produced by the endostyle and eventually conveyed to the gut. Like the salps and doliolids, their relatives, larvaceans are commonest in epipelagic waters and most diverse in the warm ocean. Larvaceans consist of a body region, usually no more than two mm in length,

191. A small siphonophore *Physophora hydrostatica*, about 9 cm tall; at the top is the small gas-filled float and below it the cluster of swimming bells and knot of tentacles, gonads and mouths

four groups can exist, at least during the summer months, in temperate and cold seas.

Highly predatory members of the zooplankton are also partly or entirely transparent. Such are siphonophores, jellyfishes and comb-jellyfishes. Apart from certain jellyfishes, which have ciliary means of collecting planktonic food (e.g. *Aurelia* and *Rhizostoma*), species of the first two groups catch their prey on tentacles armed with stinging-cells *(cnidoblasts)*. Most of the comb-jellies have tentacles bearing adhesive kinds of prey-trapping cells *(colloblasts)*, though certain without tentacles (e.g. *Beroë*), swallow quite large animals entire. Like the pelagic tunicates, these animals are relatively large and mostly composed of gelatinous tissue, a kind of hydrostatic skeleton – a skeleton supported by sea-water and moved by propulsive muscles in siphonophores and jellyfishes. Most siphonophores are moved, some quite rapidly, by swimming bells, which have a medusa-like form and special pulsing muscles that expel jets of water. In one group (Physophorida), the jets are topped by a gas-filled float (figure 191), though buoyancy mostly resides in large gelatinous appendages (bracts); members of the other group (Calycophora) have one or several swimming bells at the upper end. Below hang the tentacular, digestive and reproductive elements of the colony. Jellyfishes range in umbrella diameter from a few millimetres to the two metres of *Cyanea capillata*, the brown jellyfish. The largest individuals of *Cyanea* have been found in Arctic waters.

Drifting and pulsing fitfully, siphonophores and jellyfishes, trailing their deadly long-lines and tangles, take a heavy toll of the zooplankton and young fishes. The comb-jellies live in much the same way, but they are moved by flickering bands of ciliated plates. All these forms sweep the sea for their prey, but the arrow-worms, and they are also transparent, dart on their neighbours in the zooplankton, and seize them in their curved jaws (figure 159). Arrow-worms, ranging from about 1-10 cm in length, are distributed over most of the ocean. The largest species, *Sagitta gazellae*, lives in the Southern Ocean. The heteropod molluscs, again transparent and predatory, are mostly confined to the warmer parts of the ocean. These are a group of prosobranch snails, all with elaborate tubular eyes, moved by special muscles, perhaps for scanning their prey. *Carinaria* (figure 180) and *Pterotrachea* have elongated, gelatinous bodies, while *Atlanta* and its relatives are the most snail-like in form (figure 181). The middle part of the foot is produced as a membranous fin, which undulates and so propels these animals through the water. In *Pterotrachea* the foot is reduced to a thin muscular

to which is attached a long flattened tail. By means of special glands on the body, they secrete around them a gelatinous house, a house filled with filtering windows, filter-pipes, an oral tube and an exit. By lashing its tail, the animal draws water through the filters, so screening and concentrating minute flagellates etc., which are sucked into the mouth (figure 178). All the pelagic tunicates, including the colonial forms *(Pyrosoma)*, are thus able to feed on the very smallest kinds of plants. The latter are particularly prominent in the microflora of the warm ocean, also the headquarters of the pelagic tunicates. But members of all

192. Flying-fish at night; the lower animal is gathering speed for take-of by sculling with the lower half of its tail fin (the trail of splashes is clearly visible) and the upper animal is already airborne

flap, used as a scull, though the entire body can be thrown into propulsive waves. But whatever their way of moving, heteropods are active predators, using special teeth on the radula to impale their prey in the zooplankton.

If heteropods show how snails can evolve into lively, pelagic predators, the same is no less true of the gymnosome pteropods, which are a group of opisthobranch snails (figure 173). The gymnosomes, so-called because they have no shell, prey largely on a related group of pelagic snails, the shelled pteropods, which also swim by two wing-like extensions of the foot. But the shell-bearing pteropods are suspension-feeders, some using tracts of cilia on their wings to draw in minute plants and organic particles (figures 173, 193). Over parts of the tropical South Atlantic the shells of pteropods are prominent in the oozes of the deep-sea floor.

A proper review of oceanic zooplankton would require a full chapter, but before we turn to the nekton, the larger active swimmers, there are two final considerations. First, the epipelagic zone is not only a nursery ground for the young stages of its own inhabitants, but also for those of many kinds of deep-

sea animals (see page 223). Secondly, members of the zooplanktonic groups considered above are not confined to the epipelagic zone: they may also be caught at deeper levels, though they are generally most abundant in the surface layer.

Squids, fishes, turtles and cetaceans form the oceanic nekton. Squids are elusive animals, yet two species, at least, show themselves from time to time. The hooked squid, *Onychoteuthis banksi* and *Ommastrephes bartrami*, like their relatives, have broad triangular fins, a torpedo-shaped form and strong powers of jet propulsion. Indeed, these two squids have power enough to dart out of the sea and glide through the air, sustained by their fins. For instance, a small shark was seen to approach two *Ommastrephes*, whereupon both turned dark red and shot out of the sea to a height of about two metres and glided over some five metres. Certain other squid are also common in the epipelagic zone. All such squids are preyed on by sharks, tunny-fishes, spear-fishes and cetaceans. By taking to the air, flying squids may often elude their enemies, as do the flying-fishes, some of which can glide for a distance of 400 m (figure 192). Dolphin-fishes (*Coryphaena* spp) (figure 243) are avid pursuers of flying-fishes and so are snake-mackerel *(Gempylus serpens)*. Other epipelagic fishes include salmon, skippers (related to flying-fishes) ribbon-fishes (for example *Regalecus,* the king of the herring), breams and ocean-sunfishes (family Molidae).

Most of these fishes depend, at least to some degree, on relatively large prey. Outstanding exceptions are the whale-shark, the basking shark, flying-fishes and skippers, which take zooplankton and young fishes. The two sharks, the first growing to 20 m, the second to nearly 13 m, swim open-mouthed through the sea and amass plankton on the fine rakers attached to enormous gill arches. Salmon, basking sharks and mackerel-sharks feed in temperate or cool waters, but most epipelagic fishes have their headquarters in the warm ocean. Even so, certain species such as the great white shark, mako-sharks, blue-fin tunny, swordfish and several breams migrate into temperate waters during the productive season, when they feed and grow fat. Tunny-fishes, mackerel-sharks, mako-sharks and the great white shark, as Professor Nicol relates elsewhere (page 245), are warm blooded and able, to some extent, to regulate their body temperature. These large predators are thus able to retain their powers of quick movement when they enter cooler waters.

THE MESOPELAGIC ZONE

Nets fished by night in the epipelagic zone may contain animals that are not taken during the daytime; the most striking are certain kinds of luminescent squid and fishes. Though such active keen-eyed forms can readily avoid nets, and so might be missed by day, there is now enough evidence to show that they have come from deeper-lying mesopelagic levels. Towards nightfall many mesopelagic animals migrate up to the surface waters, where most likely they find much of their food.

In the sea, transitions from one major kind of living space to the next are rarely sharp. Concerning the lower limits of the mesopelagic zone, there are certain grounds for taking this to be the lowest reach of the sun's rays. In the underlying waters of the bathypelagic zone, where the only light is a fitful bioluminescence, the eyes of fishes, cephalopods and the larger crustaceans tend to be small or degenerate. These small-eyed animals and others form a distinct bathypelagic fauna, which is much less diverse and productive than that of the mesopelagic zone. The transition between these two faunas is around a level of 1000 m in the subtropical and tropical belts of the ocean, where live most kinds of deep-sea animals.

Besides the rapidly darkening twilight as the mesopelagic zone deepens, there is also a sharp decrease of temperature. Below the 'seasonal' thermocline of the warm ocean (see Chapter 3), the temperature falls steeply from levels near 20°C to values near 5°C at a depth of 1000 m. The level of most rapid fall, which forms the centre of a permanent thermocline, varies from depths of about 500–900 m over various parts of the Atlantic and Pacific Oceans. Thus, in migrating up from a level of 500 m to the surface waters of the tropics, a mesopelagic animal must not only have the physiological capacity to breach the seasonal thermocline (page 208), but also to withstand a total increase of temperature of 10°C or more.

While they live in dim surroundings, many animals of the mesopelagic plankton are translucent or transparent. This is true of forms ranging in size from copepods through arrow-worms to siphonophores. If these animals are transparent to the faint downwelling light, they should often escape detection by upward-looking, keen-eyed predators. Moreover, the cover of twilight is frequently lifted, particularly at night, by luminescent displays, so here is another possible advantage for animals to be 'optically-absent'. But to human eyes the most striking colour change in descending to mesopelagic levels is the prevalence of reddish and dark-coloured animals.

There are red-coloured medusae, such as the flaming-red *Agliscra ignea*; reddish nemertean worms; arrow-worms speckled or wholly suffused with red pigments *(Sagitta macrocephala* and *Eukrohnia fowleri)*; red copepods (e.g. *Gaetanus)*; red and orange ostracods *(Gigantocypris)* (figure 194); scarlet prawns *(Acanthephyra* (figure 195), *Systellaspis)*, and red cephalopods *(Mastigoteuthis, Histioteuthis)* (figure 226). The whale-fishes, certain of which may be caught at mesopelagic levels, are red or plum-coloured (figure 196), but most mesopelagic fishes are pale or translucent (e.g. *Cyclothone signata, C. braueri)*, or brown to black, or silver-sided with dark backs. Most of the dark-coloured fishes, such as the star-eaters (family Astronesthidae), dragon-fishes (family Melanostomiatidae), rat-trap fishes (family Malacosteidae) and giant-swallowers (family Chiasmodontidae), are powerful predators, armed with long, sharp teeth, some able to swallow prey as large as or even larger than themselves. Silver-sided fishes include the hatchet-fishes, certain gonostomatids (e.g. *Vinciguerria)* (figure 185), and many of the lantern-fishes (figure 209). The probable adaptive value of red and dark pigments is discussed by Professor Nicol (page 240); they reflect very little luminescent light, thus shielding their owners. Silver-sided fishes have developed mosaics of vertical mirrors, which in reflecting the stronger down-welling component of

193. Some of the bulkier pteropods such as *Cavolinia* (1.5 cm) must expend a lot of energy in swimming, for they sink very rapidly when passive

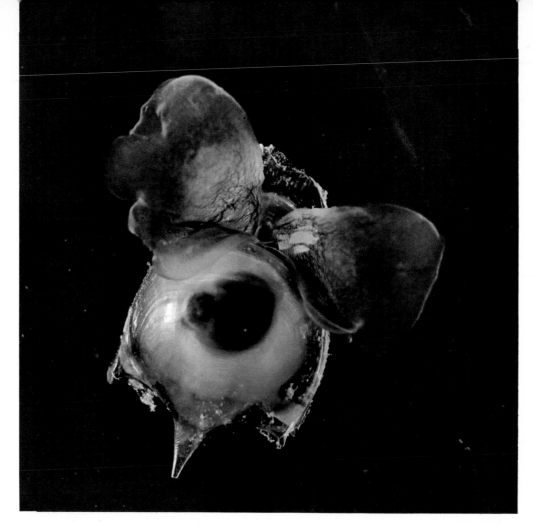

194. The bathypelagic *Gigantocypris* is the largest known ostracod, reaching 1 cm in diameter, and in this view from the front its huge reflecting eyes are very prominent

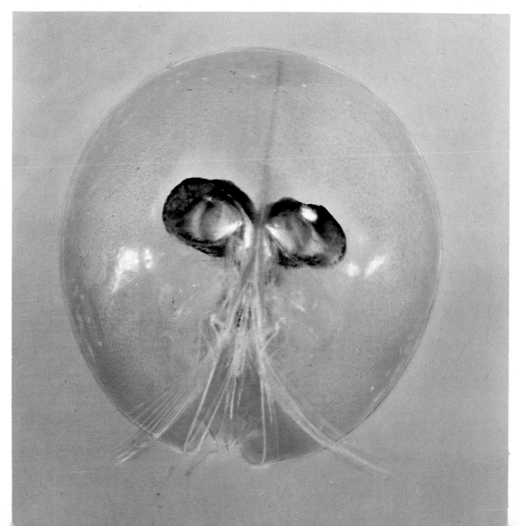

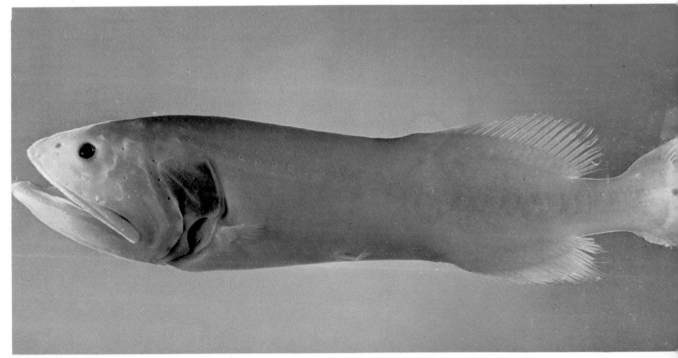

197. Cheek light organ, teeth and barbel of *Melanostomias*

sunlight may enable a fish to vanish from the view of a predator (page 239). Yet the mirrors of such fishes must often be exposed by luminescent flashes, particularly at night. If so, it is understandable that the black granules in large pigment cells on the sides of a hatchet fish *(Argyropelecus)* and its relatives should be fully spread at night.

The luminescent animals that live at mesopelagic levels include jelly-fishes, siphonophores, tunicates *(Pyrosoma* and salps), certain copepods, euphausiid shrimps, prawns, squid and fishes. Species of the last four groups develop the most elaborate light-organs (figure 197); they also have sensitive image-forming eyes (page 227).

Deep-sea animals may be seen from deep-diving submersibles, but there has not yet been time enough for a proper study of their luminescent activities. If care is taken, certain forms (e.g. prawns) may be kept alive in aquaria chilled to 'mesopelagic' tempera-

tures, so it may not be long before relevant experiments will be possible. When these times come, certain hypotheses may be tested, although we do know already that the eyes of euphausiids and mesopelagic fishes are most sensitive to blue-green light, close in wavelength to luminescent light (see page 227).

Apart from their possible use as visual aids – a natural guess for a human being – how far do constellations of large light organs serve as recognition signs by members of a particular species? For instance, there are over two hundred species of lantern-fishes, each of which has an individual pattern of pearl-button lights, the same on each side of the body. In some species the sexes also have a luminescent individuality: males bear one or more strongly luminescent plates along the upper surface of the tail, which are either on the lower surface or absent in the females. The males of other lantern fishes and certain

195. *opposite above* The bright red shrimps of which this *Acanthephyra* (8 cm) is an example are a common feature of deep-water trawl catches. The animal can produce a cloud of luminescence from glands near the mouth

196. *opposite below* Most whalefish are rather dark in colour, with a reddish tinge to the body, but the bathypelagic *Barbourisia rufa* (17 cm) is much more brightly coloured

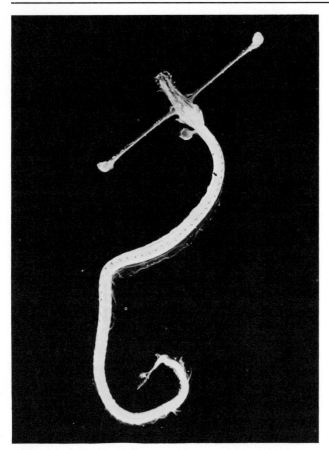

198. The remarkable larva of the fish *Idiacanthus* has its eyes on the end of long stalks; it was regarded for some time as a separate species altogether until other stages in the development series were caught

stomiatoids (e.g. *Idiacanthus* figure 198, *Photostomias*) differ from the females in developing one or more stronger lights on the head. Evidently, then, intense light signals are involved in courting and mating in a twilight world.

The larger light organs of fishes and squids are made much like those of euphausiids and certain prawns. The light-producing gland is backed by a reflecting layer and faced with one or more lenses to concentrate the light. Such similarity in design is hardly surprising, but a more remarkable convergence concerns the searsiid fishes, a squid *(Heteroteuthis* figure 210) and several prawns (e.g. *Systellaspis* and *Acanthephyra)*. These animals produce showers of luminescence; the fishes from a large gland on each shoulder, the prawns from glands near the head, and the squid from a light organ associated with the ink-sac. The most reasonable conjecture is that such a discharge confuses predators.

THE BATHYPELAGIC ZONE
The bathypelagic zone is the largest part of the oceanic space for it extends downwards from a level of about 1000 m to near-bottom waters, and we recall that the mean depth of the ocean is some 4000 m. This vast mass of water is cold (temperatures range from about 5° to 0°C), slow-moving and pitch dark—broken only by sporadic displays of living light.

The resident mass of life (expressed as biomass per cubic metre) in the bathypelagic zone is usually well below a tenth of the amount that exists at mesopelagic levels. Bathypelagic life is also less diverse: for instance, there are about 150 species of fishes, which is about a fifth of the mesopelagic fauna. There is a distinct community of animals, from copepods to fishes.

Mesopelagic fishes, cephalopods and decapod crustaceans have large, sensitive eyes. Indeed, many fishes, including (*Opisthoproctus*) (figure 186), hatchet-fishes (*Argyropelecus* spp), and giganturids (figure 186), have large tubular eyes, which are set close together in parallel to give wide-angle binocular vision, either in an upward or forward direction. Compared to normal eyes of equivalent optical and visual properties, tubular eyes are not only more sensitive but provide better means of judging the position of nearby organisms, both useful attributes in a twilight broken frequently by sparks of bio-luminescence. Even so, one group of bathypelagic fishes has evolved tubular eyes; the free-living males of one family of angler-fishes (Linophrynidae). The males of most other deep-sea angler-fishes have large, wide-open eyes, but fully grown females of all species have small or deficient eyes.

Both sexes of other bathypelagic fishes, notably the black bristlemouths (*Cyclothone*), gulper-eels and the bob-tailed snipe-eel (*Cyema*), have small, beady eyes. The eyes of whale-fishes, certain of which are bathypelagic, are regressed, which is true also of an octopod (*Cirrothauma*), and an euphausiid shrimp (*Bentheuphausia amblyops*).

If the bathypelagic zone was utterly dark, one might expect a wholesale degeneration of eyes, such as occurs in cave-dwelling animals. Concerning *Cyclothone,* a very successful genus of fishes, all but one species develop small light organs, which is presumably the main reason for their retention of eyes. The exception, *Cyclothone obscura,* a deep-living form, seems to prove the rule; for it has almost lost both light organs and eyes. But why should most male angler-fishes have highly developed eyes?

Female deep-sea anglers—and they are larger than the males—bear a special fin-ray on top of the snout which in all but two of about a hundred species ends in a luminous bulb (figures 199, 200). Inside the bulb

199. A juvenile angler fish *Himantolophus* only a few centimetres in length. Adults may weigh several pounds

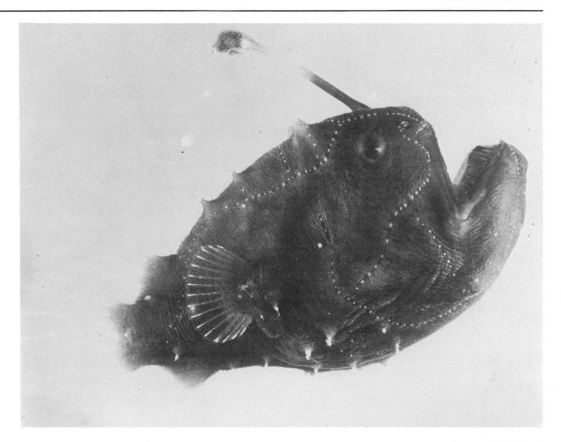

200. A very young, almost transparent, angler fish whose lure is as yet no more than a button above the mouth

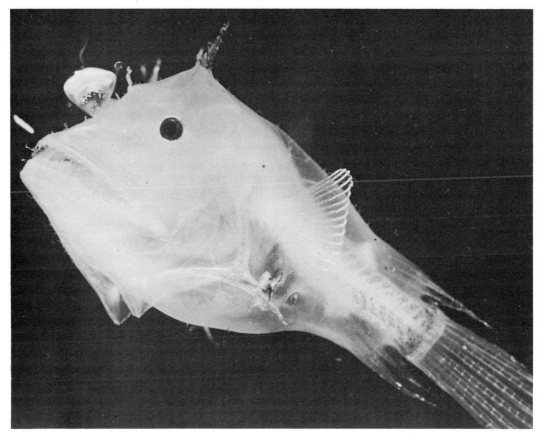

is one group – or more – of gland cells, each cell holding bacterial bodies, the probable light-producers. In a combination of features, largely of shape, luminous structure and external decoration, the bait of each species presents an individual appearance. This individuality is hardly to attract particular kinds of prey – for female anglers feed on organisms ranging from copepods to fishes – but to lure a mate, who has the right kind of eyes to locate the bait, and perhaps even to recognise the 'call' signs of his kind.

Most males also have large olfactory organs, which are regressed in the females. If the females produce specific scent-trails, they have another means of attracting mates of their species. Moreover, the males of *Cyclothone* species, again smaller than the females, also have large olfactory organs. Most probably the males of both groups mature earlier than the females and thus outnumber their partners.

Whatever the process of attraction, females of some 20 species of deep-sea anglers have been caught, each with one or more attached males, which are fused by the snout to the skin of their mates. Evidently, the male must first grip the skin of the female in his jaws, then become parasitic on her through the growing together of blood vessels and other tissues. In the end, he is little more than a resident supply of sperm. Males of other species, which have pincer-like jaw tips, may simply use them to cling to the female during the reproductive season. Life in a vast pitch-dark world has thus evolved cunning adaptations for survival.

LIFE NEAR THE DEEP-SEA FLOOR
Observers in deep submersibles and many photographs have now revealed that certain kinds of fishes swim near the ocean floor. These include black sharks, eels (e.g. *Synaphobranchus*), halosaurs (figure 201), notacanths, deep-sea cods (family Moridae), rat-tails and brotulids (figure 184). Except for the sharks, these fishes contain a large gas-filled swimbladder, and are thus likely to be neutrally buoyant. But so are the sharks, which are buoyed up by a very large liver heavily charged with squalene, a hydrocarbon oil with a low specific gravity (0.8). Certain kinds of squid also move above the oozes, as do various copepod, amphipod and isopod crustaceans. Over the continental slopes, at least, certain mesopelagic forms, such as lantern-fishes, prawns and euphausiid shrimps have been seen near the bottom.

The fishes, at any rate, form a special (benthopelagic) fauna. They are most diverse over the upper part of the slopes in the subtropical and tropical belts of the ocean. The dominant groups are deep-sea cods, rat-tails and brotulids. There are over 300 kinds of rat-tails, most of which have a projecting snout and

ventral jaws, capable of protrusion from the head. Such forms root in the oozes, which they swallow and strain through their gill-rakers, so retaining small ooze-dwelling bristle worms, bivalves, cumaceans and so forth. They also take animals that live *on* the oozes; for instance, brittle-stars and larger kinds of worms.

The swimbladder of deep-sea cods, rat-tails and brotulids is not simply a hydrostatic organ. In the first group, two forward horns of the swimbladder each make close contact with a diaphragm in the back of the corresponding auditory capsule. Fishes with such hearing-aids have sharper hearing than those not so equipped, but what do deep-sea cods hear? If they make sounds, their means of doing so are not so evident as those of rat-tails and brotulids. Most of the rat-tails, and they are slope-dwellers, have a pair of large drumming muscles on the forward part of the swimbladder, which are formed only in males. Again, the males alone of the egg-laying brotulids have a drumming mechanism, more complex than that of rat-tails. Sound signals produced by these mechanisms evidently play some part in sexual congress. There are also viviparous brotulids, the males of which introduce sperm-packets into the genital duct of the females. In these forms, so far as we know, both sexes have a drumming swimbladder. Near the bottom of the ocean, then, the dominant fishes have evolved sonic mechanisms in some way associated with reproductive activities. In the bathypelagic zone, as we saw, such activities are related to differences in olfactory organization, and the laying of scent trails somehow seems apt in pitch-dark slowly-moving seas. At mesopelagic levels the sexes of certain dominant fishes are marked by bioluminescent features; above all else, this marks the mesopelagic twilight world.

LIFE ON THE SEDIMENTS
Many deep-sea animals are attached to the sediments or exposed rocks. Most deep-sea sponges are hexactinellids with an intricate siliceous skeleton, part of which grows down into the oozes as a brush-like 'root' (figure 151). Like the deep-sea corals, their relatives the sea-fans and the sea-pens, which are most diverse in warm coastal seas, are also attached forms, the last anchored to the sediments by a bulbous holdfast. Sea-anemones grow attached to these forms or to rocks, as do many barnacles. There are also deep-sea sea-squirts, some of which, like *Culeolus* and *Styela*, are suspended on a long stalk. Stalked kinds of sea-lilies have some form of basal holdfast, while the fine chitinous tubes of the beard-worms are inserted in the sediments. The largest beard-worms are about 30 cm in length, and the tube in which they live is no more than about 2 mm in diameter.

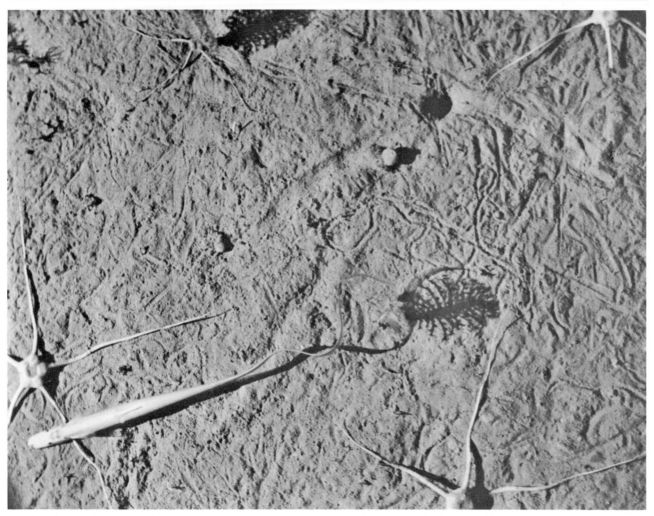

201. A halosaur of the genus *Androvandia* showing how the undulating tail is used for swimming, several large brittle-stars and a branched animal, probably a pennatulid, photo-graphed at a depth of 1500 m. Note how the soft bottom is extensively marked by the animals living in and on it

How do these fixed animals make a living? The beard-worms, which have no gut, may well use their tentacles to absorb organic substances (e.g. amino acids) dissolved in the sea. Sponges and sea-squirts must depend on currents to bring them organic particles in suspension. Deep-living sea-lilies, like their shallow sea relatives, presumably spread their arms and use podia and cilia to usher food particles to a moving tract of mucus on their ambulacral grooves. Perhaps the corals, sea-fans and sea-pens depend on living prey; or are they able, like certain sea-anemones, to collect particulate food by means of cilia on the tentacles?

Arthropods are well made to walk over the sediments. There are hermit crabs as well as true crabs, and the former, like their relatives of coastal seas, often carry one or more sea-anemones on the shell. Some of the true crabs (e.g. *Geryon*) look not unlike shore crabs; others (e.g. *Platymaia*) have long, spidery legs. The crayfish group of decapods is represented, *inter alia*, by *Polycheles*, red coloured and with degenerate eyes, and *Phoberus*, a pink lobster-like form, also with defective eyes. Most of these crustaceans presumably prey on other invertebrates, though certain hermit-crabs, at least, are able to live on fine particulate food.

There are bottom-dwelling prawns (e.g. *Nemato-carcinus*) with very long legs, and various shrimps (notably *Glyphocrangon*), which bear a heavily ar-moured skeleton. Outstanding among the long-limbed arthropods are the sea-spiders. When fully extended, the legs of *Colossendeis giganteis*, a red-coloured species, straddle over 60 cm of the deep-sea floor. Sea spiders

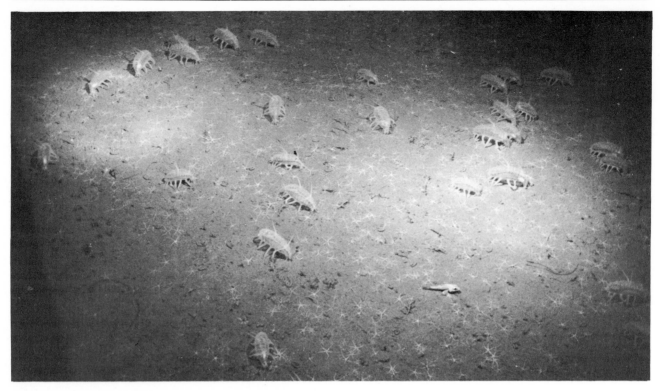

202. A large population of brittle-stars and elasipod holo-
thurians observed from the bathyscaphe *Trieste* at 1061 m in
Coronado Fan Valley

have the mouth at the end of a tubular proboscis,
through which they suck the tissues of sessile coelenterates and sea-mats (bryozoans).

Starfishes are generally carnivorous, though the
species of one successful deep-sea family, the Por-
cellanasteridae, are ooze-eaters. Deep-sea brittle stars,
which can be very numerous over the continental
slopes and rises, perhaps copy certain of their shallow-
water relatives in feeding, at least in part, on organic
particles. How sea urchins make a living in the deep
sea is uncertain; in any event there are both regular
and irregular forms. Sea-cucumbers are classic mud-
eaters, and one group (Elasipoda) is particularly
diverse and widespread over the deep-sea floor
(figure 202).

Most of the bottom-dwelling cephalopods are
octopuses. In the deep ocean there are such forms as
Opisthoteuthis, and *Bentheledone.* The former, which
is reddish, has a flattened body and large pop eyes.
The deeper dwellers, such as *Bentheledone,* have a soft,
gelatinous body, very small gills and a greatly reduced
(even absent) ink-sac. Of the benthic squids, the
sepiolids (e.g. *Rossia* and *Sepiola*), which have a
squat form and broad fins, are known from slope
regions.

Benthic deep-sea fishes, which have no swimbladder,
rest on the bottom. In the tripod-fishes (family
Bathypteroidae), the outer rays of the pelvic fins and
the lowermost rays of the caudal fin are greatly
elongated to form a tripod undercarriage, on which
the fish is poised for take-off after prey. Members of a
closely related family (Ipnopidae) have lost most parts
of the eyes except for the retinae, which are flattened
out under transparent skull bones and seem perfectly
functional. These two groups are confined to the deep
ocean, whereas, of the deep-sea fishes, species of sea-
snails (Liparidae) and eel-pouts (Zoarcidae) are representatives of groups found also in coastal seas. The
numerous intertidal species of sea-snails cling to any
suitable surface by means of a suction-cup disc
formed by the pelvic fins, but this structure tends to be
reduced in the deep-sea forms.

DEEP-SEA ECOSYSTEMS
Part of the foregoing outline of biological order has
centred on how animals are adapted to their living
spaces. This leads us to consider ecosystems (ecological
systems) which are communities of organisms in
certain surroundings. More precisely, an ecosystem
is: 'A structure of organisms and environment,

primarily arranged in terms of flow of energy through the system'. The ultimate sources of energy are, of course, the microplants, whose products either directly or indirectly sustain herbivorous and carnivorous consumers. Besides producers and consumers an ecosystem contains decomposers, largely bacteria and fungi, which in breaking down dead organisms return, *inter alia,* nutrient salts to the plants (see also Chapter 5).

In the open ocean there are two self-contained ecosystems: neustonic and epipelagic. Recent studies have shown that very close to the sea surface there is a flora of small flagellates mixed with bacteria. Neustonic herbivores, such as copepods and larvaceans, presumably graze on this film. In turn, a herbivorous copepod may fall to a larger carnivorous species: both may be seized by an arrow-worm, while all three may be the prey of a flying-fish. The flying-fish, as we saw, may form the food of a dolphin-fish or a squid. Below in the epipelagic zone, herbivorous forms, copepods, euphausiids and so forth, may be engulfed by flying-fishes and migrating mesopelagic fishes, such as lanternfishes. These fishes are the prey of gempylid fishes like the snake mackerel and squids, which in turn become the food of sharks, tuna and marlin. At the end of the food chain there may be a great white shark, able to seize smaller sharks and tuna. Thus, from microflora to great white shark there may be six links in the food chain. Since about ten per cent of a growing animal's food is used to form new tissues, 100,000 weight units of plant food are required to form 1 weight unit of growth for a great white shark (e.g. 100,000 plant units – 10,000 copepod units–1000 arrow-worm units–100 small fish units – 10 tuna units – 1 great white shark unit).

Below the epipelagic zone there are no primary producers. To exist and grow under the photosynthetic zone, plants must turn to organic food-stuffs in solution. Concerning the animals, soon after sunset the surface waters are visited by migrators from mesopelagic levels; for instance by copepods, euphausiids, prawns, pteropods, arrow-worms, squids and fishes. Certain species of the first two groups and the pteropods are largely herbivorous; the last three consist of carnivores. All such migrators stay in the surface layer during the night, but before the sun has risen they are moving down to their deeper, daytime living spaces (figure 48).

Some mesopelagic animals, such as hatchet-fishes and certain prawns, migrate upwards, but do not normally breach the 'seasonal' thermocline; others do not appear to migrate at all. Indeed, why should so many mesopelagic animals spend so much time and energy each day moving up and down through a water column of several hundred metres? If, as seems likely, the primary end is to feed in the productive zone, then the potential energy gained must be greater than the energy expended. There is little direct evidence of such gain, but one may argue–by means of mathematical models based on relevant growth equations–that vertical migrations can be advantageous, provided that the upward migration is from cool to warmer waters and that the migrators rest by day. (The first proviso certainly holds for the warm ocean, while regarding the second there is evidence that lantern-fishes are quiescent during the day). The prediction is that such migrators will gain an energy bonus, part of which can be put into their fecundity.

But the overall energy gain of mesopelagic animals must be less than that of epipelagic forms. Apart from the busy migrators, the less energetic mesopelagic residents, if they are particulate feeders, must depend on progressively degraded, down-drifting morsels of organic matter. At best a resident carnivore is sustained by the hard-won energy of migrating herbivores. It is hardly surprising, then, that the biomass of mesopelagic zooplankton, say at 500 m, is something like a fifth of the amount at epipelagic levels.

There is no certain evidence that bathypelagic animals make daily migrations. At first, one might be inclined to attribute this to their sunless surroundings, but organisms are not necessarily dependent on diurnal changes of light for regulating their diurnal activities. At all events, the biomass of bathypelagic life at 2,000 m, for instance, is something of the order of one fiftieth of the amount in the surface waters. Bathypelagic copepods tend to be somewhat smaller than their mesopelagic relatives; they are also less diverse. The latter is true of the bathypelagic euphausiids (e.g. *Bentheuphausia, Thysanopoda* spp), though they are larger than mesopelagic species. The copepods are presumably carnivores, for their limbs do not seem to be built to filter organic particles and micro-organisms. The same may well apply to the deep-living euphausiids. Of the fishes, *Cyclothone* species feed on copepods, arrow-worms, euphausiids and small fishes, while female angler-fishes lure prey ranging from copepods to fishes.

The contrasts in living at bathypelagic and mesopelagic levels may be seen in comparing a female angler-fish with a mesopelagic predator – for instance, the star-eater *(Astronesthes niger)*. The star-eater has large eyes and brain, a complex system of light organs, a swimbladder, a firm skeleton and substantial axial muscles. Moreover, the muscles contain many red fibres, needed no doubt to give the fish cruising stamina for its daily migrations. An angler-fish of the same size has small eyes, a tiny brain, a single light

203. X-rays of the shallow
Astronesthes (top) and the deep
living angler *Dolopichthys*
(bottom)

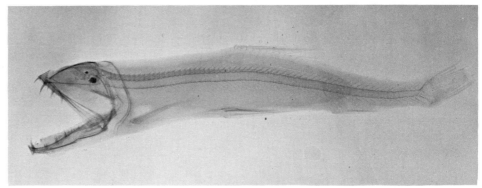

lure, no swimbladder, a lightly ossified skeleton and relatively small axial muscles with very few red fibres (figure 203). In fact, the angler-fish is a floating trapper, built very sparely to conform to its food-poor surroundings. And the evolution of dwarf males is not only a reproductive adaptation (see page 223) but also an economy in living substance, and a means of reducing competition for food between the sexes.

Thus, in descending from mesopelagic to bathypelagic levels, animals become less diverse, much less abundant and more carnivorous. The first two changes are true in descending levels of the deep-sea floor. In terms of biomass (grams per square metre of the bottom), mean values of the larger animals at depths from 0–200 m are *200;* 200–3,000 m, *20*, and over 3,000 m, *0·2*. If bottom-dwelling animals are ultimately dependent on organic matter derived from above, particularly from the surface waters, such decrease of biomass with depth is to be expected. Organic particles sink very slowly, and as they do so, some will be taken by mid-water suspension feeders. Particles eventually reaching the bottom will lose more and more of their organic content to bacteria as the depth of fall increases. Moreover, at comparable depths, the biomass of benthic deep-sea animals is greater under areas of greater productivity. Unlike the trend in the mid-waters, the number of carnivores decreases as the

ocean floor deepens. For instance, crabs and starfishes are most diverse and abundant at depths well above a level of 5,000 m, which is true also of the sea lilies, which are suspension-feeders. Most of the bivalve molluscs are also suspension-feeders, but with increasing depth the deposit feeders become dominant. In fact, mud-eaters of all kinds, and we think particularly of the elasipod sea-cucumbers, are over-whelmingly predominant in the fauna at depths below 5,000 m. Even at a depth of about 3,000 m almost half of the species, consisting mostly of bristle worms, bivalves, tanaid and amphipod crustaceans, have been judged to feed on organic particles.

REPRODUCTIVE ECOLOGY

Over the deep ocean the productive surface waters are not only feeding grounds for adult animals of the plankton and nekton, but serve also as nursery grounds for the young. Most marine animals start life as tiny larvae, which form part of the plankton. For instance, the earliest stages (nauplii) of copepods and euphausiids are usually less than one mm in length, which is true also of sergestid prawns. Most of the bristle worms, molluscs and echinoderms produce minute larvae moved by cilia and the sea. The earliest larvae of most bony fishes are just a few mm long. When they hatch, all these larval forms may be sustained for a time by yolk reserves, but very soon they must seek minute food organisms of the right kinds. Most of the microscopic plants are ideal larval food, but are necessarily concentrated in the surface waters.

To exploit the richness of the productive layer, many deep-sea animals have evolved apt life-history patterns. For instance, mesopelagic fishes, dominated by stomiatoids and lantern-fishes, presumably spawn in the depths, there releasing myriads of eggs that develop as they float towards the surface. At all events, the larval stages feed on minute plants and animals in the surface waters, and as they grow they tend to move down in the sea. By the time metamorphosis occurs, when they begin to look more like the adults, the fry have descended close to the depths of the adult living space. Bathypelagic fishes, some of which live at depths well below a level of 1,000 m, also exploit the nursery grounds of the surface. Their life-history pattern, shown very well by certain angler-fishes, is basically similar to that of the mesopelagic species. The eggs, larvae and metamorphosis stages are simply exposed to greater risks during their longer developmental migrations.

Of the bottom-dwelling fishes, halosaurs, eels (synaphobranchids) and rat-tails produce larvae that live in the upper 200 m. Since the first two groups have leptocephalus larvae (figure 185) with large jaws and teeth, they start life as predators on zooplankton; larval rat-tails, which have small jaws, seem to be restricted to a diet of copepods. But larval stages are evidently rare among the invertebrates of the deep-sea floor. Such knowledge that we have, particularly of echinoderms, molluscs and crustaceans, indicates that they produce relatively large yolky eggs, which hatch into advanced young. These, most likely, are broadcast by currents over the bottom.

The entire life-histories of nearly all oceanic copepods and euphausiids have yet to be discovered, but concerning the mesopelagic species at least some of the young stages live in more productive waters above the adults. This is true of some prawns (family Sergestidae), but not necessarily of others some of which produce large eggs, which hatch into advanced larvae, able to prey on small members of the zooplankton. The young of these crustaceans have no need of fine forms of plant food in the surface layer, nor have the newly hatched young of mid-water squids and octopods. Even if these advanced kinds of young live above the adults, they are often well below the plant-bearing waters.

How do the sexes of deep-sea animals manage to come together when they are ready to breed? Earlier, we considered how this might occur in the bathypelagic *Cyclothones* and certain angler-fishes, which have evolved dwarf males (page 218). Little is known, but whatever their means of sexual congress, male crustaceans and cephalopods supply the females with packaged stores of sperm (spermatophores); so do the males of the viviparous brotulid fishes. Concerning the bottom-dwellers, one intriguing discovery of recent years is that mature ipnopid and tripod-fishes are invariably hermaphrodites. If the sexes fail to meet, do they reproduce by self-fertilisation?

The continuing end of reproduction is the overall maintenance of a species stock of breeding individuals. Species consist of populations, but we know little of such organizations in deep-sea animals, though they certainly have immense living-spaces, some extending over all three oceans. The production of many small eggs hatching into minute planktonic larvae, which are broadcast by the seas' motion, must be very effective for exploiting, and even extending, living spaces, for the habit goes back at least as far as the seas of early Cambrian times, about 600 million years ago.

Though a deep-sea species may cover a wide area, it is not necessarily able to reproduce wherever it is. For instance, planktonic animals may drift into regions favourable for their existence but not for their reproduction. This is even true of certain deep-sea angler-fishes, which grow to a very large size in Icelandic waters, but are never sexually mature. Recently, there have been interesting studies of lantern-fishes in the North Atlantic. One species (*Lobianchia doffeini*) is widely distributed between latitudes 50° and 25°N, but only reproduces over the easterly parts of this range. The western population consists of non-breeding expatriates, evidently carried westward – first as larvae from the breeding area – by the North Atlantic currents.

DIVERSITY AND DISTRIBUTION IN THE DEEP OCEAN
We have seen already that deep-sea animals are most diverse under the warm waters of the ocean, which invest the earth between the subtropical convergences. Whatever group we consider, from Protozoa to fishes, there is a fall in the number of species in passing poleward through temperate regions. Consider fishes: off Bermuda about 300 species of mesopelagic and bathypelagic forms are known, whereas in waters south of the Antarctic Convergence the number is fifty. The entire fauna of mid-water animals off Bermuda must consist of several thousand species, probably well over five times the number in temperate or Polar regions.

The stability of ecosystems depends directly on the number of species involved. The more the diversity, the more intricate will be the food-web, which means that a decline in one species hardly disturbs the overall balance. Simple ecosystems tend to be dominated by a few very abundant species, and should one of these decline, the system will become unstable. But why should the diversity of deep-sea animals be greater under the warm surface waters of the ocean? Here, presumably, conditions must have favoured the evolution of considerable diversity over a long period of geological time, but are these conditions more favourable than those prevailing in temperate and Polar regions of the deep sea? All parts of the deep sea appear to be relatively constant in terms of physical conditions. Perhaps the main factor behind greater diversity is the production of plant food, which is more or less constant over the year in subtropical and tropical waters, but markedly seasonal in temperate and Polar seas. Marine life is founded on microplants and their consumers, and few of the latter have evolved the means of subsisting on seasonal grazing.

Evolution is a compromise between conflicting selection pressures, and the more these fluctuate the less likely it is that species will evolve to cope with such a changeable environment. Recent studies on the benthic invertebrate fauna of the continental shelf and slope in temperate America may show this very well. In a descent from shelf to slope a marked increase in the diversity of the in-fauna and epibenthic fauna occurs at a depth somewhere between 100 and 300 m. Below this depth there is little or no change in temperature throughout the year, whereas at 100 m the yearly change is $10.5°C$. Evidently the shelf fauna consists of a limited, but very adaptable fauna.

Physical conditions are so nearly constant at any level in the deep ocean that there would seem to be no barrier to the spread of animals, given time and a far-reaching circulation. Some deep-sea forms certainly live in more than one ocean. In a recent study of bathypelagic copepods, over 90 per cent of the species found in the Arabian Sea occur also in the North Atlantic, about 8,000 miles away. Bathypelagic euphausiids are also widely distributed: for instance, *Bentheuphausia amblyops* and *Thysanopoda cornuta* are cosmopolitan. The distribution of the mesopelagic and epipelagic species is more restricted. Though some live in all three oceans, they are confined to tropical or 'antitropical' regions, others are confined to one part of an ocean. Bathypelagic fishes also tend to be more widely distributed than mesopelagic forms. In his study of ceratioid angler-fishes, Dr E. Bertelsen predicted that most species would prove to be distributed in all the oceans. A good many lantern-fishes, on the other hand, have more restricted distributions, some to one area of an ocean.

Evidently, there are more barriers to wide distribution in the mesopelagic zone. One may be the high diversity of mesopelagic life itself. Species able to compete with rivals in all parts of the ocean are less likely to evolve at mesopelagic than at bathypelagic levels. Mesopelagic animals also encounter oceanographic fronts that are above the reach of bathypelagic forms. Thus, the Antarctic Convergence not only marks the northerly limit of *Euphausia superba* (krill), an epipelagic species, but also of the lantern-fish *Electrona antarctica*. Between the Antarctic and Subtropical Convergences there is a related species, *Electrona subaspera*. These and other fronts seem to form boundary conditions for the less adaptable members of the mesopelagic fauna.

Of the larger benthic invertebrates, few seem to be widely distributed. Out of an inventory of deep-sea sponges, coelenterates, barnacles, isopods, decapods, sea-spiders, echinoderms (except brittle-stars) and beard worms, only four per cent of a total of 1,031 species are known from all three oceans. Moreover, only 15 per cent have been taken in more than one ocean. Of the species trawled in the deep-sea trenches at depths greater than 6,000 m, about 25 per cent are endemic. After making these analyses, Dr Vinogradova concluded that '...the ranges of deep-sea bottom animals tend to shrink rather than to extend with increasing depth'. If this rule still holds after further exploration, it is the reverse of distributional patterns in oceanic mid-waters. To endeavour to comprehend the differences will then become a further task for deep-sea experts.

Deep-sea exploration is a slow and costly business, but there have been encouraging advances in the past twenty years or so. More than other scientists, deep-sea biologists appreciate the words of Dr G.G. Simpson: 'Scientists do tolerate uncertainty and frustration, because they must. The one thing they do not and must not tolerate is disorder'.

8 Physiological Investigations of Oceanic Animals

J. A. C. Nicol

INTRODUCTION

There was a time when the physiology of animals and their anatomy were studied independently, but now the two branches of science are fused, as indeed they should be, for structure and function are one, the parts of the animal supplying the mechanisms whereby it can operate, pullulate and survive. These things having been said, let us admit that some animals are much more difficult to study than others, the humming-bird more difficult than the pigeon, and the bat more difficult than the dog. It is with reflections of this kind in mind that we may consider the study of the physiology of marine animals.

The scope of this chapter is the physiology of the animals of the oceans, in all their diverse variety, inhabiting 1,370 million cubic km of water, from the Polar seas to the equator and from the surface to the bottom of the deepest trenches at 11,000 m. At the great shore-based stations of the world, marine animals have long been prized objects of research and, indeed, have yielded information of inestimable value to the biological and medical sciences. It is the present intention, however, to deal with the physiology of the animals of the high seas, beyond the 100 fathom line.

Here in the oceanic vastness there is a remarkable and spatially limitless environment which the physiologist has only begun to explore. Because his ancestors forsook it so long ago, it is a world most difficult for him to penetrate and understand. His physiological investigations of that world reflect the same multiplicity of interests he evinces elsehere : how man may safely submerge himself and dive into the sea (human physiology): what useful biomedical products may be harvested from the waters (pharmacology); the comparative physiology of the creatures living there. The last is the self-imposed limit of this chapter.

Of special interest to the biological oceanographer are those general and peculiar adaptations of oceanic animals by which they are enabled to endure and exploit their strange realm. The range of environmental variation daily and seasonally at any one place is small in the oceans, except for the daily changes of solar radiation near the surface. On the other hand, there are the great changes with latitude and depth, of temperature, pressure, illumination, density, viscosity, gaseous and organic content, and others to which animals are sensitive. So far as he can, the biologist desires to understand the mechanisms specially concerned with adjustment to the peculiar features of life in the deep oceans.

In planning and carrying out his investigations he has sobering difficulties to overcome and chastening experiences to endure. Frequently, instead of being allowed to watch problems open up from simple beginnings, he must conceive some research in its entirety, devise and prepare for it with great thoroughness, often at great expense, and hazard all at a single throw. Great is the joy when it succeeds, the anguish when it fails. Because the animals of the oceans are frequently so fragile, and difficult to secure in pristine living condition, severe restrictions are imposed on the investigations that he can carry out. Nevertheless, the investigations which have been made during recent years, in profundity and ingenuity, are remarkable ; in some part they are the result of the beautiful electronic equipment that has become available ; the greater credit belongs to several generations of enthusiastic scientists dedicated to the study of the oceans.

SENSING THE ENVIRONMENT

Oceanic animals each have some or all of the same senses as ourselves and possibly additional ones as well, such as echolocution. They use all senses for learning about their environment and sometimes for communicating with each other. Much has been (and more is still to be) learnt about the functioning of sense organs in aquatic animals of shallow waters, and some of this information may be extrapolated with justification to ocean animals. Particular studies of oceanic animals, however, are more meagre because of difficulties in dealing with such difficult creatures. There are novel and special features pertaining to life in deep waters which we may profitably consider. Among them are darkness, pressure, the vast spaces in an unmarked three-dimensional realm and the sparseness of the inhabitants. The most fruitful of these investigations, hitherto, has been concerned with eyes and sensitivity to light, which are dealt with more amply in the next section; by comparison, exceedingly little is known about the other sensory modalities.

Dr N.B. Marshall, from anatomical investigations, has found great variety in the sense organs of bathypelagic fishes. Some species, for example, have large and prominent lateral-line organs whereas others have diminished ones. The lateral-line is a system of canals and sensory papillae distributed over the trunk and head, and its sensory components respond to vibrations of low frequency transmitted through the water. By this means the fish is able to detect objects in its vicinity, because of the movements of another animal, or by sensing waves of its own creation reflected from other objects. Indeed, there is compensatory regulation of the degree of development of sense-organs. Among whale fish (family Cetomimidae) (figure 196), eyes are tiny or wanting, but the lateral-line organs are especially large. Again, rat-tails (figures 184 and 244) have a large sacculus – the sound detector of the inner ear – and large acoustic centres in the brain, both indicative of the important part that hearing plays in their lives. Furthermore, these fishes, although they are bathypelagic, possess large air-bladders provided with drumming muscles by which they can create sounds. The sounds they create, that they perceive, and the uses to which they put their apparatuses are still to be discovered.

The deep sea has its sounds, its callers, and its listeners, to us largely still an enigma. Submerged hydrophones are employed by research vessels for detecting underwater sounds. The R.V. *Atlantis* of the Woods Hole Oceanographic Institution has picked up calls, seemingly from a fish at 3,500 m, and the R.V. *Chain* of the same Institution has recorded calls of sperm whales on its echo-sounding gear; the sperm whale seemed to be replying, with extraordinary celerity and precision, to the output or signal of the echo-sounding machine.

There has been a remarkable interest recently in the acoustic ability and phonation of porpoises and whales, creatures which excite or should excite our curiosity, admiration and compassion by their grace and exploits.

The toothed whales (sub-order Odontoceti) – including porpoises, dolphins, pilot and sperm whales – emit a variety of sounds described as squeals, whistles, quacks and clicks (figure 228). The sounds are produced under water or in air, and observations on animals, especially captives, have revealed that they use them to communicate with each other, in the manner that land animals use calls of various kinds. There are many interesting factors involved in this behaviour, especially the physical characteristics of the sounds, their manner of production, transmission, reception, and behavioural significance (i.e. use or employment). Baleen whales (sub-order Mysticeti) also are vocal, producing sounds described as moans or screams, the significance of which is unknown. The description which follows is limited to the underwater sounds of toothed whales.

Sound-waves, like light-waves, can be characterized by frequency (equal to the velocity divided by the wavelength). The sounds of cetaceans are mostly between 0·04 and 30 kHz (kilocycles per sec). However, clicks of the bottle-nose porpoise *(Tursiops)* exceed 50 kHz, and it hears sounds in the range of 0.15 to 120 kHz. For comparison, the frequency range of the sensitive human ear is 0.15 to 20 kHz.

Dolphins emit both squeals (whistles) and clicks; sperm whales clicks only. Certainly the former, and possibly both kinds of sounds, are used by whales to communicate with each other. Clicks, moreover, have another function, namely echolocation, which is to say that dolphins emit sounds, listen to their echoes, and from the information so derived are enabled to orientate themselves, either to locate food, sense other free-swimming animals such as dolphins, or to avoid obstacles. Moreover, not only can they orientate themselves by means of sonar, but they also acquire much detailed information about their surroundings, about objects and even about their contents. For example, a captive bottle-nose dolphin, by sound alone, was able to distinguish real food from other objects of the same size and shape.

The high frequencies used by odontocetes for echolocation are ultrasonic, that is they are above the upper limit of frequencies detectable by humans. The clicks are emitted in short pulses which can be varied at will by the animal in several ways, in rate of delivery, length of pulses, frequency of clicks, etc. It is not unlikely that the animals emit clicks with individually recognizable characteristics. High-frequency pulses have high resolving power, that is, by their use small objects in the environment can be distinguished. The power output is small, however, the energy of the signal is rapidly absorbed, and this sonar system operates best over short distances.

It would take a long excursion into comparative anatomy to describe the ear of a toothed whale; anatomists believe, however, that its organization is especially well adapted for hearing under water and for detecting very high frequencies. One feature may serve as an illustration. The ears of mammals evolved to permit hearing in air. Beneath the surface of water, sound-waves set up vibrations in the skull, through which they are transmitted to the inner ear. The skull of a terrestrial mammal, when submerged, fails to insulate the two ears from each other, and directional hearing is lost. In the whales this difficulty is circumvented by a novel development. The ears (middle and inner) are surrounded by air-filled sinuses that isolate them acoustically from the skull. Consequently, sound in the water is conducted independently to each of the two inner ears.

Of touch, taste and smell, little can be said except to note that these senses are available to deep-sea animals, and most often be of great service to them by providing important information about the environment. Dr N.B. Marshall has noted that the males of most bathypelagic fishes have large olfactory organs; the females, small; and it seems reasonable to suppose that in the vast dim spaces of the sea, the males are guided by scents which emanate from the females.

LIGHT AND VISION

After light has penetrated long distances through the waters of the ocean, it becomes so attenuated that it is too dim for vision, even to the most sensitive eyes. One might therefore expect that deep-sea animals would be blind, just as are animals living in dark caves. Curiously enough, this expectation is not realized. In the dimly-lit twilight zone of the ocean, between 500 and 1,000 m, the eyes of fishes are sometimes enlarged, compared with those of fishes living near the surface, and they frequently possess large pupils and lenses. At deeper levels where no daylight reaches, some fishes have small functional eyes, others are blind. But even in very deep waters, below 1,000 m, there are fishes dwelling on or near the bottom, for example macrurids, that have very large eyes, with large pupils and lenses.

Since eyes are present in deep-sea fishes, the problem to which we should address ourselves is how such eyes function in very dim light and what special features they possess to make them more efficient in such environments.

The eyes of deep-sea fishes usually have very large pupils and lenses, compared with those of surface-dwelling species. The effect is to let more light into the pupil, forming a brighter image; a similar result is achieved by a camera having a large aperture and wide lens. The image is formed upon a light-sensitive surface, the retina, whose sensitivity varies in different animals. To return to our analogy with the camera, where sensitivity can be augmented by using faster film, the sensitivity can be and is increased by several means, structural and chemical. The light-sensitive elements, the rods, are packed much more densely in deep-sea fishes, they are longer, thus interposing more light-sensitive pigment in the path of the light rays; and they contain a visual pigment matching well the quality of submarine illumination. Indeed, in some fishes, e.g. *Bathylagus,* the rods are packed in several tiers, one above the other.

The rods contain a photosensitive pigment, affected by light, which is called visual purple or rhodopsin. At least, it has a reddish or purple colour in the retinae of inshore or coastal fishes. Absorption of light-energy by the visual pigment sets off a series of sensory and neural events which eventually culminate in a sensation of light in the brain.

If one examines a spectrum curve of such a rhodopsin, i.e. a curve obtained by plotting on a graph the density of visual pigment at each wavelength against wavelength, one observes that the highest density – where most light is absorbed – occurs at about 505 nm (nanometres, a measure of length, 1 nm being equal to 1/1,000,000,000 of a metre). In this region of the spectrum, light appears green. The rhodopsins of oceanic fishes, however, are more sensitive to blue light; they absorb maximally in the range of 485 – 495 nm (those of epipelagic fishes); or at still shorter wavelengths, down to 478 nm (bathypelagic fishes). Such pigments have a golden colour because they absorb the complementary colour, blue, and by some scientists they are called chrysopsins.

The eyes of oceanic fishes lend themselves to study in several ways. One can, as the fish fall out of the trawl, preserve the eyes for later microscopic examination. On the other hand, if one wishes to study the visual pigments, then one must arrange one's programme so that the fish are brought to the surface at night, in very dim light, no mean achievement when one considers the difficulties and hazards of raising a trawl aboard the deck of a small, usually heaving, oceanographic ship, in minimal lighting. If the eyes become exposed to white light, the visual pigment bleaches and its characteristics are less easy to determine. The retinae are removed from the eyes in dim red light and they are stored frozen in darkness until they can be examined ashore. Subsequently, the pigments are extracted and their absorption characteristics determined, a route which Dr F.W. Munz has followed with much success (figure 205). Or the density of the fresh retina is measured in the ship's laboratory, by elegant methods devised by Professor E.J. Denton (figure 204). The former method gives very precise information about the characteristics of the visual pigment, the latter about absorption in the living retina.

When we come to consider the penetration of light through sea water, we find that it is altered in several significant ways, that is to say in ways important to

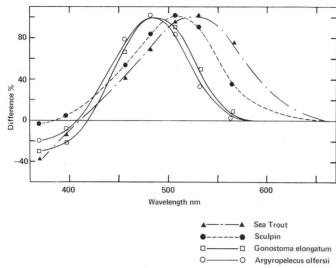

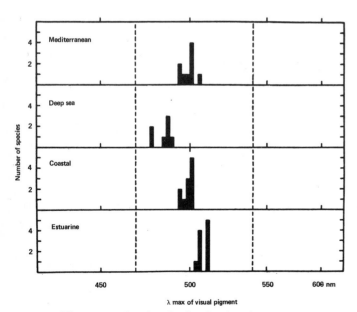

204. Curves showing how the retinae of fishes absorb light through the visible spectrum. Values to left, difference in densities between unbleached and bleached retinae; below, wavelength. The two curves to the left are for retinae of bathypelagic fishes (*Argyropelecus* and *Gonostoma*). The median curve is for a coastal fish (sculpin); the curve to the right for a sea trout (after Denton & Warren 1957)

205. Histograms showing the distribution of visual pigments of fishes from different environments and light-regimes. The blocks show the frequency of occurrence of visual pigments, which are denoted by wavelength of maximal absorption. The interrupted lines show the known limits of visual pigments of fishes, from all sources (from Lythgoe 1966)

living organisms. When light enters the water, if the rays of incident light are not perpendicular to the surface, they are bent towards the vertical because of refraction. Owing to scattering of light rays in the water, the light becomes polarized i.e. the light waves are caused to vibrate largely in one plane. As the light passes downwards its intensity decreases rapidly, a process referred to as absorption: in clear oceanic water the intensity decreases at a rate of about five per cent per metre; at 100 m it is only one hundredth of the surface value, and at 500 m one ten-millionth. Lastly, the colour composition of the light changes with increasing depth. Long and short wavelengths (i.e. red, orange, yellow and violet) are absorbed more than blue light, which passes through clear oceanic water most readily. Consequently, the light penetrating to the depths becomes restricted around a narrow band of energy centred at 485 nm (figure 206). It is to blue light of this quality that the visual pigments or chrysopsins of deep-sea fish are most sensitive, and we may adjudge with some confidence that, as fish moved outwards and downwards from their original home in coastal waters into the ocean deeps, one of the adjustments which they made individually, species by species, was to alter the character of their visual pigments so as to make them more effective for absorbing the blue light of oceanic water. Luminescent light is also of great importance to vision, a matter discussed in the next section (figure 207).

Various estimates have been made of the depths at which fish can still see by daylight. This is a most doubtful subject, at which only conjectures may be hazarded, and no direct information is available because of the, as yet, insurmountable obstacles to the study of living deep-sea fish. The estimates start with the known sensitivity of the human eye and the eye of the sunfish *Lepomis,* which is a pond fish, to very weak lights. If the eyes of deep-sea fish are at least as sensitive as those just mentioned, then they should be able just to perceive objects at depths of about 900 m. But for several reasons we may expect that their eyes are much more sensitive than those of man and the sunfish. In general, the eyes of deep-sea fishes have large apertures, and the ocular media and structures in front of the retinae are more transparent. Moreover, the retinae are much richer in visual pigments and are capable of absorbing 90 per cent or more of the light falling upon them. Consequently, it is not unlikely that these eyes are 10 to 100 times more sensitive than the eye of man or that of the sunfish. If the sensitivity to weak light be 100 times greater, then deep-sea fishes should be able just to detect daylight in very clear oceanic water at 1,000 m. At the threshold of vision the fish could hardly distinguish anything in its surround-

206. Transmission of sunlight through sea water. The curves show its spectra at various depths. Measurements were made in the Bay of Biscay (46° 29′ N, 7° 59′ W), curve A at 10.30–11.30 h; B at 13.25–14.10 h. The curves become more peaked as the depth increases because light of short wavelengths is scattered and light of long wavelengths is absorbed by water molecules and fine particles in suspension (from Boden Kampa & Snodgrass 1960)

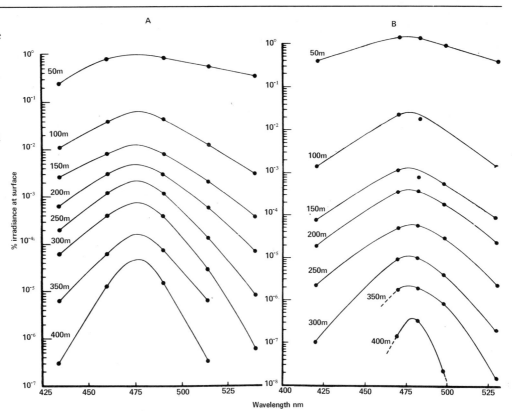

ings, but at slightly higher levels in the water column, as the light became several times brighter, it could begin to distinguish objects about it.

In the matter just reviewed we have limited our attention to pelagic fishes that spend all their lives in oceanic waters, comparing the eyes of bathypelagic and epipelagic species, and making one allusion to the visual pigments of coastal species, from which the deep-sea forms have originated. There are still other kinds of fishes that make regular migrations from rivers or coastal waters to the deep ocean, and vice versa; and because of the great interest which these migrations have aroused and the economic importance of the species involved, they have received much attention, without, however, solving all the problems posed by their metamorphoses and migrations. Adult sea lampreys and salmon breed in rivers; the progeny after a suitable interval prepare to return to the sea, the haunts of their ancestors, and in so doing they suffer a strange sea change, laying down silvery layers in their skin and acquiring a bright livery. Fresh-water eels, on the contrary, breed in the deeps of the Sargasso Sea and the young, as leptocephalus larvae, return in ocean currents to the Atlantic shores of North America and Europe. There they mature in rivers and prepare once more for a sea journey, metamorphosing, acquiring a silvery livery, enlarging their eyes, etc. Because

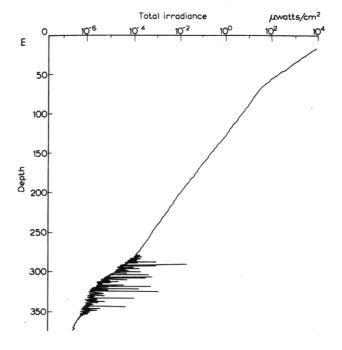

207. Measurement of the intensity of submarine light at various depths in the San Diego Trough (12.15 h). Luminescent flashes were detected between 275 and 300 m (spikes on the record) when sonic scattering was also recorded (from Kampa & Boden 1957)

208. Curves showing how the retinae of silver (sea stage) and yellow (freshwater stage) eels differ in absorption across the visible spectrum. The retina of the silver eel, containing a golden rhodopsin, absorbs at shorter wavelengths than the retina of the yellow eel (from Carlisle & Denton 1959)

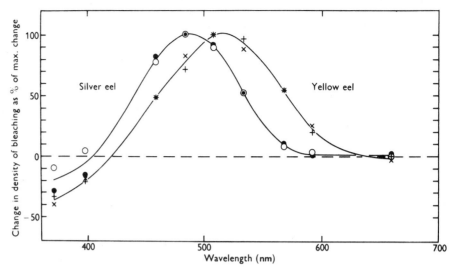

the conditions of illumination in rivers are so different from those in the deep ocean, scientists have naturally been prompted to look for correlative changes in the eye, especially in the nature of the visual pigments. Unfortunately, this work is still incomplete and parts of the story are still to be determined. It seems that lampreys essentially alter their visual pigment when they move seawards, changing over to a rhodopsin that is more sensitive to blue light than that which they possessed while living in fresh water (maximal absorption at 497 nm and 520 nm, respectively). Likewise the young salmon, preparing for a pelagic life in the surface waters of the northern oceans, resolves its several visual pigments into one absorbing in the blue-green region of the spectrum (at 507 nm). More dramatic still are the changes occurring in the eel. No one has yet undertaken the difficult task of investigating the visual pigments in the eyes of the tiny leptocephalus larva, but those of yellow and silver eels (immature and adult animals respectively) have been examined. The immature, yellow eel has a purple retina; it contains a mixture of two pigments absorbing in the blue-green and green regions of the spectrum (at 502 and 523 nm). The silver eel, on the contrary, has a golden retina, absorbing blue light at 487 nm (figure 208). Thus the eel, on making itself ready for its last journey, acquires in anticipation a golden eye pigment like that of bathypelagic fishes whose realm it is preparing to invade.

Rhodopsins occur in invertebrates and have the same chemical composition as those found in fish; they are usually much more difficult to investigate. Those of euphausiids (*Meganyctiphanes*) absorb maximally between 460–465 nm, and those of deep-sea prawns (sergestids and *Acanthephyra*), between 360 and 480 nm.

LIVING LIGHT

Luminescence is a technical term employed by scientists to denote the production of light without heat. In the process, particular molecules acquire excess energy – their electrons become excited – and when the energy is subsequently released it does so in a form which we recognize as light. When the event takes place in living organisms it is called bioluminescence, and the energy which is converted into light is provided by some chemical reaction.

It is a common belief that luminescence among marine organisms is much more common than in terrestrial ones, and that it is exceedingly rare in fresh water. Although this generalization is open to question because of the abundance and ubiquity of fireflies and glowworms, it nevertheless remains true that luminescence occurs in wonderful variety among a multitude of marine creatures, especially those of deep waters. The luminescent displays of the oceans, sometimes exceedingly spectacular, for a long time excited the curiosity of mariners and philosophers, and provoked a variety of peculiar explanations. Slowly the nature of phosphorescence of the sea was unravelled when it was discovered that jellyfishes and fishes were luminous, and that swarms of microscopic dinoflagellates (such as *Noctiluca*) could cause the surface of the sea to glow.

Luminescence is a natural phenomenon that can often be investigated successfully at sea because suitable apparatus exists which can be readily turned to that role. Extremely sensitive photosensors (light-detectors) developed for commercial and military purposes permit the measurement of the very dim lights emitted by plants and animals, and the electrical responses generated in the instruments in response to the light can be made visible by recording machines, oscilloscopes or pen writers, and permanent tracings obtained, or they can be recorded on magnetic tapes for study later when more time is available. Moreover, the photosensors can be enclosed in water-tight casings, proof against pressure, and sent down into the ocean

209. Lantern fishes such as this *Myctophum* (8 cm) are often caught at the surface at night, and can sometimes be seen in the water because their eyes shine like those of a cat in the ship's lights

210. The little squid *Heteroteuthis* (4 cm) can squirt out a luminous secretion that breaks up in the water into small spots of light

to record the luminescence occurring naturally at different times, places and depths.

Just as there is some degree of correlation between the complexity of the visual apparatus, depth, and light-penetration, so are there analogous relationships between luminescence and the depth preferences of animals. Luminescent phytoplankton is limited, of course, to the surface waters; photosynthesis demands it. On the other hand, luminescent animals occur at all depths (including those that live in the hadal zone, i.e. in the deep ocean trenches). Among the zooplankton, lacking visual organs, i.e. eyes capable of forming images of objects, luminescence is present equally in epipelagic species that could only employ it usefully at night, and in bathypelagic species that live in the continual darkness except for the fitful gleams which they themselves create, e.g. radiolarian protozoa, jellyfish and copepods. A variety of benthic animals in deep waters are luminous, e.g. sea pens and sea spiders. It is among the nektonic squid, decapod crustaceans and fishes, however, that luminous organs attain their greatest variety and profusion; these animals have true visual organs.

Before proceeding further with this matter we should recognize the three kinds of light-production obtaining among animals. They are intracellular, whereby the animal itself generates light inside special cells, usually gathered together into light-organs or lanterns; extracellular, by which the animal discharges a luminous secretion or glowing cloud into the sea water; and bacterial, by which the light is generated by luminous bacteria which the animal harbours and cultivates within special luminescent glands.

In the mesopelagic realm between 200 and 1,000 m, where residual daylight suffices for vision, are multitudes of small fishes bearing batteries of light-organs, e.g. lantern and hatchet fishes, round-mouths (*Cyclothone*), and others such as *Gonostoma* and stomiatoid fishes unburdened by popular names (figure 185). Light-organs are aligned in rows along the lower surfaces of the head and trunk, and they cast their light downwards. Often, like a bull's-eye lantern, they have a lens in front; enveloping the light-generating tissue is a reflector and, behind that, a black screen that prevents light from penetrating into the animal. No investigation as yet has been made of the biochemistry of the light-organs of deep-sea fishes but it is not unlikely that the luminescent reaction may prove to be like that which has been worked out for the midshipman *Porichthys*, a fish of coastal waters. The luminescent material or luciferin of *Porichthys*, of the fishes *Apogon* and *Parapriacanthus*, and of an ostracod crustacean *Cypridina*, is believed to be identical. It is a complex organic molecule; in the presence of a specific enzyme,

luciferase, it undergoes oxidation with emission of light:

$$LH_2 \ (luciferin) + \tfrac{1}{2}O_2 \ \xrightarrow{\text{luciferase}} \ L + H_2O + light$$

The luminescence of these fishes is intracellular, that of *Cypridina* extracellular.

The light-generating cells of pelagic fishes, examined with the electron microscope by Dr Bassot, were found to contain large numbers of secretory granules which in all likelihood give rise to the luminescence. These granules are very small, slightly below the resolution of the light-microscope. In addition, the wall of the light-organ usually possesses a shiny reflector containing stacks of flat crystals, which are referred to later.

In the bathypelagic realm, below the limit of visible light, the arrays of downward pointing photophores generally disappear. In this place, some bathypelagic fishes have luminous glands lying in their bellies, such as *Opisthoproctus* (figure 186), and rat-tails. It is likely that the light of these glands is produced by symbiotic bacteria.

And lastly we may take cognizance of those bathypelagic fishes that bear luminous bulbs or lures at the end of tentacles, especially the deep-sea angler-fishes, the stomiatoid *Chauliodus* (figure 185), and many others. Again, by the aid of electron microscopy, it has been possible to distinguish the luminous bacteria responsible for producing light in the tentacular bulb of the angler-fishes.

Corresponding problems arise from, and similar approaches have been made to, the luminescent systems of deep-sea squid, shrimp and euphausiids (the planktonic crustaceans generally known as krill, which are very abundant in high latitudes). The shrimps have a luciferin-luciferase system requiring only oxygen for consummation; the luciferin, however, is dissimilar from that of ostracods. Two species have been investigated, viz. *Acanthephyra* (figure 195) and *Oplophorus*. As with fishes, studies of the luminescent systems of shrimps have just begun, and a happy coincidence of suitable animals and interested investigator will yield most interesting results.

The luminescent system of euphausiids is one that is unusual among animals. The light-emitting molecule or substrate itself is a protein, referred to as a photoprotein. The system contains a smaller organic molecule, which is not a protein and is fluorescent, i.e. it emits light for a short time after it has been illuminated by ultra-violet light. The latter catalyzes or facilitates the oxidation of the luminescent substrate, whereupon light is emitted. We may recall that in the systems described earlier in the light-emitter was a small molecule, the oxidation of which was catalyzed by an enzyme, which is a protein.

211. *opposite* Among the clumps of Sargassum weed are to be found an array of exquisitely camouflaged animals, like this little trigger fish (6 cm) whose body is covered with weed-like tufts of skin. An anemone and some hydroids are attached to the weed at top left

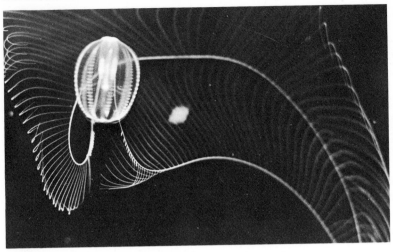

212. Luminous ctenophore *Pleurobrachia* (1 cm diameter) with tentacles retracted and extended

If we are to consider luminescence as having biological significance by stimulating the photoreceptors of other animals, then we must concern ourselves with the physical aspect of the light emitted. Luminescent lights of all colours have been described, from violet to red, but generally the colours of the lights of marine animals are blue or green, that is, they lie in the spectrum region to which the eyes of marine animals are most sensitive. The spectrum curves show that the light of each animal is spread over a fairly wide band, and it has a maximum at a wavelength (λ) characteristic of each luminescent system. Unfortunately, the data are rather sparse; some values are shown in Table 10.

The lights of marine animals are very weak and the irradiance, the energy from the light falling on a unit area, rarely equals moonlight. The intensity of luminescence is usually expressed in units of energy or power, as microwatts (μW) falling on a square centimetre (cm^2) of a receptor surface at a given distance, usually one metre. Some measurements of the output of luminous animals gave these results:

Radiolaria *circa* 10^{-9} μW/cm^2 at 1 m
Jellyfish, siphonophores 10^{-7}
Ctenophores (figure 214) 10^{-5}
Copepods 10^{-8} to 10^{-4}
Euphausiids 2×10^{-7}
Lantern-fish 0.5×10^{-7}

Intensities of sunlight and moonlight, for comparison, are 10^5 and 10^{-1} μW/cm^2 respectively.

The width of the spectrum band of luminescence is usually rather less than that of the absorption band of rhodopsin, i.e. the light-output lies in a narrower spectrum range than the sensitivity of the eye. Spectrum and sensitivity curves overlap, without exact matching of λ max, which may differ by 15 or 20 nm. Nevertheless, luminous efficiencies, in terms of luminescent light absorbed by the photosensitive pigment of the eye, among euphausiids or lantern-fish, for example, are high, 70 per cent or more.

The emission of luminescence being restricted to fairly narrow spectrum bands, it seems likely that the region of maximal emission, in the range 460 to 510 nm, represents an adaptation to the visual sensitivities of animals (figure 214).

TABLE 10
Colour and emission maxima (λ max) of bioluminescence

Animal	Colour	Wavelength of maximum emission (λ max)
Protozoa		
Radiolaria	Blue	
Dinoflagellate *Gonyaulax*	Blue	478 nm
Coelenterata		
Jellyfish *Atolla*	Blue	ca 470
Ctenophore		
Mnemiopsis	Blue–green	490
Crustacea		
Euphausiids	Blue	476
Protochordata		
Pyrosoma [figure 215]	Blue	480
Fish		
Myctophum punctatum	Blue	470
Bacteria from macrurid fishes	Green	510–520

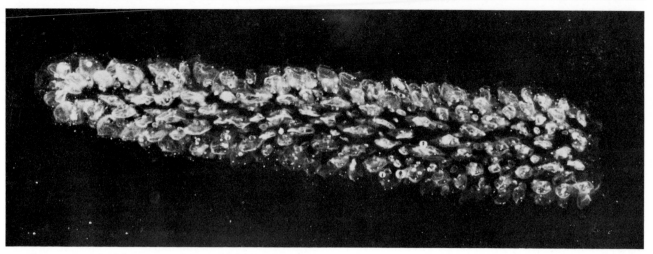

213. *Pyrosoma,* a colonial pelagic tunicate, which is luminescent

It is a nice question whether, among deep-sea animals, the spectrum sensitivity of the eye is adapted to the composition of luminescent light, or vice versa. One inclines to believe that the eyes of deep-sea fishes have developed rhodopsins well matched to the colour of daylight having λ max (wavelength maximum) at 485 nm, which penetrates ocean water most readily. Secondarily, they have acquired luminescent systems the colour of which approximates that of submarine daylight and corresponds to the sensitivity of the eye. Although the spectra of luminescence and of visual pigments do show such variation as to suggest that the

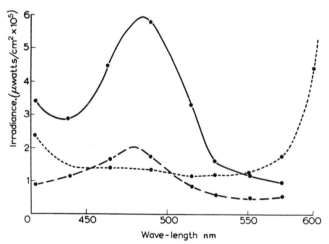

214. Spectrum composition of luminescence in the sea. The measurements were made by photometer submerged at 25 to 60 m in the San Diego Trough. The solid line represents brightest flashes; the broken line, average intensity of luminescence; the dotted line, light at the sea surface. Time, 02.40–03.50 h. (from Kampa & Boden 1957)

chemical systems may be readily alterable, still there are probably finite limits to such ability, determined by the nature of the chemical compounds involved, limits which restrict the closeness of correspondence which can be achieved. Indications of this nature already exist for visual pigments, and further work may well clarify the matter.

Another factor which is of importance is the transmission of luminescent light through the sea, and therein several factors are involved. Propagation of light from a small source, such as the light-organs of a fish, is governed by distance, transparency of the water, scattering and the geometry of the light-beam. We have some reason to believe that there are often dioptric structures in the photophore – lens and reflectors – that focus or concentrate the light-beam, causing it to be propagated through a narrow path. If this be not so, and the light-organ be a point source radiating freely in all directions, then its energy decreases with distance inversely as the square of the distance and in conformity with the absorptive characteristics of the medium. Because clear oceanic water transmits blue light (at 485 nm) more readily than it does other wavelengths, luminescent light of this quality is propagated with least diminution.

The distances at which luminescent lights can be perceived depend not only on attenuation in passage but also upon the intensity of the source (radiance), the sensitivity of the receptor, the level of background light, and the amount of scattering of the image. Under ideal conditions a very sensitive eye could just see a bright emitter such as a comb jelly at a distance of 40 to 120 m; in practice, weak luminescent lights probably become indistinguishable at much shorter distances than this, at 10 m or so. Luminescence is usually emitted in short flashes which causes them to be detected more readily, because the eye is usually very sensitive to small flashing lights.

LUMINESCENCE, PHOSPHORESCENCE OF THE SEA AND THE DEEP-SEA SCATTERING LAYER

When Professor G.L.Clarke was measuring the transmissive characteristics of ocean water, by lowering a very sensitive photosensor into the sea, he discovered that below certain depths during the daytime irregular flashes began to show on his recording apparatus. These were the productions of luminescent animals in the ocean depths. Further studies showed that in all the many regions in which he worked, some bioluminescence was recorded either at night or at deeper levels during the day. This was true at all depths, with the proviso that bioluminescence would be detected during the daytime only at depths below which attenuated daylight did not obscure the weak flashes of the animals. The intensity of the flashes, their duration and frequency, vary greatly from place to place, at different depths and times. Measurements made over the continental slope south-east of New York at night revealed luminescent flashes occurring at a rate of 160 per minute at 100 m, 90 flashes per min at 900 m and 1 flash per min at 3,750 m. The irradiance from the brightest flashes was greater than $10^{-3}\mu W/cm^2$. At some places and depths the bioluminescent flashes became so frequent that they provided a level of continuous background light. Luminescence began to exceed the intensity of background illumination at depths of 500 m and ambient light levels of $10^{-3}\mu W/cm^2$. This latter intensity is one 100 millionth of bright sunlight at the surface of the sea. There is little doubt that luminescence is widely used by deep-sea animals as light by which to see.

Marine biologists and oceanographers, analyzing their net collections, have found that many animals in the upper waters, 1,000 m or so, of the sea execute great diurnal movements. In brief, they move towards the surface in the evening, leave the surface and move about randomly around midnight, return to the surface at dawn, and descend into the depths at sunrise. Such movements are known as vertical migrations. Prawns and euphausiids, for example, move through distances of 400 m or more. It is not to be thought that all species come to the surface at dark; rather, the animals are stratified in depth, some species occurring deeper during the daytime than others, and moving upwards but still keeping well below the surface in the evening. The factors which regulate vertical distribution, determine its limits, and regulate its progress are of

great interest to oceanographers and physiologists alike.

A second method of investigating the diurnal movement of the oceanic animals (plankton and nekton) is by means of echo-traces, a technique which has also been exploited with great success by commercial fishermen to locate shoals of food fishes. The sound signal from an echo-sounding machine is reflected from the bottom and from aggregations of marine animals in the water column. The latter appear on the records as mid-water traces; often the traces move up and down daily because the animals producing them are making such movements (figure 48).

There is general agreement that the principal environmental factor determining the timing of these movements is light. The concept is that the migratory organisms tend to congregate in a zone of optimal light-intensity and, seeking it or conforming to it, they move downwards in the early morning to the level where it occurs, and follow it upwards in the evening as the light weakens and it moves towards the surface.

Vertical movements may be arrested, however, at levels where there are abrupt temperature changes, owing to warmer water above or colder water below.

The concept of an optimal light-intensity, although useful, may be an over-simplification of the complex factors regulating vertical migration. It has been observed that the deep scattering layer sometimes ascends so quickly during the evening that the animals reach levels when the light ephemerally is many times brighter than the 'optimal' intensity of the level where they hovered during the daytime. It is conceivable that animals are sensitive and respond to the rate at which the light intensity changes; furthermore, they probably have inherent rhythms, leading them to carry out movements at certain times each day (so-called circadian rhythms). The exact timing and phasing of the internal rhythm may need to be set by external physical events, such as nightfall or dawn, but it is capable of marching independently and ordering internal events, even in the absence of external change, for some time.

The importance of natural illumination on vertical migration has been demonstrated by the technically brilliant investigations of Dr Boden and Dr Kampa. Employing a submersible photosensor, they followed the descent and ascent of one particular scattering layer in the deep water near the Canary Islands. They measured the light-intensity at the horizontal level favoured by the animals at the beginning of their descent or ascent, and they determined the change of vertical position of that intensity throughout the migration. Marrying their results, they found that their particular sonic scattering layer kept very close to the isolume of $5 \times 10^{-4}\mu W/cm^2$ (at 474 nm), that is, it

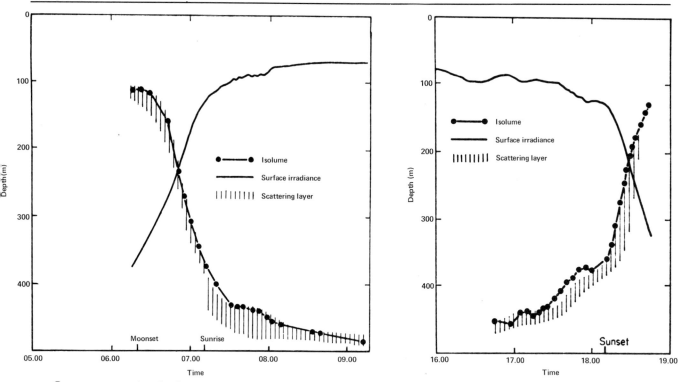

215. Curves representing the downward and upward movements of a migratory sonic scattering layer near the Canaries, at dawn and dusk, respectively. The linked points show the movement of an isolume ($5 \times 10^{-4} \mu$ W/cm² of 474 nm) (an isolume is a line connecting points of equal light-intensity) (from Kampa & Boden 1967)

followed a level $5 \times 10^{-4} \mu$W/cm² up or down as it progressed through the water column (figure 215). Others have found that it is possible to force down the sonic scattering layer or delay its upward movement by turning on bright submerged lights.

Many of the organisms carrying out vertical migrations are luminous. Dr Boden and Dr Kampa have measured luminescence closely associated with a sonic scattering layer off Hawaii; they observed, moreover, an increase of luminescent flashing when the layer was migrating at twilight and at night. Luminous intensity was $10^{-4} \mu$W/cm², and λ max was near 478 nm.

No one submarine instrument provides all the information that the biologist needs, and neither bathyphotometer (photosensor) nor echo-sounder reveals what animal or animals are migrating and flashing. Concomitant net-hauls at the levels where the phenomena originated have made it likely that lantern-fishes, euphausiids, and siphonophores are responsible, and the deep-sea submersible vessel *Alvin* has in one case enabled scientists to catch and observe the animals that do the scattering: in this instance it was a small lantern-fish (*Ceratoscopelus maderensis*) (figures 71–72).

The great majority of luminescent animals shine not continuously but in short or long flashes. Fish and some squid shine spontaneously, of their own volition, when schooling or during their breeding activities. Others shine only when excited by an external stimulus. Comb-jellies and scale worms flash rhythmically when they are jostled and disturbed, and so do the luminous unicellular algae. Some of the great luminous displays observed at sea are produced by dense concentrations of such organisms in surface waters, where they are stimulated by turbulence and wave-action. The luminescence sometimes becomes inhibited when the animal is illuminated, normally during the daytime, for example comb-jellies. Or it may be evoked by a brief flash, one dark *Pyrosoma* replying by a flash to the glowing of another (figure 213). Luminescence is a fascinating field of research, but it is also a very difficult one requiring great patience and persistence.

COLOURS AND PIGMENTS

In that excellent book, *The Depths of the Ocean*, the two oceanographers Murray and Hjort commented upon the coloration of pelagic animals and observed a general tendency for their colours to change with depth. They noticed that many small animals living at or very close to the surface are generally transparent or blue in colour. Colourless animals indeed occur in deeper waters, but down to 500 m, in the mesopelagic zone, silvery or light-coloured fishes are abundant.

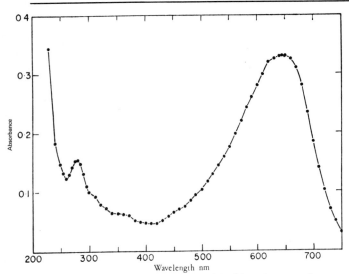

216. Curve showing the absorption of the blue pigment of a copepod *Pontella fera*, a surface-living species (see also figure 187) (from Herring 1967)

At still deeper levels many of the creatures are red, brown or black, e.g. reddish brown jellyfishes (figure 117), red nemertines (figure 159), arrow worms, crustaceans (figure 195) and squid (figure 226), and brown and black fishes. Especially striking are the red prawns of the middle layers, and the uniform dark tones of abyssal fishes. There is a host of problems posed by these and similar observations and generalizations, few of them easy to investigate or explain. Among them may be mentioned the nature of the pigments, their variation, functional significance, and the effects, if any, of environmental conditions upon them.

Dr P.J.Herring has made a special study of the blue pigments of animals that occur in the extreme surface layer of the ocean. This zooplankton is now known as neuston and, especially in warm tropical waters, many of the animals are characterized by blue colours. He has found that most often the blue pigments are carotenoproteins, containing the carotenoid astaxanthin combined with protein. Carotenes are that class of yellow or reddish coloured fat soluble substances that confer a reddish colour upon shrimp, boiled lobster and carrots. In the integument of the lobster, the carotenoid, in that animal astaxanthin, is combined with protein, in which form it appears blue; boiling destroys the bond whereupon the red astaxanthin, being released, stands revealed in its natural colour. The blue carotenoproteins of animals of the zooplankton are very similar to that of the lobster; they preferentially absorb light of long wavelengths, above 550 nm (figure 216).

Among pelagic molluscs of the neuston there are other, as yet unidentified, pigments that are blue or purple in colour, for example in the gastropods *Janthina, Atlanta* and *Creseis.*

More rarely the blue colour may be of structural origin, i.e. result from the interplay of light at the surfaces of thin films (interference) as in mother-of-pearl, or be due to scattering of light-rays, like that which takes place in the blue iris of humans and in blue feathers. A structural basis is certainly responsible for the blue colour of the backs of epipelagic fishes and a blue squid *Onychia*; and interference from thin crystalline plates gives rise to the brilliant iridescent colours of a small copepod *Sapphirina,* which flashes in the water like a cut jewel.

Various explanations have been offered as to the role of these blue pigments in animals of the neuston. At the surface of the sea, animals may remain inconspicuous either by being transparent, but this is not always practicable because of their structure and habits, or by taking on a blue tint like that of the surrounding water. A noteworthy feature of this environment is the high level of light intensity and the abundance of ultra-violet rays, especially in low latitudes. Therefore, it is possible that the blue pigments may sometimes have a role in protecting the organism against the deleterious effect of intense radiation. Alternatively, the carotenoid or protein moiety may have functional importance in metabolism, and the carotenoprotein may be more stable than either component in the free state under conditions of high temperature and illumination.

There is another organ in which screening pigment has a protective function, and that is the lens of the eye. In surface fishes and cephalopods the lens contains a yellow pigment that cuts off far blue and near ultra-violet light to which they are exposed in the upper reaches of the sea. Ultra-violet light is rapidly absorbed by sea-water; deep-sea fishes are not exposed to it and, correspondingly, their lenses lack screening pigment.

In deeper waters, between 150 and 500 m some fishes are certainly very silvery, especially lantern-fishes and hatchet-fishes, and even black stomiatoid fishes such as *Chauliodus* and *Stomias* exhibit a silvery or bronze sheen along their flanks. Before considering these fish in particular we may consider the general nature of reflecting layers in marine animals.

The iridescent or silvery coats of marine animals are composed of piles of very thin flat platelets. If these are sufficiently thin, of a composition to retard light-rays, and properly spaced, they cause the light-rays reflected from their consecutive surfaces to be in phase and the result is constructive interference (figure 217). In one version of this system, the platelets and the spaces between them each have an optical thickness of a quarter of a wavelength. Light-rays reflected from the

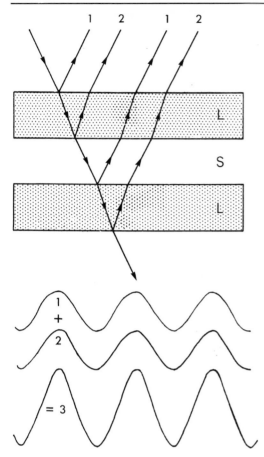

217. Reflexion from thin plates. Platelets (L) are one quarter of a wavelength thick, as are the spaces (S) between them. The total path-length within the platelets is one half wavelength. The ray (1) reflected from the surface of the platelet (L) suffers a phase shift of one half wavelength. Consequently, reflected rays 1 and 2 are in phase with one another (crests and troughs coincide in space and time), and the effects are additive. Moreover, similar effects take place throughout the pile of lamellae

many superposed platelets are additive, and a high degree of reflection is achieved; the resulting brightness of the surface may resemble that of polished metal. Places where constructive interference is utilized are silvery exteriors of fishes and squid, tapeta lucida or reflecting layer of eyes, and reflectors of photophores.

Because the reflecting surfaces, being plane, act like little mirrors, the orientation of the platelets becomes of importance, and animals have hit upon the trick of employing them in various ways, to diverse ends. In the eyes of fishes the platelets are so aligned that they lie mostly normal to the light rays reaching them. Hence they reflect light rays back into the same sensory elements (the rods) which the light traversed in its first passage through the retina. In photophores they

are oriented so as to cast the light through the aperture of the organ. In the skin of fishes and squid they occur in reflecting cells called iridophores which give rise to brilliant metallic blue and green colours.

The beautiful silvery fishes of the sea owe their metallic lustre to layers or sheets of reflecting cells called guanophores. In these cells the reflecting platelets are thin crystals of an organic substance, guanine. This material, because of its high refractive index (n = 1·80), strongly retards light, and very thin lamellae made of it, only some 70 nm in thickness, act as plates of a quarter wavelength. It is interesting that this substance should be used so widely by aquatic animals to achieve interference effects. For constructive interference to occur there must be an appreciable difference between the refractive index of the reflecting lamella and that of the bounding medium. In the case of guanine this value is about 0.46 (being the difference between guanine and protoplasm). On the other hand terrestrial animals, such as insects and birds, to achieve interference effects generally make use of organic lamellae separated by air spaces; again, the difference in refractive indices is about 0.50, the organic materials having refractive indices n = 1.50, and air n = 1.0. Being highly insoluble in water, the platelets of guanine are very stable entities. Guanine is a constituent of nucleic acid, an essential element of living cells. Animals usually possess a complete series of enzymes capable of degrading excess guanine to urea or ammonia, in which form it is excreted. On the other hand, those that employ it for structural coloration tap off some of this material as required, so to speak. It must be carried to the reflecting cells in some soluble form, and then laid down under enzymatic control, in precise planes defined with great rigidity.

Some pelagic fishes having bodies circular in cross-section, notably the great sharks, make use of counter-shading to render themselves inconspicuous. Their upper surfaces are dark grey or blue and pigmented, whereas the lower surfaces are dull white and reflect diffusely. The arrangement is such as to make the animal somewhat uniform in tone. The dark upper surface of the fish, viewed against the dark water-mass below, is difficult to distinguish. Another system is employed by silvery fishes which have reflecting cells aligned more or less vertically. The platelets in these cells act as fine mirrors. Consequently, from whatever direction the fish is viewed, it tends to appear about as bright as the water-mass behind it, against which it is seen. These fishes usually have dark pigmented backs, which reflect little light and are no brighter than the dim upwelling light of the ocean depths. The platelets of the skin, reflecting light from an angle of inclination equal to that of the background illumination, cause the

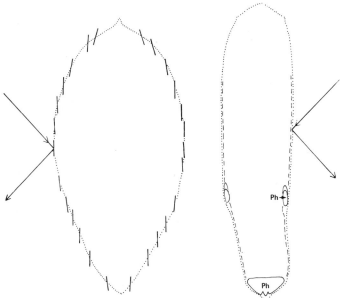

218. Orientation of reflecting platelets in the skin of two silvery fish, a herring (left) and a hatchet fish (right). Strokes represent reflecting platelets. Ph, photophore (light-organ)

219. Orientation of the reflecting plates in the eye of a deep-sea shark *Deania calcea* (figure 184). Reflecting plates are depicted in their several orientations about the wall of the eyeball, and light-rays are shown entering the eye on the right. Wavefronts, refracted by the lens, are mostly perpendicular to the surface of the plates (from Denton & Nicol 1964)

fish to melt into the background. To achieve these effects, fishes with rounded sides have the platelets set at considerable angles to the surface of the body in the upper and lower flanks, e.g. saury pike *Scomberesox*, lantern-fish *Myctophum* (figure 209). In thin fish with vertical sides, such as the hatchet fish *Argyropelecus* (figure 227), the platelets everywhere lie parallel to the surface of the body. Either arrangement brings the platelets to the vertical position (figure 218). The agency controlling the precise orientation of the platelets is not known but it is probable that the reflecting cell is oriented first, and the platelets then are deposited with reference to predetermined cell axes. The salmon parr, while it is still in its home stream, before it has laid down silver, contains oriented guanophores (cells destined to produce guanine) beneath its scales, already anticipating the changes which will occur. Then, when the time comes to become a smolt and go down to the ocean, the guanine platelets are deposited in their pre-ordinated places and a silver sheen grows over the surface of the fish.

Below the mesopelagic realm where daylight weakens and fails, silvery coats are lost and fishes become drab and uniformly dark. Because the little light present is non-directional, silvery sides are of no avail. All light of the bathypelagic realm is that which is generated by

luminescence, feeble fitful gleams, randomly distributed. Yet, as we have discovered, these lights are frequent and ubiquitous, and can serve to reveal animals one to another. Therefore, to remain inconspicuous, bathypelagic animals often assume dark coverings that reflect very little light. The reflectance of the skin of dark *Cyclothone* is very low, only some four per cent. Or they are reddish in colour, e.g. prawns, the integument of which contains a red carotenoid astaxanthin. Now, the reflectance of blue and green light in the spectrum region 400 to 560 nm by red prawns is very low, around three per cent. This is the region of the spectrum containing most of the energy of submarine luminescence, and to all intents and purposes these creatures are virtually black in the light by which they are normally illuminated (figure 220). From a consideration of these various facts, and the low level of luminescence, it is unlikely that dark or red bathypelagic animals can be seen at distances of more than a half metre from the source of illumination. That is, an animal using its light organs as a torch would not see another fish or shrimp if it were distant by more than half a metre.

Analyses made of silvery fishes of the upper well-lit regions of the oceans have shown that they contain up to a milligram (1/1000 gram) of guanine in a square cm of skin. A silvery fish about 20 cm long might contain

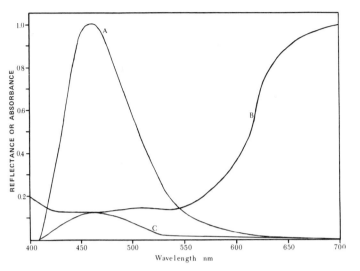

220. Spectrum curve of luminescence of *Cypridina*, A, reflectance of a red prawn *Acanthephyra*, B, and a product curve, C, which shows the quality of the light reflected by the red prawn from any such luminescence. The red surface of the prawn reflects blue luminescence poorly (A and B are normalised at maximum wavelength)

up to one gram of guanine in its silvery layers. This may seem to be very small quantity to produce the striking effects which are manifest, but then we should remember that the guanine is laid down in very thin platelets in a layer of microscopic thickness. Guanine, as might be expected, disappears from the skin of bathypelagic fishes, but it reappears again in another role in bathybenthic fishes, viz. in the tapeta lucida of deep-sea sharks, rays and rabbit fishes. The tapetum of the eyes of these fishes is like burnished aluminium, it is rich in guanine, and it reflects over 85 per cent of incident light. The fish have golden retinae, rich in visual pigments which absorb much of the incident light (some 70 per cent). That which passes through is reflected back by the tapetum, giving the retina an opportunity to absorb more of the light on its second passage (figure 219). As a result of this concatenation of events the eye is able to absorb about 90 per cent of the light falling upon the retina. The tapetum therefore confers an advantage by increasing the light-absorbing power of the retina by about 20 per cent. These sharks and their allies of which we are speaking live at great depths, below the level of effective light-penetration, and the illumination which they are utilizing for vision must be derived from the luminescence of other animals (or from themselves, because several bathypelagic sharks carry their own light-organs).

It is interesting to compare the eyes of these deep-sea sharks with those of bony fishes from the same environment. There is no reflecting layer in the eyes of the latter, but their retinal pigments are far more dense, absorbing 90 per cent or more of the incident light. Because the shark's eye has an internal reflector, it is almost as efficient in absorbing light as the eye of a comparable bony fish, although the eye of the latter has a more deeply pigmented retina.

A further example of promising lines of biochemical investigations which are centred on marine animals is the investigation of vitamin A distribution. One carotenoid, ß-carotene, is the principal source of vitamin A. ß-carotene is synthesized by plants and transmitted along the food-chain from herbivores to carnivores. Animals have a capacity for transforming ß-carotene to vitamin A in the intestine. During the course of investigations of carotenoids and vitamins of pelagic animals, it was discovered that euphausiids, especially, and decapod crustaceans contain large amounts of both astaxanthin and vitamin A. So rich are the krill (euphausiids) in vitamin A that they can supply enough of that vitamin to baleen whales without the necessity of conversion. Most of the vitamin A and much of the astaxanthin occur in the eyes of krill. A small fraction of the vitamin A aldehyde gives rise to the visual pigment rhodopsin, essential for vision, but a large part of it must have other as yet unappreciated functions.

FLOTATION AND ACHIEVEMENT OF NEUTRAL BUOYANCY

The specific gravity (s.g.) of sea water ranges from 1·028 to 1·021. (Specific gravity is the ratio of the weight of a substance to that of an equal volume of pure water at 4°C.) Protoplasm is heavier (s.g. 1·03 to 1·10), skeletal material even more so (s.g. 1·50); consequently, without compensatory devices animals will sink in sea-water. To stay at a certain level, counteracting the pull of gravity, animals can swim actively, a process requiring the expenditure of energy, and this is what strong swimmers such as euphausiid and acanthephyrid shrimps do. These animals have good muscles and are strong swimmers.

Another way of reducing the pull of gravity is by diminishing the density of tissues. Many deep-sea animals have very gelatinous tissues e.g. nemertines, squid, octopods, salps and fish. Jellyfish are 95 per cent water, and contain less than one per cent of organic matter and have a salt concentration equivalent to that of the surrounding sea-water. Moreover, many deep-sea fishes, lacking air-bladders, are fragile and lightly built, their muscles are slight and their bones fragile. The low level of protein and diminished hard parts

221. Buoyancy balance sheets of the bathypelagic fish (without swim bladder) *Gonostoma* and a coastal fish *Ctenolabrus* (with swim bladder). Positive values (+) are for components heavier, negative values (−) for components lighter than sea-water that they displace (weights per 100 g of fish). Lighter components tend to float the fish, heavier components sink it. *Dil. Flu.*, dilute body fluids; *Sk + C*, skeleton, etc. (from Denton & Marshall 1958)

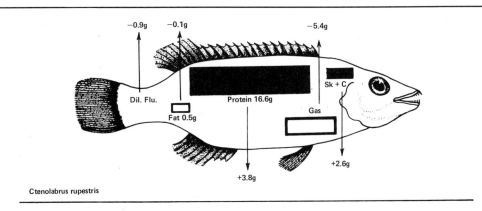

Ctenolabrus rupestris

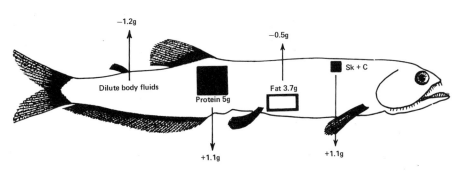

Gonostoma elongatum

depress the specific gravity (figures 203, 221). An animal cannot by this means diminish its specific gravity until it is equal to that of the sea-water, it can only reduce the rate at which it sinks. Of course, it also loses strength, cannot exert sustained effort and becomes passive.

The third way is to introduce a compensatory factor to make the creature neutrally buoyant. Substances that have specific gravity less than protein or sea-water are accumulated; in sufficient concentration they reduce the specific gravity of the whole animal to that of sea-water, i.e. it becomes neutrally buoyant. The means by which it does so are referred to as flotation mechanisms.

Sea-water is a complex mixture of salts some of which (the divalent cations and anions calcium, magnesium and sulphate) are heavier than others (sodium, potassium and chloride). All marine animals have ionic regulatory mechanisms whereby they maintain the body salts at concentrations different from that of the sea-water about them, although total osmotic concentrations inside and outside generally are the same. By reducing the heavy ions, especially sulphate, replacing them by lighter ones, gelatinous animals achieve neutral buoyancy, notably jellyfishes, comb-jellies, planktonic gastropod molluscs and salps.

Animal fats have specific gravities below unity (cod liver oil is 0.93), and the deposition or accumulation of fats can contribute to achieving buoyancy. Oil droplets in unicellular algae and deposition of fat in fishes such as mackerel, besides serving as energy stores, lower

the specific gravity slightly. Noteworthy are deep-sea sharks *(Centroscymnus)* which have enormous livers, four-fifths of which is a hydrocarbon oil, squalene, having a specific gravity of only 0.86; the enormous oil store serves to counterbalance the weight in excess of sea-water of other tissues. These bathypelagic sharks probably live near, but spend much time swimming freely off, the bottom. Sperm whales also store large amounts of low-density fats (esters of long-chain aliphatic alcohols and fatty acids) which can be important during deep dives when the whale's air spaces are compressed and buoyancy reduced.

Cranchiid (figure 176) and other squid are neutrally buoyant. They contain large cavities filled with fluids of low specific gravity (about 1.01). The low density is secured by accumulating ammonia to substitute for heavier ions.

Various animals have air pockets for flotation. There are various siphonophores which have gas-filled bladders enabling them to stay neutrally buoyant, neither ascending nor descending (*Nanomia*), or float at the surface (*Physalia*). That of the Portuguese man-of-war (*Physalia*), projecting above the surface, also acts as a sail. The physiology of these floats has invited much attention because it has been discovered that they contain a high proportion of carbon monoxide, 90 per cent in *Nanomia* and 11 to 21 per cent in *Physalia*. The carbon monoxide is secreted by a gas gland in the floor of the float (figure 222). It is peculiar that a toxic gas should be employed for this purpose; an overriding advantage seems to be that it is far less soluble in water and therefore diffuses less readily than carbon dioxide.

Two pelagic cephalopods have chambered shells containing gases, namely the pearly *Nautilus* and *Spirula* (figure 174). Both these animals are neutrally buoyant. The successive chambers are separate and fully enclosed, and a tube runs through the centre of them in a spiral course. The first chamber contains fluid; the others, gas at a pressure of nine-tenths of an atmosphere. When a new chamber is formed it is, at first, filled with fluid; the salt and water are pumped out and gas diffuses in from body fluids through permeable regions of the central tube.

Fishes having air-bladders are neutrally buoyant; the air-bladder is lost in benthic and many deep-sea species. When the fish is in equilibrium with the pressure of the environment, the volume of gas in the bladder displaces sufficient fluid to counter-balance its weight in excess of sea-water. As it rises in the water and the pressure of the water decreases its air bladder expands, and it must needs absorb gas from the bladder to reduce volume and buoyancy; and conversely, when the fish descends.

Gas is secreted into the bladder by a gas gland in the wall. The gland is richly vascularized and has a *rete mirabile* or network of parallel arterial and venous capillaries, which makes use of the counter-current principle to maintain a high concentration of oxygen in the gland. By this principle, oxygen adventitiously absorbed from the gland by the venous blood, which would be lost into the blood system, diffuses back into the arterial blood passing to the gland. Two salient problems connected with air-bladders of deep-sea fishes are the efficiency of the secretory mechanism under great pressures, and the speed of volume-adjustment. Mesopelagic fishes having air bladders occur down to depths of 500 m, at which the pressure is 50 atmospheres, against which gas is secreted to bring about expansion of the organ. Bathybenthic and bathypelagic fishes, macrurids and brotulids, at even greater depths, down to 4,000 m, have air-bladders. Lantern-fish move upwards and downwards at rates of about two metres per minute, which gives some idea of the speed of adjustment required.

EFFECT OF PRESSURE ON ORGANISMS

Pressure and temperature are two factors which change with depth, in opposite directions. There is an increase of pressure of about one atmosphere (atm) for each increment of depth of 10 m. Great changes of pressure affect the viability of organisms and the performance of tissues, and there has been some question whether

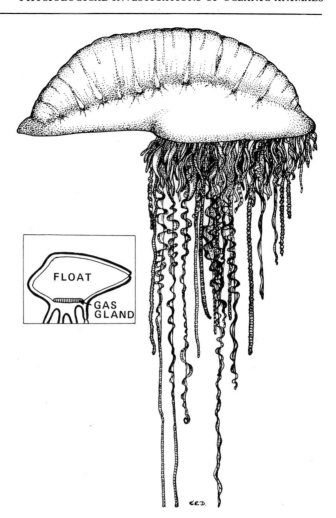

222. The siphonophore *Physalia,* the Portuguese man-of-war, may have a float up to 20 cm long, and tentacles that can inflict a very painful sting hang below it. In the inset is shown the position of the gas-producing gland (after Copeland 1968)

changes of pressure or of temperature cause the death of animals hauled up from deep water. The effects of pressure are investigated for the light they cast on motility and contractile processes, and for the changes they produce in molecules. There is interest in the different abilities of animals to tolerate changes of pressure, and in pressure-sensitivity, whereby an animal can appreciate depth and adjust its activities accordingly. Pressures in the ocean range from one atm at sea level to 1,000 atm in the deep trenches.

Examples of work in this field are experiments on deep-sea prawns *Systellaspis* (figure 169) which consume more oxygen when they are subjected to more pressure, an increase from 33 to 100 atm. Raised pressures reduce the degree of gelation of protoplasm and affect cell-movements and cell-division. Small increases of pressure, of 10 to 20 atm, applied to animals originating in shallow water cause a temporary increase of activities (of heart, body movements, cilia). But high pressures cause muscles to contract, respiration to decline; they finally cause dissolution of the cells.

Subjected to high pressures, solids and liquids are only slightly compressed, the volume of water decreasing by five per cent at 1,000 atm. Probably a major effect of raised pressure is on very large molecules, especially proteins, brought about by a decrease of molecular volume. Many chemical reactions are accompanied by an alteration of molecular volume, e.g. reproduction of protein molecules, and the union of enzyme and the substrate upon which it acts. The cell is an extremely complex structure containing a host of disparate enzymes and delicate protein membranes, working in concert and interrelated in function. Disturbances of the structure and specific properties of these systems, say by opposing the unfolding of protein molecules (a process accompanied by an increase of volume), can lead to blockage of essential metabolic processes.

Effects of changes of pressure upon animals from great depths have not been well studied as yet. On the other hand, it is interesting that when the pressure-resistance of animals from shallow water is tested, those from groups that are spread over great vertical ranges prove to be the most pressure-resistant.

One of the surprising discoveries relating to the behaviour of marine animals in recent years is their pressure-sensitivity. Many small benthic and plank-tonic creatures react to small changes of pressure by altered movements. Small crustaceans are sensitive to changes as small as 5 cm of water (0.005 atm); for a fish without an air-bladder the threshold sensitivity is 25 cm of water, and for one having an air-bladder, 0.5 cm. When the pressure increases the animals may respond by moving upwards; when it decreases, downwards; and there are many variations of this basic pattern revolving about other environmental features, such as light and water currents. It is likely that some planktonic animals maintain themselves within a suitable vertical range by this means. Stimulated by the increase of pressure encountered on sinking they become more active and swim upwards; and *vice versa*. Although changing light-intensities are the major factor regulating vertical migration in the open sea, it may be that pressure-sensitivity acts to keep the migrants within certain limits.

RETURN TO THE SEA

Many groups of higher vertebrates, from crab-eating frogs to whales, have made the return journey to the sea, reversing the course pursued by their distant ancestors. Because they are air-breathers and because some of them are warm-blooded, they have encountered special problems and difficulties in adapting themselves to an aquatic existence. Air-breathing vertebrates which have adopted a pelagic existence wholly or in some part of their yearly cycle are sea turtles, sea snakes, oceanic birds (fulmars, petrels, shearwaters, albatrosses, penguins and kittiwake gulls), seals and whales. Some, notably birds, live above or at the surface; others, especially whales, dive deeply and for long periods. Functional adaptations of special interest to physiologists are: resistance to pressure; breathing and respiratory arrest whilst diving; thermal regulation; water balance and water conservation; signalling and perception underwater. Some of these topics are reviewed in the paragraphs following immediately and elsewhere in this chapter.

In most vertebrates the salt concentration of the blood and of the tissues is only about half that of sea-water. Since they have no access to fresh-water, and they are constantly taking in salt, either in food or by swallowing water, they suffer from a real water-shortage. To get water they must needs pump out salt, or as much as is not needed, and retain the right amount of water for their needs. Marine reptiles and birds do just that by means of a nasal salt gland that secretes a fluid very rich in salt (sodium chloride). Indeed, so efficient is this mechanism that when wholly oceanic birds are held captive, they require salt water to stave off salt depletion. Marine mammals presumably have salt pumps in the kidneys. Man does poorly when limited to sea-water only as water supply.

Diving mammals, seals, porpoises and whales are capable of staying underwater for long periods, 15 minutes is normal, yet they carry only a slightly larger oxygen-store than non-divers do. When air-breathing vertebrates dive beneath the surface, not only do they cut themselves off from an external supply of air, but they also hazard the vitality of several internal organs which are highly dependent upon a steady supply of oxygen. Diving mammals and birds, when they dive, slow the heart-rate to about one-tenth normal whilst they maintain the arterial blood-pressure constant. At the same time, by differential constriction of blood vessels, blood is restrained from circulating in some organs, especially the muscles, gut and kidneys, which are tolerant of oxygen-lack, and the slow heart continues to supply blood to the heart muscle and brain, which are especially sensitive to anoxia. A form of anaerobic respiration continues in the other tissues (partial utilization of energy-stores in the absence of oxygen), oxygen debt is built up, which is paid back by heightened respiration for a period after the animal emerges.

Phenomena of this nature have been observed in diving mammals, porpoises, seals, in birds and reptiles. Indeed, similar responses accompany asphyxial conditions in fish when out of water, e.g. codfish, and flying fish when they soar.

Whales not only dive long but also deep, down to 1,000 m on occasion. When man dives deeply, provided with gas under pressure, he is liable to caisson disease (bends) if he decompresses when returning to the surface too quickly. Caisson disease is brought about by bubbles of gas coming out of solution in the small blood vessels. Cetaceans when diving seem to be spared this difficulty. They carry no excess supply of gas, and no troublesome nitrogen in excess of atmospheric pressure. The gas in their internal spaces is compressed, but the amount of nitrogen present in solution is insufficient to become damaging to their tissues as they ascend to the surface.

Cetaceans are found in all seas from tropical to polar regions, and their size spans the range from the small dolphin to the mighty blue whale. They are, of course, warm-blooded animals and, because water is such a good conductor of heat, they have the problem of keeping warm in water which is always colder than they are themselves. The conservation of heat is achieved by laying down a thick layer of blubber or fat. This reaches 45 per cent of body weight in the porpoise, and 25 per cent in the whale. Heat loss is also a factor of size; large whales have a larger volume relative to surface area than do porpoises; therefore they lose the heat produced less quickly through their skin. It has been calculated that whales actually carry twice the amount of blubber needed for insulation. On the other hand the small porpoise, in cold waters, must use up more energy to keep warm, and it has a higher metabolic rate than the whale. The desired condition is not one of merely conserving heat, but of maintaining a steady body temperature. Heat is lost across the skin and in expired gases. There are mechanisms in the cetacean skin, tail and flippers for regulating heat exchange, and when a porpoise becomes cold it shivers to increase its heat production.

Having noticed certain features of thermal regulation of marine mammals, we may take this to be a convenient place to consider heat conservation in two groups of fishes – universally acknowledged to be cold-blooded.

Heat conservation and thermal regulation have been achieved in some of the larger marine fishes, with resultant increases in muscular performances. The body temperatures of mackerel sharks (mako and porbeagles) are maintained at 7 to 10°C above those of the surrounding water. These are large sharks, the porbeagle reaching about 2·5 m and the mako about

4 m; they are swift active fish capable of great bursts of speed. Now, the power required for swimming increases rapidly with speed, and the power available from the swimming muscles increases as the temperature rises (an increase of 10°C increases power threefold). By conserving metabolic heat the sharks are enabled to obtain the additional power necessary for swimming at high speed. The various tunas are large oceanic species (Table 11), the blue-fin reaching a length of 3 m and a weight of over 400 kg. They maintain their body temperatures at 25-30°C, within 5°C, over a range of water temperatures of 10-30°C. The warmest regions of the body are areas of red muscle in either flank, which are the muscles used for continuous swimming at cruising speed (figure 223). Tuna swim constantly to respire, they are swift and

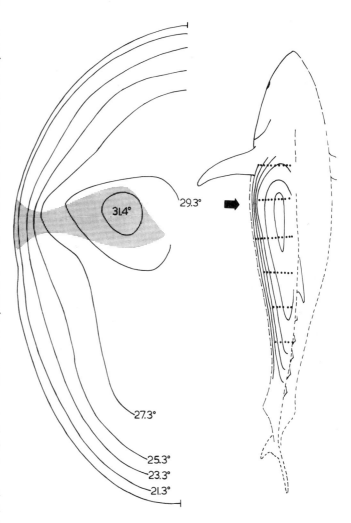

223. Temperature-distribution in the body of a blue-fin tuna. Left, section through the body at the level of the arrow, showing temperatures at several levels. Dark stipple is the red muscle mass. Lines on the right hand figure are isotherms (lines of equal temperature) at intervals of 2° C. (from Carey & Teal 1969)

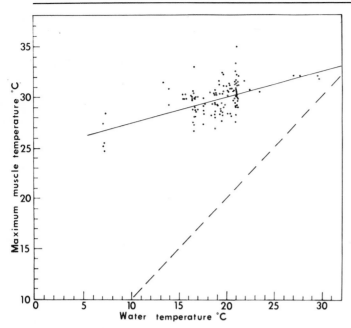

224. Muscle temperatures of blue-fin tuna caught at different water temperatures (solid line). The broken line shows what the temperatures of the fish would be if it were not capable of conserving metabolic heat (from Carey & Teal 1969)

they make long migrations. Temperature regulation permits them to range so widely over cold and warm oceans (figure 224). Heat conservation is favoured by a special arrangement of blood vessels in the lateral body wall, in the vicinity of an area of red muscle. Major vessels supplying this area lie beneath the skin, and a complex network of parallel small vessels extends deeply into the flank. This network, called a *rete mirabile*, consists of many small arteries and veins lying in parallel. The anatomical arrangement conforms to a counter-current system, warmer venous blood proceeding outwards in the vessels passing heat into the colder arterial blood passing inwards; the exchange causes heat to be conserved, elevating body temperature.

VENOMOUS COELENTERATES

There is much interest today in the pharmacology of marine organisms, either because of their adverse effects upon man or as sources of unusual biochemicals of service to medicine and biomedical research. Rarely are the animals examined oceanic, particular exceptions being pelagic coelenterates, especially large jellyfishes and Portuguese men-of-war.

Coelenterates contain batteries of stinging cells or *cnidoblasts*, which are small sac-like structures containing an involuted thread. When the cell is stimulated some change, as yet poorly understood, occurs in the cell which causes the thread to be discharged. In this process the thread extends itself inside out, somewhat as if one drew out and extended the turned-in finger of a rubber glove. The thread is hollow and open at the tip, the sac of the cell contains poison or venom and, if the thread succeeds in penetrating something – the surface of another creature – when it is ejected, venom is squeezed into the wound.

Cnidoblasts are aggregated densely on tentacles and are employed for offence and defence. Especially powerful and venomous cells are borne by sea wasps (figure 156) and *Physalia* (Portuguese men-of-war). The tentacles of these animals can inflict very painful wounds upon man, and they immobilize and kill fishes and other small animals. The tentacles thereupon shorten, bend, and convey the prey to the mouth.

It is generally believed that the cnidoblasts are independent effectors, that is, they respond directly to a stimulus by discharging their threads, without the mediation of a nervous system conveying signals from one point to another. The cell is sensitive to touch and to chemical agents, lipids or fatty materials adsorbed on proteins. This is a combination of stimuli the like of which the jellyfish encounters in its prey.

Being very small, it is not easy to isolate cnidoblasts in quantity for chemical analysis. The toxin contains a variety of substances – quaternary ammonium compounds, especially tetramine, 5-hydroxytryptamine, histamine and histamine-releasers, and several toxic proteins. The latter are responsible for the paralysing and lethal effect of the toxin, which acts directly upon nerve cells (those producing the transmitting substance acetylcholine) at their endings. 5-Hydroxytryptamine is a strong pain-producer; the effects of the other substances in the toxins, in terms of the concentrations in which they occur, are still open to clarification.

In the preceding pages certain aspects of physiological investigations of marine animals are described. They should be regarded as examples, not as exhaustive listing. They show, after a fashion, the kind of problems and ventures that capture the attention of zoologists desirous of understanding how the functional systems of deep-sea animals operate. Because of the brevity of the accounts and the complexity of many of the matters described, the descriptions do scant justice to the efforts and the publications of the capable scientists who have carried them out.

9 The Animal Crop

Extension of Commercial Fisheries into the Deep Ocean *C. T. Macer*

Since the Second World War, the total annual world catch of marine fish has been increasing steadily by about seven per cent each year (faster than the growth of the world's population) and is now approaching 50 million tons (figure 225). This quantity may seem very large but when one takes account of the vastness of the oceans, it works out at only about 0.13 tons (286 lb) per sq km of sea surface or 0·03 tons (66 lb) per cubic km. These figures give a rather misleading impression, however, because well over 90 per cent of the catch is taken over the relatively shallow shelf areas which border the land masses, and these shallow regions comprise only about eight per cent of the oceans (figure 7). Thus, over 92 per cent of the oceans contributes very little to the world catch, and indeed the tuna are virtually the only truly oceanic fish heavily exploited.

These facts and figures pose two obvious questions – what are the reasons for the restricted distribution of present-day fisheries, and are there vast untapped living resources in the deep oceans?

The close association of the world catch with the continental shelf is due to a number of factors. An obvious one is that it is more economical to fish close to one's own coast, because of the reduced steaming time to and from the grounds, and because relatively small boats with low running costs can be used. But although this may have been an important factor in the past, the modern fishing fleets of some of the major fishing nations are very far-ranging and able to exploit almost any part of the ocean. Another factor is that some fish (the so-called *demersal* species) often depend upon the sea bed for food or shelter. Roundfish such as cod or haddock often eat invertebrates (worms, molluscs and brittle stars, for example) which live on or in the sea bed. Flatfish such as plaice and sole also have a diet of bottom animals and, in addition, they use particles from the sea floor to cover themselves as a form of camouflage. Other species such as the weever fish and sand eel or lance actually bury themselves more or less completely.

In the deep ocean, the sea bed cannot support these species because (a) food organisms are not so plentiful, (b) oxygen concentration is low in the muddy substrates, (c) constant darkness interferes with feeding, in which vision plays an important part, and (d) pressure itself may have some effect at great depths. These factors do not affect the so-called *pelagic* species, such as herrings, anchovies, sardines and jack mackerels, which spend much of their life in the upper water levels. One might therefore expect these species to be abundant in all areas of the ocean; yet, with the notable exception of the jack mackerel, they also are largely confined to the shelf and slope areas. As figure 225 shows, it is the pelagic species which comprise the major part of the world catch and are largely responsible for its upward trend.

The deep ocean thus appears to be rather sparsely populated compared with shelf areas, though it must be admitted that present-day fishing techniques may not be entirely suited to the demands of this environment and so tend to under-estimate potential fish resources. In general, the fish which inhabit the deep oceans are quite different from the shelf species. They include pelagic fish such as tuna, bonito, sharks, flying fish, saurys and jack mackerels, and species such as rat-tails, lantern fish and hatchet fish, which show special adaptations to life in deep water.

The qualitative and quantitative differences between the fish inhabiting shallow and deep waters are probably connected with basic differences between the two habitats, and one such important factor lies in the nature of the productive cycle. Production by phytoplankton, that is the amount of living plant material produced per unit volume per unit time, is usually much less in oceanic regions than in the shelf areas, but because the euphotic zone (the region in which light is sufficient for photosynthesis) extends to greater depths in the former than in the comparatively turbid waters of the latter, the difference in production per unit *area* is not as marked, being of the order of

225. The world catch of marine fish. The herring-like group of fishes pre-dominates and is largely responsible for the up-ward trend. (Modified from Gulland 1968)

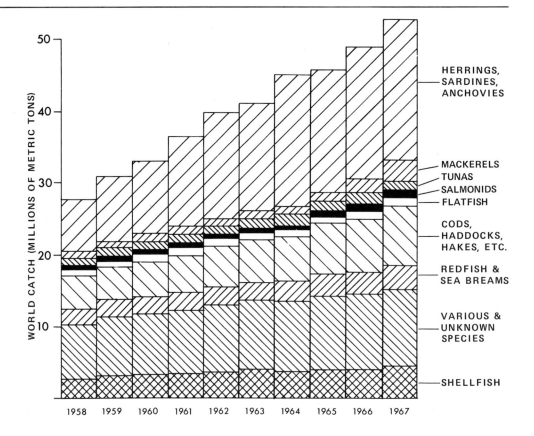

HERRINGS,
SARDINES,
ANCHOVIES

MACKERELS
TUNAS
SALMONIDS
FLATFISH

CODS,
HADDOCKS,
HAKES, ETC.

REDFISH &
SEA BREAMS

VARIOUS &
UNKNOWN
SPECIES

SHELLFISH

WORLD CATCH (MILLIONS OF METRIC TONS)

1958 1959 1960 1961 1962 1963 1964 1965 1966 1967

four to six times and sometimes less. The differences are probably related to the extent to which the water-masses mix vertically. In the shelf areas, the water is mixed from surface to bottom, at least in winter, by the action of strong winds, tides and winter cooling. This ensures that the productive euphotic zone does not become depleted of essential nutrients and, in addition, nutrients are gained by run-off from the land, particu-larly at river estuaries.

By contrast there is little exchange between bottom and surface waters in the deep ocean. The dead organic matter, from which some of the nutrients are released by bacteria, tends to sink below the euphotic zone and so to be lost to the productive cycle. So, although nutrients are plentiful in the ocean depths, they cannot be utilized by the phytoplankton because of the absence of light. In certain areas where the process of upwelling brings deep water to the surface, production is extremely high and this gives rise to very productive fisheries. The best-known example is along the coast of Peru, where, in the upwelling areas of the Peru current, the most productive fishery in the world occurs (figure 241). In 1968, about 11 million tons of anchoveta were caught, but the total fish production in this area is probably con-siderably in excess of this figure. Probably about two million tons are eaten annually by the birds which

produce guano, and there are other fish species in the area, at present little exploited.

It is possible that the production of the deep oceans could be markedly increased if more of the deep waters could be brought to the surface by the creation of artificial upwelling. Several schemes to accomplish this have been proposed, including the use of nuclear power. However, in the immediate future the prospects for maintaining the growth in supplies of marine fish seem to depend on three other possible methods. These are: better management of stocks already exploited, rearing fish in artificial environments, and the exploitation of new resources.

Better management often involves such measures as control of fishing effort and mesh size regulation, and marginal increases in catch could be obtained in a few fisheries, for example with cod and haddock in the

226. *opposite above* The soft body of *Histioteuthis* (6 cm) is buoyed up with ammonium salts and the animal is remarkable in having one large eye and one small eye. Each dark spot on the animal's body is a light organ.

227. *below* One of the commonest and most characteristic of mesopelagic fishes is the silvery hatchet fish *Argyropelecus* (6 cm), whose tubular eyes look upwards while its ventral light organs point downwards. The reduction of pressure on being brought to the surface has caused the stomach of this specimen to balloon out of its mouth

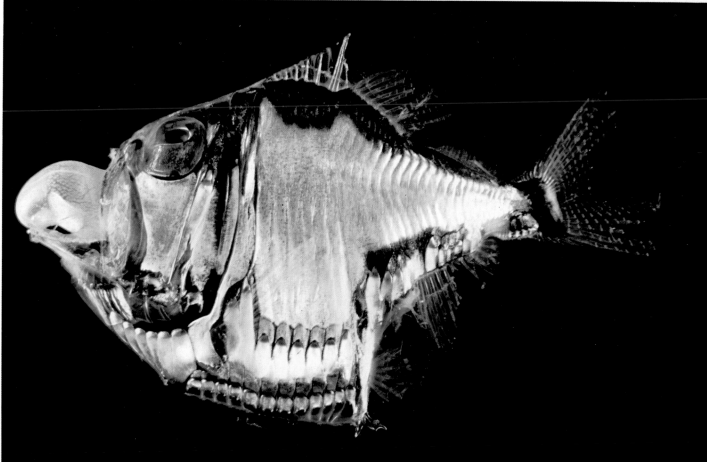

228. Dolphins often swim at the bows of ships of all sizes in all oceans. These animals were part of a school of some hundreds encountered in the Eastern North Atlantic

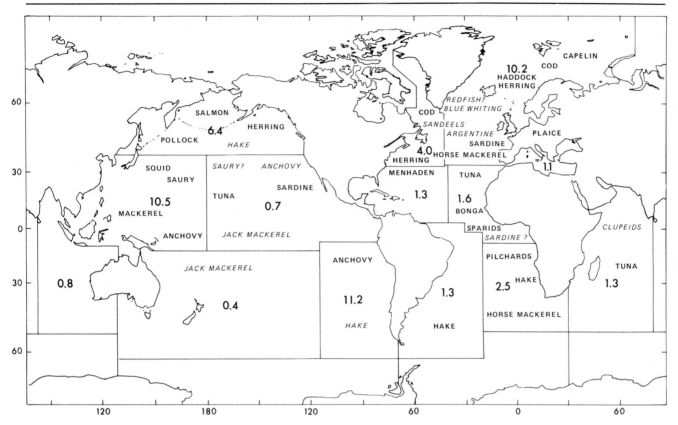

229. The principal marine fisheries of the world and, those in italics, those which are at present under-exploited. Also shown is the world catch in 1967 by regions, in millions of tonnes. (Modified from *FAO Yearbook of Fishery Statistics*, Vol. 24)

north-east Arctic. But it should be stressed that benefits from optimum fishing effort will be small.

Rearing of marine fish, or fish farming as it is sometimes called, is practised on a large scale only in Japan, where, for example, about 20,000 tons of yellowtails are produced annually. Elsewhere, economic and other factors are presenting problems to the large-scale development of fish farming.

The exploitation of new resources, which means using either unfamiliar species or presently-exploited species in unfamiliar areas, seems to offer the best prospect for increasing fisheries production at the moment. Some difficulties will have to be overcome because some of the species may not be acceptable for direct human consumption. In such cases, the fish will have to be processed into fish meal for feeding to farm animals or into a form which is palatable enough to be eaten directly by humans, perhaps mixed with other foods. Such a product is FPC (fish protein concentrate), and work is in progress to make this both palatable and cheap. So let us now briefly survey the world's fisheries and consider the possibilities of further expansion.

As can be seen from figure 229, the southern hemisphere as a whole contributes much less to the world catch than does the northern hemisphere. However, the area off the west coast of South America is the most productive in the world, and were it not for the catch in this region the difference between the two hemispheres would be even more marked. But more about this area later.

Considering firstly the Atlantic, the northern part of this ocean supports some of the longest-established fisheries. There is heavy exploitation over most of the area and some of the stocks are overfished, with the result that their present yields are below the maximum. Any increase in the catch of fish from this area would come mostly from fish not presently exploited, rather than from increased effort on familiar species. Two unexploited species which occur at the edge of the shelf are the blue whiting and the argentine. The eggs and larvae of the former are extremely abundant to the west of the British Isles and a substantial stock of adult fish must be present. The size of the stock has been estimated as one million tons. The greater silver smelt or argentine, though not so abundant as the blue

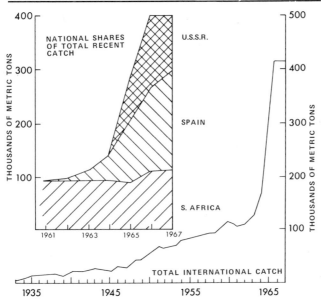

230. Landings of Cape hake, showing how they increased dramatically after 1963. (Modified from Jones & van Eck 1967)

whiting, also exists in this and other areas in substantial quantities.

There are indications that two fish species which are currently exploited could yield greater catches. The sand eel (or lance) supports a fishery of about 200,000 tons in the North Sea and Skagerak, the fish being converted into meal and oil. There is, however, no fishery for them in the western Atlantic, although it is known from adult and larval data that large stocks exist off Greenland, Labrador and Newfoundland. The other species is the redfish, of which about 400,000 tons are landed from the shelf areas. Recently, larvae have been discovered over deeper areas and it may be that an oceanic stock exists. Redfish are very slow-growing and long-lived (apparently reaching 50 or more years), and the stocks are not likely to withstand heavy fishing.

Passing southwards towards the equatorial Atlantic, the abundant demersal species such as cod and plaice disappear, and pelagic species become predominant – sardine (pilchard) and horse mackerel in the east and menhaden in the west. As the tropics are reached, a great diversity of species is found, including breams, sparids, hake, bonga, sardines and tuna. The sardines may be capable of supporting increased catches over the continental slope, and this possibility is at present being assessed by various African countries and by the United Nations.

In the South Atlantic, there were until recently two relatively unexploited stocks of hake, one in the east off

South Africa (the Cape hake) and one in the west off Argentina (the Patagonian hake). Both provide good examples of how quickly a stock can be exploited by present-day fleets, particularly that of the USSR. The fishery for the Cape hake was an almost exclusively South African one up to about 1960; when, however, Spain and the USSR joined in, landings increased very sharply (figure 230). The stock is now thought to be fully exploited, though extension of fishing to the north and into continental slope waters may yield additional supplies. The other main species caught off South Africa are pilchard and horse mackerel, but no great increase in the catch of these species can be expected.

The fish stocks around the Falkland Islands were surveyed by British research vessels between 1926 and 1932. The promising stocks of hake which were found remained virtually unexploited until 1967, when Russian vessels caught about half a million tons. This probably represents a catch of an accumulated stock, and future sustainable catches may be lower than this quantity. It would thus appear that all the hake stocks in the Atlantic are now fully exploited or nearly so, since the northern stocks (the European hake and the silver hake) are also heavily fished.

Turning now to the Pacific Ocean, the main fisheries in the north are those for pollock, cod, hake, herring and salmon. There are also important fisheries for shellfish, notably king crab. In the eastern Pacific, off California and Mexico, there was formerly a productive fishery of several hundred thousand tons per year for sardine but this has now greatly declined to a level of 20,000-30,000 tons. By contrast the anchovy stock, which is virtually unexploited, has greatly increased to a size estimated at about four million tons. It is possible that the great reduction in numbers of one species has allowed the other to increase and fill the 'ecological space'. It is interesting that a similar phenomenon may be taking place off South Africa, where there are signs that the local anchovy has increased with the reduction in stock size of the pilchard.

The two most productive regions in the Pacific are the coastal regions in the western central area and off Peru. In the former area, much of the catch is taken by Japan, one of the world's major fishing nations; the most important species taken are squid, anchovy, mackerel and saury. The fishery off the coast of Peru is certainly the largest in the world, based on the Peruvian anchovy. The catch started to increase dramatically in the mid-1950s and in 1968 it reached a record of over 11 million tons. Thus, Peru is now the world's major producer of fish meal and oil, into which most of the fish are processed.

In addition to the northern anchovy, there are

several other species which could help to expand the catch of fish in the Pacific. There are two main groups of hake in this ocean, the North Pacific hake and the Chilean hake. Current landings are between 200,000 and 100,000 tons, but catches could probably be increased. In view of the tremendous productivity off the South American coast, the stocks of hake are probably considerably more abundant than the present landings indicate.

Judging by the distribution of eggs and larvae, there are large quantities of two truly oceanic species in the Pacific, namely the saury and the jack mackerel. Estimates of stock size are lacking but it is clear that the jack mackerel in particular is very widespread and abundant over the Pacific Ocean. Further research, both in terms of the biology of the fish and methods of catching it, is needed before this resource can be properly exploited.

Knowledge of the potential of the Indian Ocean is extremely limited. Its contribution to the world catch is very small at present and much of the fishing is carried out by small, unmechanised boats close inshore. The chief resources are probably fish of the sardine type, and the most promising areas seem to be the Gulf of Aden and off the coasts of Somalia and Southern Arabia, where upwelling occurs.

Three other important resources have not been mentioned because they are dealt with later in this chapter; these are the Antarctic krill, the 'deep scattering layer', and the tuna.

What are the prospects for a continued expansion of the world fish catch from the sea? It is fairly certain that the fish living on the continental shelf and slope will be fully exploited soon, say within the next 10 or 20 years, by which time the world catch may be around 100 million tons, according to United Nations estimates. In the deep ocean, fish resources are abundant but are rather thinly spread and mostly consist of rather unpalatable fish. Exploitation may therefore be costly and will involve new methods of catching and processing. Estimates of the total annual fish production in the sea vary widely, as might be expected. One estimate is 3,000 million tons of fish such as herring and 450 million tons of fish such as cod. The amount which could be harvested is much less than these figures, of course.

The deep ocean may be compared with the deserts on land; both yield very little to man at present but could be made productive. Experiments on increasing the fertility of the deserts are well under way and perhaps an equal amount of effort ought to be put into a similar study of the deep ocean.

Pelagic Oceanic Fisheries *N. R. Merrett*

The exploitation of the deep sea beyond the continental shelf is associated mainly with one group of pelagic fish, the scombroids. The most important representatives of this group are the tunas (family Scombridae), the billfish (family Istiophoridae), and the swordfish (family Xiphiidae). All are large, swift predators with typically streamlined bodies. The tunas are among the most streamlined of fish; perfected by smooth lines which are accentuated by bullet-shaped heads with closely fitting gill-covers, and grooved inlets for the pectoral, pelvic, and spinous dorsal fins. The billfish and swordfish are equally powerful, although their bodies are more elongated and laterally flattened. The upper jaw is extended forwards as the bill or sword referred to in the common name. This is almost round in section in the billfish, and flattened in the swordfish.

The body form of the members of these families is reflected in the swimming speeds these fish attain. Measurements based upon the rate at which a hooked fish strips line from a reel show that yellowfin tuna can reach 41 knots, while sailfish are reputed to attain 60 knot bursts of speed. On the other hand, observations of the swimming speeds of yellowfin, albacore, and bluefin (tuna) by echo-sounder under natural conditions indicate a maximum of only 2-3 knots. Nevertheless, the ability to produce bursts of high speed is an advantageous factor for catching the active squid and fish which form the major proportion of the diet of these species.

The commercially most important scombroid species are shown in Table 11 together with an indication of the maximum size they attain. Shown also is a group of secondary importance, the pelagic oceanic sharks, which are also caught in the fisheries directed at scombroids. Once again these species have a powerful streamlined form, and most are large and voracious. They cause considerable damage to longline-caught tuna, which often results in their final capture. Similar predation is not uncommon in billfish. In the experience of the writer, albacore up to 40 lbs are the most often attacked, while the most remarkable example was a 3·4 m black marlin in whose stomach was an 82 lb yellowfin with the longline hook still in its jaws.

Table II. Maximum sizes of the larger scombroids and predatory sharks

Family	Genus	Species	Common name	Approx. max. wt.	Age at max. wt. (approx) years.
	Thunnus	*albacares*	yellowfin	300 lbs	10+
	T.	*alalunga*	albacore	90	9+
SCOMBRIDAE	*T.*	*obesus*	bigeye	450	9–10
(Tunas)	*T.*	*thynnus*	bluefin	1000	17–18
	T.	*atlanticus*	blackfin	30	5+
	Euthynnus	*pelamis*	skipjack	40	8+
	Istiophorus	*platypterus*	sailfish	140	3+
ISTIOPHORIDAE	*Tetrapterus*	*audax*	striped marlin	500	—
(Billfish)	*T.*	*albidus*	white marlin	200	—
	Makaira	*nigricans*	blue marlin	2000	—
	M.	*indica*	black marlin	2000	—
XIPHIIDAE (Swordfish)	*Xiphias*	*gladius*	broadbill swordfish	2000	—
ALOPIIDAE	*Alopias*	*vulpinus*	common thresher	1000	—
(Thresher shark)	*A.*	*superciliosus*	bigeye thresher	1000	—
LAMNIDAE (Mackerel shark)	*Isurus*	*oxyrinchus, alatus*	mako	1000	—
SPHYRNIDAE (Hammerhead shark)	*Sphyrna*	*zygaena, lewini* and *mokkaran*	hammerhead	1000	—
CARCHARHINIDAE (Requiem shark)	*Galeocerdo*	*cuvier*	tiger	1500	—
,,	*Prionace*	*glauca*	blue	500	—
	Carcharhinus	*longimanus*	white-tip	13	—
	C.	*falciformis*	silky	10	—

The geographical distribution of the groups represented in the catches of the pelagic oceanic fisheries is very widespread. It is best known for tuna, the most sought after of these fish, which are found throughout tropical, sub-tropical, and temperate seas. The early juveniles of all species are limited to tropical and sub-tropical waters. Their distribution is closely related to the surface water temperature. However, only in the temperate zone are adult tuna similarly controlled by temperature and by specific oceanographic conditions at temperature barriers. Bluefin and albacore adults are most abundant in temperate waters, while bigeye and skipjack migrate from tropical and subtropical to temperature regions. Only the yellowfin is entirely confined to tropical and sub-tropical seas. Its distribution is related to areas of high basic productivity resulting from upwelling and divergence.

The migration of these large fish is known from the seasonal changes of the area in which peak catches are taken and from the recapture of fish previously caught and 'tagged'. The latter method provides the most accurate information, but mortality caused by tagging is high and the areas for recapture are extremely large. Much more knowledge is required to understand fully the migration of these species. Most is known about tuna; in general, those species which move into high latitudes do so in the summer months for feeding purposes, and return in winter to the warmer waters where spawning takes place. However, this behaviour may change with age in certain species. Tagged fish have provided considerable evidence for migration, particularly of the bluefin; the young are known to move from the east coast of the United States across the Atlantic to the Bay of Biscay. Two large bluefin tagged near the Bahamas in 1961 were recaptured northwest of Bergen on the Norway coast 118 and 119 days later. They covered a direct distance of more than 4,000 miles during this time (over 34 miles per day) but the following year a third tagged fish covered this distance in only 50 days (over 80 miles per day).

A dominant feature in the behaviour of tuna is that they are concentrated in the upper 200 m, often being associated with the thermocline. This depth preference is exploited to good advantage by specific fishing methods. Of the food stimuli, taste, brightness, size, and movement affect the intensity of response. This has led fishermen engaged in capturing fish with baited hook and lines to increased catch rates. The aggregation of tuna into schools during certain periods of their life cycle renders them vulnerable to purse-seining and pole-and-line fishing. Fishermen are assisted in the location of such schools by the knowledge that they are frequently associated with dolphins and whales, floating logs and other flotsam. The geographical distribution of billfish and swordfish is less well known than for tuna, and even less is known of shark distribution. However, in general the billfish have an overall distribution similar to tuna, and range between 40°N and 40°S in all oceans. The swordfish penetrates colder water, and the capture of adults is greatest between 30°N and 45°N.

The truly oceanic species of billfish are the striped marlin and blue marlin, while the black marlin and sailfish are found in waters adjacent to land. The white marlin is an Atlantic species, found in large numbers near land during the last part of the year. Although less information is available, billfish share a similar environment to tuna and are probably influenced by the same factors that determine the vertical and horizontal distribution of tuna. The billfish, swordfish, and shark do not exhibit a strong schooling behaviour, although at times they will congregate into small groups.

Accurate age determination of these fish is not easy. Estimates of age in some species have been made by counting growth rings visible in the scales and vertebrae. Once again, this feature has been most thoroughly studied in tuna. The estimated age at the suggested maximum weight for each species is shown in Table 11.

Because Japan catches more than half the world yield of tuna it is not surprising that there is considerable Japanese influence in the methods and gear used in these fisheries. By far the most important method is *tuna longlining,* which originated many years ago in the inshore waters of Japan. As the name suggests, the primary purpose of this method is the exploitation of tuna stocks. However, this form of longlining is successful to a greater or lesser extent in catching all the commercially useful species mentioned earlier (Table 11). The larger individuals of the tuna species, with the exception of albacore and skipjack, rarely school at the surface, but swim in small groups to a depth of about 200 m. Billfish, swordfish, and shark

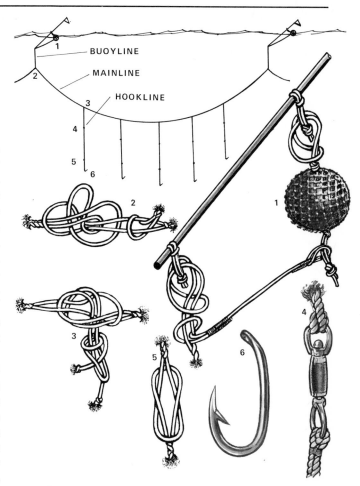

231. One basket of tuna longline in the fishing position, showing the components and the knots used in joining them (see text for explanation)

are also distributed within this upper layer, and consequently all of them are vulnerable to tuna longlining.

Basically the tuna longline is a drifting horizontal line of considerable length, from which single-hooked branch lines hang down at regular intervals. The unit of gear is known as a 'basket'. The term originates from the early days of the fishery when each length of line was stored in a separate basket. One basket of gear is shown in figure 231, and comprises a 300 m synthetic fibre line of approximately 400 lbs. breaking strain, bearing five hooklines, and a 30 m buoyline at one end. The main line is in 50 m sections and the hooklines are knotted into it at each intersection. Typically the hooklines are 22.5 m in length, being made up of a length of mainline incorporating a weighted swivel, looped to a length of twisted multi-strand wire served with light line, which in turn is looped to a wire trace bearing the hook. Each basket is suspended from the sea surface by a glass or PVC float looped on to the

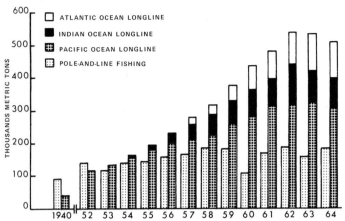

232. Development of the Japanese tuna longline fishery in the three major oceans, compared with the annual catches of the Japanese skipjack pole and line fishery in the Pacific Ocean (from Kamenaga 1967)

buoyline and attached to a bamboo pole and flag. When the longline is being put out, or shot, the free end of one basket is knotted to the intersection of the buoyline and mainline of the next. In this way any length of line may be fished. In normal conditions a Japanese tuna longliner will fish a line of 350–400 baskets daily. This gives a stretched length of 120 km or 65 nautical miles – an amazing length of line to shoot and haul regularly every 24 hours.

A certain degree of elasticity is required in a line of such length to enable it to equilibrate with the forces of wind and current without breaking. This is achieved by laying the mainline loosely. A downward curve in the mainline thus forms between buoys, as shown in figure 231, contracting the total length of the line by about one third. Under normal conditions, therefore, a tuna longline extends approximately 45 nautical miles in the open sea. The tension of laying will also affect the depth reached by each hook (in the basket). When the line reaches equilibrium with the prevailing sea conditions, the sag in the mainline between buoys causes the middle hook to lie deepest, the two on either side are shallower, while the outer two are shallower still. Under optimum sea conditions and with the dimensions of the line given above, the hooks have been calculated to lie at about 160 m, 140 m, and 100 m respectively. Apart from altering the distance between buoys when shooting, the hook depths may also be varied by lengthening or shortening the buoy line. The number and length of the hooklines are also varied to the personal preference of the fishermen. Up to twelve hooklines per basket can be fished for albacore, but as few as one for yellowfin, due to the difference in size

and swimming depth between the species. Both buoylines and hooklines are often shortened for billfish swimming close to the surface.

A single fishing operation is carried out in three phases, shooting the line, a short waiting period, and hauling in the line. Considering the length of line involved and the size of the fish caught, it is remarkable that the whole cycle is achieved in about 24 hours. It is an even more notable fact that daily operations are sustained for periods of three months and longer during commercial cruises. Shooting normally begins at 4 a.m. and continues for approximately four hours, during which time 400 baskets are paid out over the stern of the vessel, which steams ahead at eight to ten knots. A large dhan buoy is tied to each end of the line, which drifts freely. The ship is then kept close to the dhan buoy shot last, until about noon, and the majority of the crew rests. The longline is hauled by a mechanical line hauler which retrieves the line at three to four knots continuously lifting it from the water as the vessel steams at the same speed back along its length. The catch is gaffed and brought aboard through a door on the starboard side of the vessel, a little way behind the hauler. All species of tuna, except albacore, are gutted; billfish, swordfish, and shark are filleted, ready for freezing. Much of the hauling process takes place after dark, when work is continued under powerful floodlights until completed at about 2 a.m.

The most numerous and efficient tuna longliners are vessels of 250–300 gross tons. They operate independently on cruises of up to six months duration out of Japanese ports or foreign bases. Larger vessels, around 1000 gross tons, either carry or tow one or more 20 ton (40–50 ft) catcher boats for use in multi-vessel operations. The catcher boats are built solely to fish, and depend on the mother ship both for transport to and from the fishing grounds, and for other support. Each has a small fish hold but the majority of the space is taken up with the storage of fishing gear, since even these tiny ships operate 200 – 250 baskets of line (about 21 – 27 nautical miles immersed) per day. The mother ship and catcher boats shoot in a pattern radiating outwards from a central point, and each hauls in the line while steaming back towards it. There they rendezvous at the end of each operation, and the mother ship takes the catches aboard for storage.

Japanese tuna longlining has developed to its present scale only since the Second World War. Previously, less efficient pole-and-line fishing, the next method to be discussed, predominated. Figure 232 shows how pole-and-line fishing has been gradually replaced by tuna longlining in the Japanese tuna fishery. It also shows the relative growth of Japanese tuna longlining in the Pacific, Indian and Atlantic

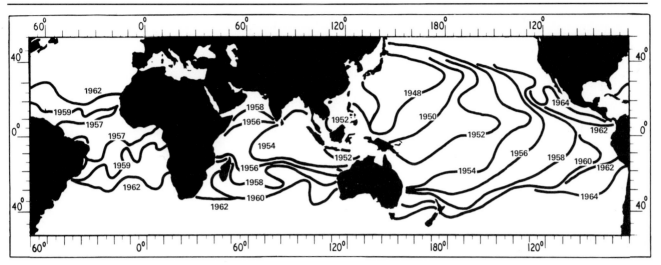

233. Chart showing the expansion of the areas fished by Japanese tuna longline vessels since 1948 (after Kamenaga 1967)

Oceans. This rapid spread is more readily seen from figure 233. By 1962 the Japanese fleets were fishing the breadth of the Pacific and Indian Oceans. Meanwhile, in 1957, Las Palmas (Canary Islands) was made available as a Japanese base port in the Atlantic Ocean. Again rapid expansion followed, and in the same year as expansion across the Pacific and Indian Oceans was completed, all the main tuna fishing areas of the Atlantic were also being exploited.

Despite the change-over by the Japanese to tuna longlining, *pole-and-lining* from a drifting or slowly moving boat is still probably the most widespread method of tuna catching. It is exclusively a day fishery and exploits the surface shoals of skipjack and young tuna of all species. It is carried out in many parts of the world on varying scales, although the techniques used are very similar. Pole-and-lining is notably more successful in areas where a shallow thermocline (20–70 m deep) exists, in contrast to longlining, for which a deep thermocline is required. Approximately one-third of the catch of albacore in America, the majority of the Japanese albacore fishery and, until recently, 80 per cent of the other tuna caught in tropical American waters, are the results of pole-and-line fishing utilising live bait. To some extent this has been superseded by the method of purse-seining, but pole-and-lining is still regarded as the most important method for catching skipjack.

Pole-and-line fishing requires the availability of live-bait, usually anchovy or sardine. This is used to lure schooling tuna into a feeding frenzy close enough to the ship's side for them to be hooked by a rod (=pole) and line. It is a near-shore method, as the live-bait are caught inshore by encircling nets. Often pole-and-line vessels purchase bait from dealers instead of wasting potential fishing time. Most of the Japanese vessels are between 30 and 80 gross tons, and may be converted to longliners during the off season for pole-and-line fishing. Their long high bowsprit is characteristic, with fishing 'racks' which extend around the vessel. The modern tuna clippers of the Pacific coast of America fish mainly for yellowfin, albacore, and skipjack. They extend up to 150 feet in length and have a high bow and foredeck, with a very low main deck aft, on which are several raised live bait wells. Steel fishing racks are. hung outboard from the bulwarks on the port side aft and around the stern (figure 234).

The behavioural characteristics of tuna are used to locate schools on the fishing grounds, the association with bird flocks, dolphins, and floating timber being well known. The vessel steams towards a school at slow speed and upon reaching it the live bait is scattered or 'chummed'. When the school rises to the bait, the vessel stops with the fishing side, usually the port, to leeward, and fishing begins. The gear consists of a bamboo pole 3–6 m in length, on to which is tied a length of line slightly shorter than the pole, with a hook or jig at the end. Jigs are made of barbless hooks attached to whale bone, feathers, or fish skin. They are used when the fish are in a frenzy and biting readily. When biting slackens, live bait is used on barbless hooks. Fishermen use single poles for fish up to 20–30 lb, but for 30–50 lb fish a two-man line is used, which consists of two poles with lines attached to a single hook (figure 234).

The fishermen stand in the fishing racks casting the

234. Two-man pole and lining of tuna in the Pacific

lines into the water which are taken or 'struck' immediately by the fish. When hooked, the fish are jerked out of the water, over the fishermen's heads, into the boat. A sea-water spray system is used in conjunction with live bait chumming whilst fishing. The spray ruffles the sea surface, and is said to improve the catches. It is suggested that the broken surface simulates the presence of large numbers of small fish, and also serves to obscure the vessel and fishermen from the view of the schooling fish.

Pole-and-line fishing for tuna in the United States was used extensively until the latter part of 1958 at which time there was a mass conversion of the Californian bait boat fleet to purse-seiners. Although the *purse-seining* method began in the United States after the First World War, such a rapid gain in popularity was caused by the introduction of an all-nylon net and a powerblock for recovering the seine. By 1961 nearly all the suitable pole-and-line vessels were converted.

Nowadays the tuna purse-seiners are the most mechanized American fishing vessels.

Basically the purse-seine is a wall of netting which is used to encircle a school of fish (figure 235). Tuna seines typically measure 700–900 m in length by 70–90 m in depth. The nylon nets used today are handled more easily, need less care, and do not deteriorate as rapidly as the cotton ones previously used. Floats are attached along the top of the net and weights along the bottom. Rings are also attached to the bottom of the net. A pursing wire is threaded through the rings which acts as a drawstring, enabling the fishermen to close off or 'purse' the bottom of the net. The catch is thus trapped in an inverted umbrella-shaped enclosure (figure 235). To surround a school of tuna, in the American fishery, one end of the net is attached to a skiff which is dropped overboard, and the net is paid out over the stern of the seiner as it swiftly encircles the school. This is a modification of the older

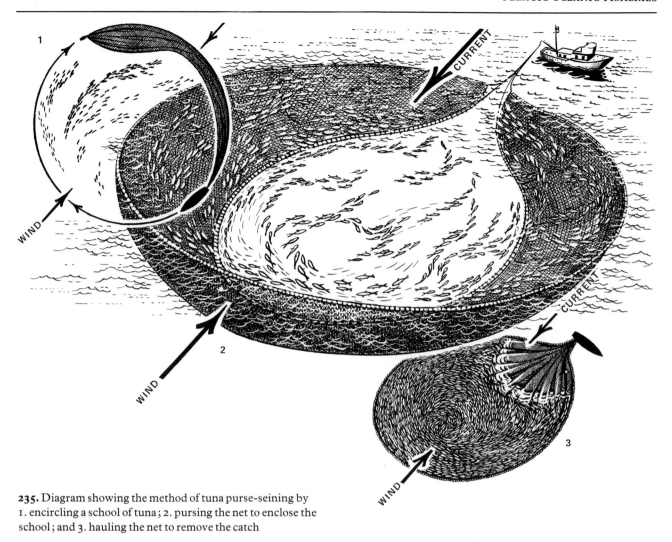

235. Diagram showing the method of tuna purse-seining by 1. encircling a school of tuna; 2. pursing the net to enclose the school; and 3. hauling the net to remove the catch

system of using two equal-sized vessels: each carries half the net and runs in an approximate semi-circle around the school. Once surrounded, the net is immediately pursed. In the one boat system, the size of the enclosure is reduced by pulling the net aboard the seiner from one end. In the two boat system the net is hauled evenly from each end on to both vessels. As about 50 per cent of the sets are unsuccessful, the introduction of the power block for hauling the net saves a considerable amount of time. When no fish are caught, a set is now recovered in one hour instead of two. When the size of the enclosure is sufficiently reduced to concentrate the fish, the latter are lifted aboard in a 'brailing' net used to scoop the fish out of the water. During the pursing operation, in the one boat system, there is a likelihood that the seiner may drift over the net. This is prevented by the skiff keeping the seiner off by means of a towline.

During seining, schools of tuna are located by utilizing the knowledge of their behavioural characteristics. Recently, this knowledge has caused purse-seine fishermen to modify their technique. When dolphins are sighted, fast skiffs chase after them, being directed by radio from the crow's nest on the seiner. The chasers run ahead of the school, herding it into a tight circle. The seiner then circles the dolphin school, releasing the net. Schools of 1000 or more dolphins are encircled and caught, and with them, the tuna. Unless a large portion of the dolphin school is netted, the set is unsuccessful for tuna. Even a few escaping dolphin can lead the tuna out of the net with them. Every effort is made to release the dolphin when the set has been completed. Statistics for the yellowfin catches in the eastern tropical Pacific show that this seemingly needless capture of dolphins is a reliable technique for catching tuna. In 1966, about 62 per cent of the seine-caught fish were captured from schools associated with dolphins. The reasons for this associa-

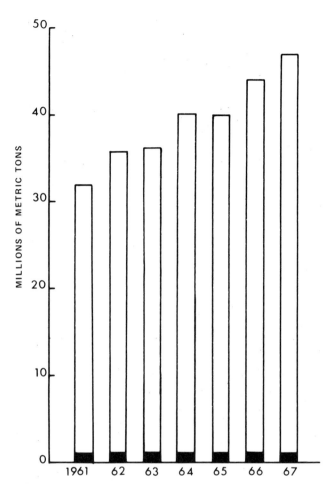

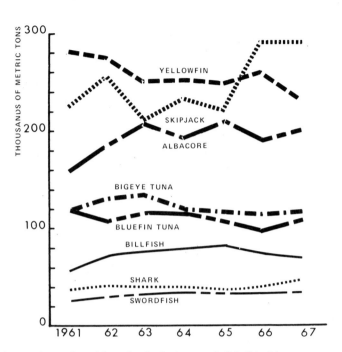

236. Annual world catch of pelagic oceanic fish (black) compared with the total world marine catch 1961-1967 (left)
Annual world catch of pelagic oceanic fish 1961-1967 (centre)
Mean percentage composition by countries of the world catch of pelagic oceanic fish 1961-1967 (right)

tion are by no means fully understood, and the economic implications have prompted a current study.

One of the oldest methods of catching all pelagic oceanic species is *trolling*. In its simplest form, a line with a single lure, natural or artificial, is pulled behind a moderately fast-moving boat, near or on the surface of the water (figure 243). Commercially, vessels troll with multiple lines mounted on 8–18 m (tangons), which are hinged masts swung outboard to approximately 45°. The outer lines are longest and the inner shortest, being about 10–33 m in length. Shock absorbers mounted midway along each line and at its point of attachment to the tangons take the initial drag of any fish which bites. The lines are towed at four to eight knots, and hooked fish are hauled aboard while the vessel is trolling. This is often the most efficient method for catching fish from loosely-knit schools in widely scattered groups, as in the albacore fisheries off California and in the Bay of Biscay. Trolling has been brought to perfection by sport fishermen angling for

large billfish, tuna and shark.

Finally, *harpooning* is a fishing method used to capture swordfish, and a growing fishery exists in Nova Scotia based upon this method. The equipment is extremely simple. A pike pole of wood or aluminium, 4.5 m long, is shod with a 60 cm rod, which fits into a socket on the back of a metal dart. The latter is a double-headed arrow with broad flanges set at a slight angle to the main axis. A line of 80–150 m is attached to the dart, and tied lightly along the pike pole with slip knots to hold the dart in place. The other end of the line is attached to a buoy. The pike pole is held by a 'striker', a fisherman on a platform at the end of the bowsprit of the fishing vessel. Lookout men perched in the mainmast rigging search for swordfish basking on the surface on calm sunny days. When a fish is reached, the striker drives the dart into its back. The pole is retrieved by an additional line while the swordfish rushes off towing the buoy tied to the line. A dory is then lowered, and the doryman picks up the

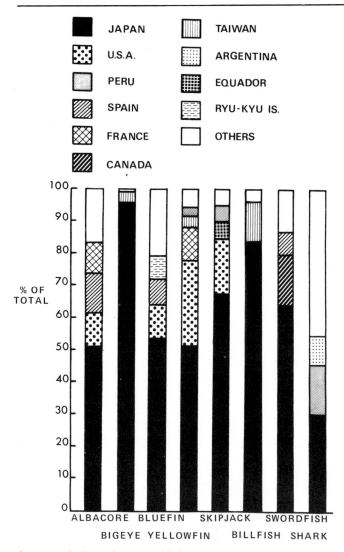

JAPAN | TAIWAN
U.S.A. | ARGENTINA
PERU | EQUADOR
SPAIN | RYU-KYU IS.
FRANCE | OTHERS
CANADA

% OF TOTAL

100
90
80
70
60
50
40
30
20
10
0

ALBACORE BLUEFIN SKIPJACK SWORDFISH

BIGEYE YELLOWFIN BILLFISH SHARK

Figure 236 shows the annual world production of the pelagic oceanic fisheries from 1961–7, indicating the relative importance of the component species or groups of species. The species of tuna are most important. The rise in the catches of skipjack is noticeable, while, apart from the decline of yellowfin catches, those of the remaining species have been approximately maintained. Whereas the tuna catches are sufficiently large to be computed by individual species, the billfish statistics represent the total catch of all the species in the family. Further, in the case of the shark, statistics represent not only the total catches of all the species of the several families of pelagic oceanic sharks, but also includes certain groups from other habitats. Therefore the true numbers of oceanic shark are somewhat lower than the total numbers shown here.

Considering that the waters available to the fisheries described above cover 70 per cent of the world's surface, the number of countries actively involved is small. Those countries contributing more than three per cent of the world catch for any species or group are shown in figure 236, and number only ten. The most striking feature is that apart from the shark catch (already shown to be exaggerated in the present context), Japan catches more than 50 per cent of the world production in all species or groups.

It has been pointed out earlier in this chapter that tuna longlining has gradually replaced pole-and-line fishing in Japan, but the latter is still an important industry (figure 232). Pole-and-line fishing has also been superseded in the United States, but by purse-seining, which is by far the most important method in the American tuna fishery. Longlining is not an economically viable technique for countries like the United States, but where wages are low and fishermen are prepared to spend long periods at sea, it is commercially worthwhile. Consequently Taiwan, South Korea, and China have growing longline fleets, while Japan is beginning to experience problems in this industry from rising labour costs.

The post-war increase in the world's pelagic oceanic fishery was made possible by the development of modern freezing techniques which enabled the industry to produce super-quality frozen fish meat. This has met the growing demand from the United States and European markets for canned tuna, and to a lesser extent swordfish. It has also provided the home markets in the Oriental countries with frozen tuna, billfish and shark, for raw fish dishes and fish sausages. Such an increase in fishing effort into hitherto largely untapped resources has inevitably been followed by a decline in catches in some species. However, as this is an oceanic fishery on a world-wide scale, the effect on the stocks is difficult to assess accurately. Whether

buoy and plays the swordfish out by hand.

Having discussed the species and methods involved in the pelagic ocean fisheries, their importance among the total catches of marine fish can now be considered. The world's marine environment is broadly divisible into two main regions; the shallow sea covering the continental shelf areas from the coast to a nominal depth of 200 m, and the deep sea stretching downwards from the continental slope. The large proportion of the world's surface area covered by deep sea is shown in figure 7, together with the percentage catch of marine fish taken from both regions. It is startling to see that the vast majority of the world marine catch comes from the shelf areas. During the past six years the yield of the extensive oceanic areas has only varied between 1.0 and 1.1 million metric tons. However, due to the expanding world production this represents a steady percentage drop from 3.1 to 2.3 of the total world marine catch (figure 236).

237. Large commercial whales. From the top, humpback (50 ft), sei (50 ft), sperm (55 ft), southern right (55 ft), fin (70 ft) and blue (85 ft). The African elephant is drawn to the same scale

the current yields are sustainable, or control is necessary for the recuperation of the overall stocks, are

questions the expanding research in this field is required to answer.

Whales and Whaling *S.G.Brown*

There are some ninety species of whales (Cetacea) living today. They range in size from porpoises and dolphins four or five feet long and weighing around 120 lbs, to the great blue whale, largest of all living animals with a length of up to 100 feet and weighing as much as 120 tons, equivalent to 24 elephants. Although whales resemble fish in having a streamlined body with fins, swimming by means of their tail and spending their entire lives in water, they are mammals, warm-blooded with lungs so that they must surface to breathe air at regular intervals, and they feed their calves on milk.

Existing whales are divided into two main groups. The toothed whales, sub-order Odontoceti, comprise some 80 species and include the familiar bottle-nosed porpoise or dolphin and the pilot whale of oceanaria. The sperm whale is the largest member of the group, with males reaching a length of 60 feet, but the majority of toothed whales are less than 20 feet long. The number of teeth varies from two in the narwhal, in which both normally remain concealed in the gum in the female and only one grows to form the tusk in the male, to over 200 in the La Plata dolphin. Toothed whales feed mainly on fish, squid, and to a lesser extent on crustaceans. The killer whale sometimes takes seals, penguins and porpoises. Most of the toothed whales live in the sea but some frequent estuaries and rivers and a few are found exclusively in fresh water.

The baleen or whalebone whales, the second group, sub-order Mysticeti, are all marine and include the largest whales. They have no teeth and use a sieve of whalebone plates (baleen) which hang down from the roof of the mouth, to filter their food (plankton or shoaling fish) from the sea. There are 12 species including the right, blue, fin, humpback, sei and gray whales (figure 237).

Baleen whales undertake extensive annual migrations from subtropical and warm temperate waters where they breed in the winter months, to cold temperate and polar waters where they feed in the summer months. Since they feed little if at all on the breeding grounds there is an alternation of non-feeding and feeding periods during the year. It is when they are concentrated on the feeding grounds and laying down a thick layer of fat known as blubber, under the skin, that they are hunted by the whaling fleets. Sperm whales have a less well-defined migration; they are mainly confined to warmer waters and are hunted from land stations there, but some of the males apparently migrate regularly to colder waters and are caught with the baleen whales.

In many species of whales the individuals swim together in groups known as 'schools', which in some dolphins may number hundreds of animals. Schools may include adults of both sexes and calves, but in some species, such as the sperm whale, the sexes and different age groups are segregated at certain times. Dolphins swimming in schools apparently communicate with each other by whistling. They also produce other sounds which are supersonic (Chapter 8), and in the bottle-nosed porpoise these have been found to be used in combination with its acute powers of hearing as a kind of sonar system to locate food and other objects under water.

The life cycles of a few species have been worked out, mainly those of commercial importance. The large baleen whales are believed to become sexually mature at around ten years and to have a life span of at least 50 years. The gestation period is between 10 and 12 months and the calf suckles for at least six months, so that the females produce one calf (rarely twins) at most every two years. In the sperm whale gestation lasts between 14 and 16 months and lactation possibly as long as two years.

Whales are a valuable source of raw materials. Stranded whales on the shore first provided man with meat and blubber as a source of food, light and heat, and with bones for making tools. Until the mid-nineteenth century, baleen whale oil and sperm whale oil were both used for lighting and the baleen or whalebone for corsets, sieves, carriage springs, etc. Today, baleen whale oil, which is edible, is hardened and processed to produce margarine and cooking fats. Sperm whale oil is of different chemical composition and is not edible. It has a variety of uses in industry, e.g. in the making of steel, dressing of leather and textiles, as a lubricant and in the production of soap and cosmetics. Spermaceti, a waxy solid from the head of the sperm whale, is used in cosmetics and for candles. Meat meal is a constituent of foods for livestock, bone meal is used in fertilizers, and the frozen whale meat provides human and animal food.

Whaling has a long history reaching back over the

238. Whale catcher in the Antarctic. Note the long catwalk between the bridge and forward gun platform, and the look-out's barrel on the mast

centuries. Some of the smaller species including porpoises and pilot whales were hunted in the Stone Age of northern Europe, and the Eskimoes and Indians of North America also pursued whales at an early date. Of the larger whales, the North Atlantic right whale was hunted in the Dark Ages in Europe, and it was from the Basque fishery for this species on the shores of the Bay of Biscay that whaling on the high seas

developed in the fifteenth century. The Greenland or Northern whale fishery began at Spitsbergen in 1611 and lasted for 300 years. In 1712 the great American sperm whale fishery started, and in the nineteenth century hundreds of American and small numbers of British, French and German whalers scoured the world's oceans for sperm and right whales. This open-boat whaling with hand harpoons was for species which

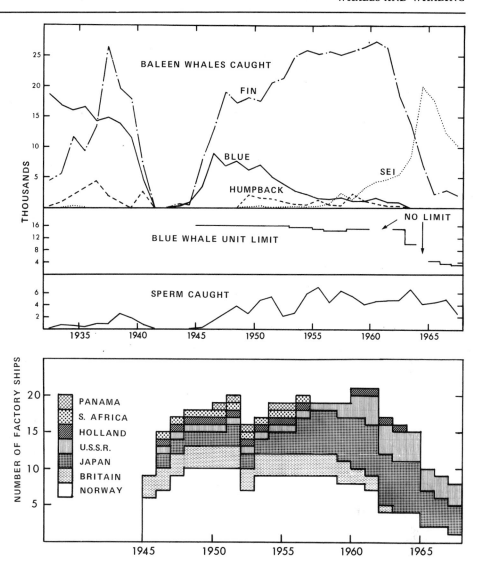

239. *Top,* catches and blue whale unit limits in Antarctic pelagic whaling. *Bottom,* numbers of factory ships operating in the Antarctic

swim slowly and could therefore be chased under oars or sail, and which usually float when dead. Modern-style whaling began in the 1860s with the invention by the Norwegian, Svend Foyn, of the steam whale catcher equipped with a harpoon cannon and firing an explosive harpoon. The fast-swimming rorquals (the blue, fin and sei whales) were now open to hunting. They could be run down only by a steam vessel and they usually sank when killed, so that compressed air had to be pumped into the carcass to keep it afloat. A rapid development of whaling from shore stations followed; the catching boats operated within a hundred miles or less of the station and took blue, fin, humpback and sei whales with the occasional sperm whale. From northern Norway, land-based whaling spread widely in the North Atlantic and North Pacific and in 1904 it reached the Antarctic with the opening of a station on the island of South Georgia. More stations followed,

and whaling was also conducted from factory ships moored in harbours among the islands off the Antarctic Peninsula.

Modern whaling on the high seas, known as pelagic whaling to distinguish it from shore-based whaling, now carried on within a range of up to three hundred miles from the station, is today confined to the Southern Ocean around the Antarctic continent, and to the North Pacific. It began in the Antarctic in the 1925–6 whaling season, when whales were first hauled up a stern slipway on to the deck of the factory ship *Lancing* for processing, thereby ushering in a new whaling era. Fin, sei and sperm whales are hunted and until very recently blue and humpback whales were also captured, but hunting of these two species has now been stopped to prevent their extinction.

An Antarctic whaling expedition consists of a factory ship, catching boats and one or more refrigerator

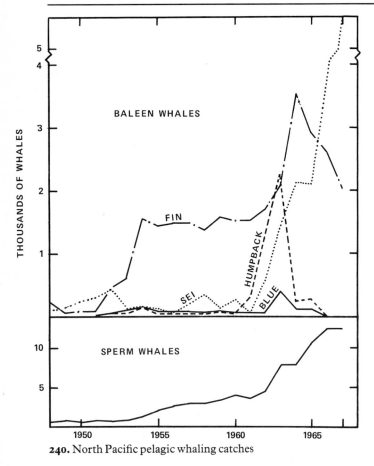

240. North Pacific pelagic whaling catches

winched to the side of the catcher and inflated with compressed air to prevent it from sinking. It is then marked with a flag, radio buoy and radar reflector to enable it to be found easily and is cut adrift. Its position is radioed to the factory ship from which a buoy boat is sent out to collect the flagged whale, while the catcher goes off in search of more whales.

On arrival at the factory ship the carcass is hauled up the stern slipway on to the after deck and 'flensed' or stripped of the layer of blubber, this being peeled off in three strips with winches (figure 242). The strips are cut into pieces and dropped into boilers below deck where they are cooked under pressure to extract the oil. The flensed carcass is hauled forward to the foredeck for 'lemming' or butchering, during which the meat is stripped from the bones which are then sawn up, using steam saws. The best meat is transported to the refrigerator ship for freezing, other meat goes to make meat meal and the remainder is cooked to extract the oil. Oil is also extracted from the bones. A fin whale 70 feet long and weighing around 50 tons will be completely worked up and have disappeared from the deck within half an hour, and as soon as one carcass has been flensed another takes its place.

Within five years of the first factory ship being used, 41 factory ships were at work in the Antarctic Ocean. They produced nearly 3,500,000 barrels (570,000 tons) of oil from some 37,500 whales. This enormous production brought a collapse of the world market for whale oil in 1931, and since the 1932–3 season first voluntary production agreements and later international agreements have regulated catches in the Antarctic.

The development of pelagic whaling in the Antarctic from the 1932–3 season onwards can be traced in figure 239. This illustrates the catches of the four baleen whale species, blue, fin, humpback and sei whales, the level of the pelagic catch limit set by the International Whaling Commission, the catch of sperm whales in the Antarctic and the number of factory ships of different nationalities operating in each post-war whaling season.

Considering first the baleen whale catches, it can be seen that in the pre-war years there was a gradual decline in the catch of blue whales, the most valuable species, and a corresponding increase in the fin whale catch. The humpback whale catch fluctuated and the catch of sei whales was negligible. Fears that the Antarctic whale stocks might be depleted by overfishing, in the same way as whaling history showed other stocks to have been earlier reduced by excessive catches, were expressed as early as the 1920s. In 1929 the Bureau of International Whaling Statistics was set up in Norway to collect and publish statistics of whaling

ships with attendant cold storage meat transporters. During the course of the whaling season, an oil tanker may visit the expedition to refuel the factory ship and take off the whale oil. The factory ship is the mother ship of the expedition. Of about 22,000 gross tons, she is a complete floating factory for processing whales and producing whale oil from baleen whales, sperm whale oil, meat and bone meal, meat extract and other by-products. Frozen whale meat is processed on the refrigerator ship and stored in the transporters.

From 10–15 whale catcher boats accompany the factory ship. These are diesel-engined vessels of about 750 gross tons, developing some 3,200 h.p. and fast enough at around 17 knots to run down any large whale (figure 238). They may operate up to 200 miles from the factory ship, searching for whales whose 'blows' (expelled breath at the sea surface) are sighted as a puff of vapour by the lookout. At closer range, an echo whale-finder may help the catcher to follow the movements of the whale below the surface until it comes within range of the harpoon gun mounted on the bow. This fires a heavy harpoon attached to a rope whale line. The grenade head of the harpoon explodes within the whale's body, killing it; the carcass is then

241. *opposite* The waters off the coast of Peru (with the Andes mountains to the right) seen from Gemini IX at a height of 150 miles. These waters are one of the most productive fishing grounds in the world

242. *top* Sei whale on the deck of a factory ship

243. *above* The spectacular ocean blue colour of the dolphin-fish *Coryphaena* disappears within minutes of capture, to be replaced by a kaleidoscope of different colours ending with yellow and silver ; this warm-water species makes excellent eating, and is usually caught by trolling

throughout the world. The League of Nations introduced an International Convention for the Regulation of Whaling in 1931 which provided protection for the much reduced right whales and forbade the catching of female whales with calves. In 1937 a new international agreement additionally introduced minimum length limits for the different species, limited the Antarctic season to three months and closed to pelagic whaling all oceans north of 40° south latitude except for the North Pacific. In the last pre-war season 1938–9, 34 factory ships from Norway, Britain, Japan, Germany, Panama and the United States caught 36,600 whales yielding 2,700,000 barrels of oil. Antarctic whaling was greatly reduced during the Second World War but the respite from hunting was too short to allow the whale stocks to recover very much.

In 1946 a most important conference in Washington set up the International Whaling Commission which regulates whaling in most parts of the world, with the aims of safeguarding the whale stocks and securing their optimum exploitation. It held its 22nd annual meeting in London in 1970. A scientific committee reports on relevant biological research and on the state of the stocks and, with the technical committee, advises on conservation measures. The Commission decides which species and categories of whales need protection, the length of the whaling season, the minimum length limit for legal catches and, most important, sets an annual total catch limit for Antarctic pelagic whaling. This is defined in terms of 'Blue Whale Units', one BWU equalling 1 blue whale or 2 fin whales, or $2\frac{1}{2}$ humpbacks, or 6 sei whales. Sperm whales are not included in the limit. Contracting governments have the right to object to decisions made by the Commission, and it cannot limit the number or nationality of factory ships or land stations operating, nor allot specific catch quotas to them. This has proved to be a weakness but the countries concerned have since 1962 reached agreements on catch quotas outside the Commission.

In the post-war period there was a decline in the catch of blue whales as a result of overfishing. With the catch down to 20 whales, complete protection in the Antarctic was introduced in 1965. Protection for humpback whales throughout the Southern Hemisphere came in the 1963–4 season. With the decline in numbers of blue whales, the catch of fin whales increased to around 26,000 whales annually from the 1954–5 to 1961–2 seasons. A steep drop in the catch followed, with a corresponding rise in the catch of sei whales to a peak of 19,800 whales in 1964–5.

In the mid-1950s signs of a decline in the fin whale stock appeared, but not until 1963–4 was the Commission able to introduce a substantial reduction in the

catch limit to 10,000 BWU. This was a result of new assessments of the stock sizes of the different species of Antarctic baleen whales, arising from the adoption of methods similar to those used in modern fisheries research to calculate the size of fish stocks. These studies are now carried out annually by the scientific committee of the Commission with the collaboration of experts from FAO and elsewhere. They involve the collection of biological data relating to the ages of the whales and the number of mature and pregnant animals in the catch, together with statistics of the numbers, lengths, sex ratios, etc. of the different species caught, and details of the numbers of catcher boats and their employment provided by the Bureau of International Whaling Statistics.

A further substantial reduction of the catch was necessary to halt the decline in the fin whale stocks but the Commission was not able to agree to this until the 1965–6 season when a limit of 4,500 BWU was set. An additional proviso was that there should be reductions in the limit for the next two seasons to make the catch in 1967–8 less than the sustainable yields of the fin and sei whale stocks. The sustainable yield is the number of whales which can be caught without reducing the size of the stock or allowing it to increase. With the catch limit for 1967–8 set at 3,200 BWU and the sustainable yields of the two stocks combined calculated at between 3,100 and 3,600 BWU this aim was finally achieved. Provided that the catch limit is kept below the level of sustainable yield of the stocks for some years to come, the depleted stocks will be able to increase gradually to a level at which much larger catches will again be possible.

The catch of sperm whales in the Antarctic is confined to males. It has fluctuated in the post-war period but shows a gradual increase up to the 1963–4 season, when there were signs that the catches were affecting the stock. Some additional sperm whales are caught in warmer waters by factory ships on their voyages to and from the Antarctic.

After a rapid rise, the number of factory ships operating in post-war years remained at around 20 until the 1962–3 season (figure 239) but the composition of the fleets altered gradually with Panama and South Africa ceasing pelagic whaling, a slight decline in the number of Norwegian and British factories, and with Japan and the Soviet Union increasing their fleets. More rapid changes followed as whaling became uneconomic for Norwegian, British and Dutch companies, who sold their factory ships to Japan. Lower catch limits then led to a reduction in the fleets of the three remaining Antarctic whaling nations, and in the 1967–8 season only one Norwegian, four Japanese and three Russian factory ships operated. Since then

244. Rat-tails (macrurids)
round a bait can on the
bottom at 5860 m

245. A mass of hag fish around
a baited wire basket at 1450 m
in the Gulf of California

the last Norwegian factory has been withdrawn, and with the present reduced catches necessary to rebuild the stocks it is unlikely that Antarctic whaling will prove profitable to any countries other than Japan and the Soviet Union.

Pelagic whaling catches in the North Pacific since 1948 are shown in figure 240. The picture is similar to that for the Antarctic. The catch of fin whales rose to a peak in 1964 and then fell, the decline being accompanied by a steep rise in the sei whale catch. The numbers of blue and humpback whales taken remained at a low level (except for 1962 and 1963) until protection for both species was introduced in 1966. The catch of sperm whales shows a steady increase to a peak of 12,400 in 1966 and 1967.

The increasing importance of North Pacific pelagic whaling is seen in the 1967 total catch of 19,900 whales, only 300 fewer than the Antarctic catch in the 1966–7 season. Three factory ships with nine catchers were employed in 1948 compared with six factories (three Japanese and three Russian) and 58 catchers in 1969. Studies of the North Pacific stocks indicate that the blue and humpback whale stocks are at present very small and in need of continued protection. The fin whale catch is above the present sustainable yield, and the state of the sei and sperm whale stocks is uncertain. Encouragingly, the countries whaling in the North Pacific agreed to limit the fin and sei whale catches in 1969.

According to Mackintosh, the gross value of products from pelagic whaling and land stations throughout the world in 1959 was probably around £50-60 million. This is about one-twentieth of the total value at that time of all other marine fisheries. With the recent decline in Antarctic whaling the present value of whaling products will be smaller but not insignificant. If present catches can be kept at levels which will allow the various stocks to rebuild their numbers over the years to a position at which large catches can again be taken without fear of over-fishing, the whaling industry may once more become an important world source of valuable food and other products.

It seems unlikely that present whaling methods employing factory ships or expensive shore installa-tions can be easily adapted to the hunting of the smaller species. The minke or lesser rorqual, a small baleen whale growing to a length of some 30 feet, is the basis of a separate Norwegian whaling industry employing small catching boats in the North Atlantic and Arctic, and it has been taken by factory ships in the Antarctic on occasion. To support a factory ship's operations solely on small whales, however, very large sustained catches would be necessary and would require different catching methods from those at present in use.

The large baleen whales have suffered heavily from whaling over the years. The stocks of right whales and the gray whale were reduced to a mere remnant of their former size by whaling in the nineteenth and early twentieth centuries, and even after many years of protection are only now slowly recovering their numbers. More recently the populations of blue and humpback whales have been depleted by pelagic and shore-based whaling to a level at which possible ex-tinction has been feared. It is calculated that in 1961–2, before they were protected, the stock of blue whales in the southern hemisphere was somewhere between 930 and 2790 animals, and it is unlikely that much larger numbers were to be found in the North Atlantic and North Pacific. While there is apparently no case of a whale becoming extinct because of overfishing in the past, it is possible that this could occur now without special protection for individual species. Today's highly efficient whaling industry might, for example, while supporting itself economically on catches of fin and sei whales, have reduced blue whales to such low numbers in the southern hemisphere that extinction there was possible. This, however, would not neces-sarily mean the final extinction of the species since some blue whales would remain in the northern hemisphere.

With our increasing awareness of man's responsibi-lity to the world's wildlife and to future generations, it is to be hoped that no species of these magnificent and beautifully adapted animals will ever again be placed in danger of extinction through mismanagement of this unique resource. Equally it would be sad, after its long and often romantic history, to see the end of whaling itself as a way of life for man.

Other Resources of the Deep Sea *Malcolm R. Clarke and John D. Isaacs*

Application of laws of exponential growth to human population studies has led both to alarming forecasts about overpopulation and vigorous speculation on new sources of food. The deep ocean often figures in such speculation, and its potential as a source of food is frequently exaggerated. Let us attempt to put the living resources of the deep sea into perspective.

As mentioned elsewhere, the production of living matter is limited by the total amount of sunlight falling upon the earth's surface. Calculations suggest that the oceans produce two or three times the total production of the land (page 162). A large proportion of this living matter is in the form of small plants and animals, the plankton. To filter plankton from water is a very expensive procedure. If we imagine a large ship able to travel at its economic cruising speed and magically catching *all* the plankton in the path of its hull we are envisaging the cheapest possible means (though impossible in practice, of course) of collecting plankton. If such a ship were to traverse the open ocean it is easy to show that the cost of the journey would exceed the value of the weight of food even supposing the food to be prime fish and not a low grade meal. Fortunately, this food is concentrated in one of two ways: it may be collected by larger animals that eat it and build it into their own substance or it may tend to form dense patches, shoals, or schools due to the influence of various environmental or internal factors. Economics dictate that any exploitation must be of the concentrated food, either larger animals, or dense shoals of smaller ones. The present-day aim is to detect concentrations, discover the reasons for their existence, develop means of forecasting their appearance, and generally to improve on their acquisition. Such concentration takes place at certain physical boundaries such as regions of sudden change of sea temperature, may result from seasonal changes in behaviour of some animals, or for other reasons. In the future, artificial means to encourage such concentration may prove possible.

Exploitation of herbivorous or small carnivorous plankton is attractive from a purely theoretical point of view because at each stage in a food chain there is a wastage of substance (page 221). The total weight of small planktonic animals is many times greater than the total weight of, for example, fish or whales. By catching the smaller animal, the supply from the sea would be greater. The baleen whales in the Southern Hemisphere tap the food concentrations in the Antarctic during summer at which time quantities of the shrimp *Euphausia superba* (figure 169) form large patches in the surface waters. Rough calculations suggest that many million tons of this 'krill' are consumed by whales, seals and sea birds each year. Speculations concerning the effect of the reduction of whale stocks (page 267) could hardly avoid the supposition that as the whales drop in number the krill stocks should increase.

Recent Russian and Japanese expeditions have investigated the possibility of commercial exploitation of krill. The animal is rich in protein and mineral salts and its fat reaches 7.5 per cent of dry weight by the end of the summer. Preliminary work has shown that krill is unsuitable for salting down but may be canned or used to produce a 'fish' meal. This meal has proved

very good for fattening pigs. Hydrological studies have reduced the problem of finding the krill patches and several catching methods have been tried; the most useful nets are those that are towed on booms from the foredeck with a frame to support the mouth and with which a pump is employed for continuous removal of the catch from the cod end into the hold of the vessel.

In other regions, large concentrations of euphausiid shrimps, copepods, and amphipods are sometimes seen near the surface, but their sudden appearance cannot be forecast at present so that exploitation would not now be profitable.

As mentioned elsewhere (pages 55 and 56), echo sounders often show the presence of sound scattering organisms. These fall broadly into two types; diffuse layers of varying density (scattering layers) and more discrete patches (scattering groups). The layering is, to some extent, a product of the particular frequency of the echo sounder; different frequencies show layers at different depths. However, there is now little doubt that these layers are caused by echoes from small animals, sometimes fish, sometimes such animals as siphonophores. The animals that cause such a layer normally contain a bubble of gas and are excellent sound reflectors but may be very diffuse; only one small fish, 2.5 cm in length, need occur in every 300 cubic metres for a distinct acoustic layer to be recorded on an echo sounder. Some of these layers are caused by myriads of tiny fish called *Cyclothone*, which are certainly among the most numerous vertebrates on earth (figure 185). Scattering groups are caused by fish in shoals. Usually these indicate such things as snipe fish or horse mackerel, which occur over deep water near islands, but occasionally such patches may show aggregations of less well known and more typically deep sea fish such as myctophids. An example of a seasonal occurrence is *Ceratoscopelus maderensis* off the eastern United States. In 1967 dense patches observed on an echogram were found to be this myctophid fish by simultaneous sampling and observation from a submersible (figure 71). These patches, consisting of fish about five to eight cm long, extended above the continental slope from Cape Cod to Cape Hatteras. Observation from the submersible and catches of some fish in trawls permitted the total weight of fish to be estimated at between one and ten million tons. This is of the same order as the world annual catch of *Anchoveta* or herring. Other examples of seasonal concentrations of myctophid fish are known. Among other groups, one of the commonest fishes is the small gonostomatid *Vinciguerria* (figure 185). The Opah or Moonfish *(Lampus guttatus)* reaching two metres in length, the sun fish *(Mola mola)* reaching three metres in length, and hammerhead

246. A large shark, with a prawn on the top of its head, and several smaller fish round a bait at 1890 m off Hawaii

sharks *(Sphyrna)* reaching five metres, all occur in dense concentrations over large areas from time to time.

Flying fish, noticeable by day over large areas of the warmer waters, are the basis of a special fishery at Barbados. At night, a ship invariably attracts squids to its lights unless it is in high latitudes. The most frequently seen surface squids all belong to the Ommastrephidae, to which family the common Japanese squid belongs. This squid, *Todarodes sloanii,* is a very important source of protein in the Far East and the recorded annual catch in Japan alone sometimes exceeds 300,000 metric tons (1967). Its nearest relative, *Todarodes sagittatus* (figures 24–25), which seasonally occurs in very large numbers off Iceland and Norway and extends in subsurface water throughout the North and South Atlantic, is not caught for food although small quantities are utilized for long line bait by the Russians. Yet another squid, *Illex illecebrosus,* forms the bait of the Newfoundland Bank line fishery, and in years of its failure has caused economic hardships, before the advent of large-scale refrigeration which allowed storage of bait.

Other medium-sized invertebrates, less rich in protein, perhaps, but sometimes observed in large numbers, are the small so-called 'red crab' *Pleuroncodes planipes* which forms vast shoals off lower California;

the stomatopods, which occurred in enormous numbers in the western Indian Ocean during 1967; and the larvae of the decapod *Munida gregaria,* which swarm on the Patagonia shelf but are also found over deep water at times.

What of the subsurface, out-of-sight animals? If baited cameras are lowered into *midwater,* by far the most frequent animals photographed are squids of the Ommastrephidae (figures 21–25). If lowered to the *bottom* a vast new world of fishes is disclosed. By using cameras which take photographs at frequent intervals, many interesting facts come to light. Because the first photograph often includes at least one fish it is clear they cannot be all that rare on the bottom. Subsequent frames often contain large numbers, and rough estimates of population are possible. Rat-tails (figure 244) and hag fish (figure 245) are extremely numerous. Sharks with heads as much as four feet between the eyes are sometimes photographed (figure 246). The sable fish *(Anopaploma fimbriata)* occurs off the California coast at 800–2000 m in apparent densities of 0.5 kgm/sq m but is only fished much further north in cold water where it occurs nearer the surface. A flat fish, *Bathyichthia maculatus,* has a shallow limit of 1400 m below which it occurs in large numbers. Invertebrates on the bottom include the Tanner crab

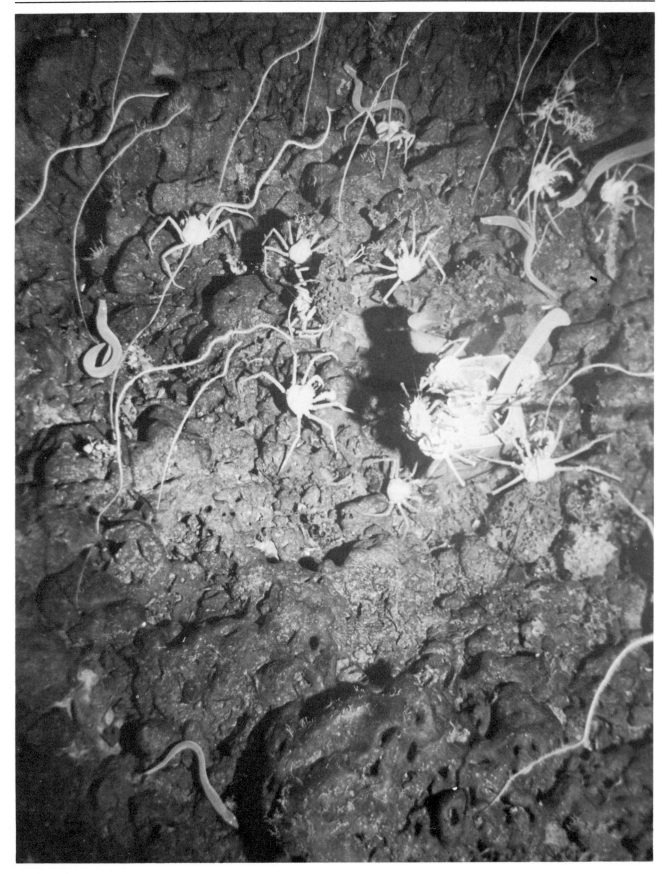

247. Hag fish and Tanner crabs attracted to the bait at 620 m
off Baia, California

248. Numbers of a penaeid prawn on the bottom at 1890 m off Hawaii

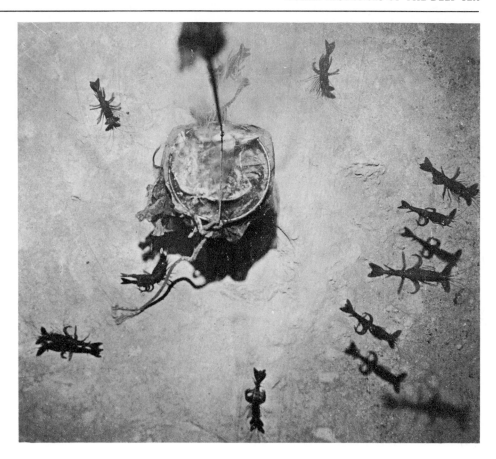

(Lithodidae) – common below 1000 m in some regions, and a penaeid prawn (Penaeidae – figure 248) which was first observed by deep water cameras and is now fished experimentally off Hawaii.

Fishing with drop lines or bottom lines on the continental slopes fully supports the evidence of cameras. Such lines in the Bay of Biscay, off the Canary Islands and in the Indian Ocean, at depths greater than 500 m, often catch fish on 20–25 per cent of the hooks (cf. average of three to five per cent on tuna longline). In the North Atlantic the majority of the fish caught are deep sea sharks (*Centrophorus, Centroscymnus, Squalus*) but rabbit fish (*Chimaera*), rays, rat-tails (*Macruridae*), and eels (*Synaphobranchus*) are common. A recent Royal Society expedition to Indian Ocean islands caught quantities of a large (up to 60 pounds) edible fish (*Etelis marshi*), previously little known in that ocean but highly prized for food in the Far East. The catch rate (9 per cent of hooks) certainly suggests that a local fishery might be worthwhile.

Line fishing in midwater is carried out from various islands. The scabbard fish (*Aphanopus carbo*) is a metre long black fish which is caught in considerable numbers off Madeira, but it is only recently that its occurrence off Newfoundland and the Bay of Biscay at about 1000 m depth has been discovered.

Fish traps used experimentally in the Pacific nearly always catch fish in depths of 500–2000 m and often come up literally packed with fish (figure 245).

Another source of evidence regarding bottom and midwater fauna is the stomach contents of bottom fish and deep-diving sperm whales. From these it appears that squids form an extremely important part of the animal population on the slope bottom in spite of their absence from photographs and rare occurrence on deep lines.

It would be naïve to suppose that the only pre-requisite of feeding the ever growing human population with protein from the deep sea is the presence of the protein in large quantities. Evidence that there are large quantities is often more subjective than quantitative but evidence there surely is. However, for there to be utilization there must be consumers. Consumers are dietarily conservative, thus any new food must be made palatable, attractive, and acceptable, and, for the business of commercial exploitation to succeed, the consumer must already exist. The would-be exploiter would thus best employ an intermediary; by converting unpopular fish to meal and using this to enrich the food of domestic farm animals, the businessman is more assured of a market.

275

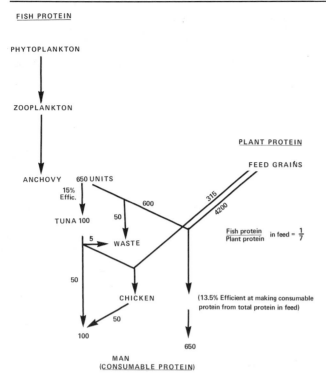

249. Potential pathways of fish protein utilisation by chickens, and protection of consumable protein for man (figure and text explanation provided by Professor W. L. Schmitt)

This route of utilizing unpopular marine species via conventional livestock products would also serve to shorten the food chain to man, as touched on earlier, and consequently make possible a much greater harvest of the sea's productivity. Consider, as an example, two ways in which anchovies are employed in man's nutrition (figure 249).

In the purely marine way, anchovies serve as substrate for tuna, and their protein is depreciated by a factor of about 13 for food chain and processing losses. If tuna offal is turned into meal for chickens, another 1/13 can be recaptured for man's benefit. If, however, the meal is made from the anchovies and used to upgrade chicken feed, seemingly *all* of the anchovy protein reaches man. This apparent miracle is the result of the complementary spectra of amino acids in the proteins of fish and in plant grains and from the assumptions of an unlimiting supply of plant protein and of fully beneficial assimilation of the fish protein. The first assumption is fairly defensible and as for the second it must apply for the essential amino acids that are lacking in plant protein.

Even at the low level of fish meal used to supplement chicken feed (three to five per cent) it is barely competitive with plant-base protein preparations. Cottonseed, soybean, sunflower, and peanut are some of these bases, and their relative composition in mixtures are now computer determined. In times of abundant and hence cheap crops the fish meal is often displaced by them. Chicken farms, though they are among the most efficient agricultural enterprises to the point that they undersell popular fish products, add to the cost of animal protein, and one can envisage yet a third way for anchovy utilization, namely via fish flour or FPC (fish protein concentrate), the fish meal equivalent for direct human consumption. Here losses would be slight (about 1/7), and though costs would be higher than in meal because of higher sanitary standards, the product would be cheaper than in the meal-chicken route.

In North America FPC has problems. The opposition of dairy interests allegedly persuaded the US Food and Drug Administration (FDA) first to outlaw it as an 'impure product', 'the Administration' then rescinded this but is now limiting commercial production by restricting it to hake and to one pound packaging. Thus, a small plant in New Bedford, Massachusetts, under contract to the Agency for International Development (AID) is only in partial operation. The Bureau of Commercial Fisheries will be testing another extraction process in Aberdeen, Washington. A larger plant with yet another process will go on stream before long in Nova Scotia, Canada, in order to bypass the US restrictions. It is, however, much too early to tell whether any of these FPC's will achieve the market acceptance and penetration their promoters claim.

10 Beneath Oceanic Windows

R.H.Belderson and A.H.Stride

BENEATH OCEANIC WINDOWS

The floor of the ocean can no longer be regarded as dead ground where little happens, for despite the great age of the earth its evolution continues apace, and its face will continue to be remoulded for many millions of years into the future. In this outline the emphasis will be on motion; that of submarine landslides and catastrophic flows of muddy water, vast outpourings of lavas, repeated vertical upheavals of large parts of the earth's surface, the break up and sideways migration of whole continents, the consequent birth and growth of new oceans and the disappearance of older ones. Many of these changes are only possible because the rocks of the earth are not as strong as they seem; given time and sufficient pressure the toughest rock will creep at up to a few cm per year.

If the earth were filmed from outer space with a time lapse camera shooting but a single frame every few thousand years, the movement of the continents might be observed directly. Although this obviously cannot be done, it has been possible, by the patient assembly of fragmentary data from all over the earth, to infer that continental drift has taken place. At the present time we are learning most from the windows between the continents – the ocean floors where the more deep-seated and fundamental processes of the earth's evolution can be observed most directly.

THE SCENERY

Is the scenery of the sea floor anything like that which we know on the lands, or is it totally different in aspect and in scale? This question must have been asked by many travellers across the broad waters separating shore from shore. Today the stage has been reached where enough has been learned to enable us to imagine what this hidden two-thirds of the earth's surface really looks like. During its exploration, first by laborious plumb-line and then by echo-sounder (Chapter 2), many unexpected features have been revealed. For example, the oceans are not deepest in the middle, as might be expected. Instead, the deepest parts tend to be close alongside the continents and can take the form of narrow, arcuate trenches up to twice the depth of the average deep floor and reaching as much as 11,000 m below sea level, while the middle of the oceans tends to be occupied by the highest, broadest and longest mountain range on earth, which locally even reaches above the surface as mid-ocean islands (figure 250).

The scenery of the land results from the interaction of the rival processes of erosion, deposition, uplift and subsidence. If erosion and deposition on land were to proceed unimpeded for long enough, the continents would be worn down to a smooth, featureless surface, partly depositional and partly erosional in origin, and standing just above sea level. With the continued deposition of sediments on the ocean floor, sea level would slowly rise and the waves would beat upon and wear away the perimeters of these plainlands until all were submerged and the earth's surface became a single, unobstructed ocean. That this is not already so means that the forces of construction are yet active, so that they periodically squash up a narrow belt of sediments to form mountains, usually at the edge of a continent, or else broadly uplift the continents to give them a new lease on life. Such long continued, massive erosion of the continents should make the oceans the sites of deposition, the dumping grounds for the bulk of the products of mechanical and chemical breakdown of the continental rocks. The ocean floor ought, then, to consist of a featureless blanket of sediment accumulated over the time span of earth history, or at least since the oceans existed. This is only partly true. The study of its shape and composition has shown that constructional activity is abundant and that the keynote in this is volcanism. In fact, the ocean floor may be considered as an immense cauldron of congealed volcanism, the surface of which has only a partial cover of sediments, at most a kilometre or so thick.

The surface of the solid earth stands at two main levels, much of the continents being at about 500 m above sea level and much of the ocean floors standing at about 4000 m below sea level. These two basic levels

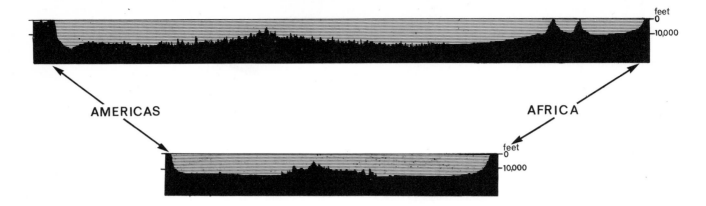

250. Two depth profiles of the Atlantic Ocean, from the West Indies to the Cape Verde Islands and north-western Africa (*top*), and from South America to equatorial Africa (*bottom*). These show the mountainous topography of the mid-Atlantic Ridge, the gently sloping continental rise leading down to the flat abyssal plains, and in one case a trench (after Heezen & others 1959)

reflect the difference in density of the materials underlying them, so that the relatively light granite raft of the continents, up to about 40 km thick, 'floats' upon a floor of denser basaltic and ultrabasic rocks. Surrounding each continent is a flat, shallow apron of variable width called the continental shelf. This is really an extension of the continent beneath the sea, the depth of which is determined by the upward or downward movements of sea level and of the continent itself, as well as by the sea's erosional and depositional activities. Thus, after any vertical movements of the continent, the sea will either tend to cut or build up a new shelf until a balance between erosion and deposition is once again achieved. When this happens the shelf can no longer build upwards, so that sediment supplied from the land must go towards the outbuilding of its oceanward edge. Here there is a fairly abrupt change in slope known as the shelf break, occurring at an average depth of about 140 m. Below this, the floor plunges away 3000 m or more, in what is known as the continental slope. This impressive incline, stretching for tens of kilometres, joins the two fundamental levels of the earth's surface and defines the outer limit of the continents. Its steepness varies a lot. If it has been strongly bent down or fractured during major earth movements, the gradient may be 20° or more, whereas if it is a depositional slope representing building out of the continental edge, the gradient is more gentle, about 3° or less. The latter kind of continental slope is often not the smooth feature which would be in keeping with its origin as the outer dumping ground of the continents. Instead, its surface is much gouged and scarred by erosional features of all sizes (figure 251), from incipient gullies to canyons whose dimensions rival the largest found on land. The discovery of numerous canyons and then the attempts to explain their origin was one of the exciting phases through which modern marine geological research has passed. It was only with the proposal and then laboratory demonstration of the turbidity current phenomenon that the riddle was to a great extent solved. Today it is generally accepted that although some canyons have been cut along faults, where the rock has been broken and can be easily eroded, and that the shallower portions of some canyons are drowned river valleys, the dominant process responsible for their formation is turbidity current erosion. Turbidity currents have even now retained a certain aura of mystery in that no naturally occurring and fully developed one has yet been seen. Nor is it likely that anyone will seek to observe one directly because of the danger involved. Their existence has been deduced from the distinctive deposits they leave on the deep-sea floor (see page 283), from torn telephone cables, sometimes broken in downslope sequence, from analogy with related 'density currents' which occur in certain lakes, and from small-scale demonstrations in laboratory models. The basic idea is this: if a quantity of sediment is put into suspension, as by a muddy undersea 'landslide', then the effective density of the liquid will be increased according to the amount of mud in suspension. If this muddy water overlies a slope, then the sediment-water mixture will inevitably flow down it. Thus, gravity is the driving force, and the continental slope the necessary incline. Because it is such a long slope it is the ideal locality for the full development of a turbidity current, which speeds up, picks up more sediment, becomes larger and of greater density until eventually it is moving downslope at a speed of perhaps 30 m.p.h. That

251. A block diagram of canyon heads at the top of the continental slope, based on a side-scan Asdic record (7 kms long by 1 km wide). There are numerous gullies in the canyon walls and a large slump below the ragged cliff

such a violent commotion could now and again be taking place in what were supposed to be the eternally still depths of the ocean came as rather a surprise. But at least it helped to enlighten the mystified telegraph operator as his line was parted somewhere on the sea floor beneath a hitherto undreamed-of onslaught of mud-laden water – as well as providing the powerful erosive force needed to explain the large and numerous canyons.

At the base of the present continental slope the gradient commonly decreases gradually to about 1:1000, within a rather ill-defined region known as the continental rise. This feature probably owes its origin largely to deposition of the coarser constituents of the turbidity currents where they begin to slow down at the base of the slope. Below the foot of major canyons great depositional 'fans' have been built up, and, where they coalesce side by side, the more general feature known as the continental rise develops (figure 252). The turbidity currents pass over these along shallow channel systems towards the deepest portion of sea floor accessible to them. The persistence of turbidity currents is indeed remarkable, for the larger ones may travel many hundreds of miles of low

gradient before reaching their death bed in the abyssal plains. These great ponds of sediment are the flattest regions on earth, often uninterrupted by hills for hundreds of miles, with a slope of less than 1:1000 and in some areas as little as 1:7000. They mask an older and much rougher buried topography which, towards the further side of the monotonous abyssal plain and nearer to the mid-ocean ridge, begins to protrude in the form of scattered, low abyssal hills. Locally there are shallow channels extending along the abyssal plains and sometimes connecting adjacent plains. They must mark the routes of turbidity currents seeking the deepest available sites for deposition. Further on, the plain abuts against and interfingers with the foothills of the mid-ocean ridge, where in many of the isolated hollows there are flat floors of ponded sediment built up by turbidity currents originating from sediment sliding on the steep slopes of sea-mounts.

The mid-ocean ridge system dominates the ocean floor (figure 253). It has only recently been mapped as a single continuous feature which straddles the globe, a mountainous chain 50,000 km in length and occupying a third of the ocean floor or nearly a quarter

252. Diagrammatic view of submarine canyons gouged into the older sedimentary rocks of the continental slope by the passage of younger material. The coarser fractions are deposited in the fans at the base of the slope while the finer material in the turbidity currents is carried out to the abyssal plain

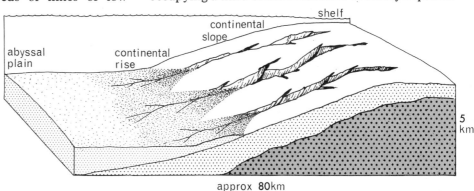

abyssal plain

continental rise

continental slope

shelf

5 km

approx **80**km

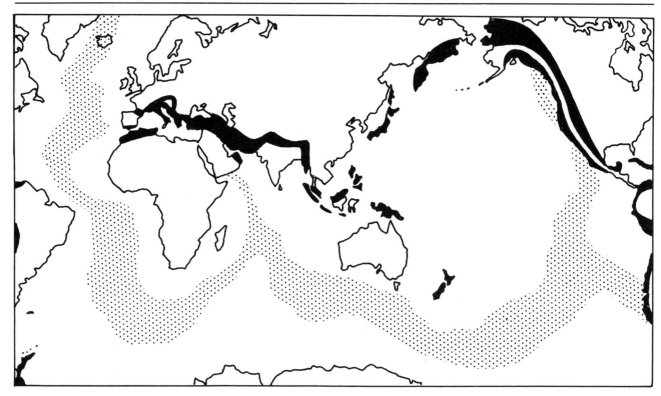

253. The outer limits of the exposed mid-ocean ridge system (dotted) and the young fold mountains of the land (black)

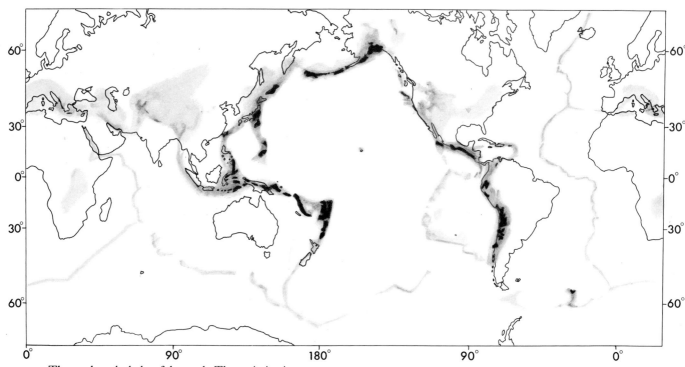

254. The earthquake belts of the earth. The variation in tone gives an indication of their relative frequency

of the earth's surface. In detail, the ridge seems to consist of rugged, somewhat parallel ranges of coalesced sea-mounts, some of which rise several kilometres and reach above the sea as volcanic islands. The trend of the ranges is frequently aligned parallel with the mid-ocean ridge as a whole, while along its crest there is a rift valley, so named because of its supposed faulted boundaries, resembling those known on land.

Continuing the crossing of the ocean basin, the topography repeats itself in reverse across to the shore of the opposite continent. This symmetrical type of arrangement is typical of much of the Atlantic and Indian Oceans, although even here it is really more complicated. A further important feature of the ocean floor is the common occurrence of linear fracture zones, sometimes displacing the axis of the ridge and sometimes cutting for thousands of kilometres across the ocean basins. These lineaments mark the lines along which there has been a relative horizontal displacement, so that points which originally were adjacent can now be several hundred kilometres or more apart.

The Pacific Ocean, apart from its great size, also has other distinguishing characteristics. It is almost encircled by a narrow zone of large and frequent earthquakes (figure 254) and by a series of deep trenches, generally curved in plan view and paralleled by curved chains of volcanic islands atop sub-sea ridges (figure 270). These features, known as island arcs, also occur in the Atlantic and Indian oceans, but are far more distinctive features of the Pacific. Another difference is in its 'mid-ocean ridge', which is not in the middle and not such a pronounced ridge, but rather a broad swell lacking a median rift valley and known as the East Pacific Rise. Abyssal plains are not common in the Pacific since much of the land-derived sediment is trapped in its marginal trenches and thus cannot get out into the further reaches, at least not by way of turbidity currents. Many thousands of sea-mounts are scattered about the basin as well as the liberal sprinkling of islands and atolls which reach above the sea at present. In some localities these sea-mounts are grouped in lines, and many have flat tops. The latter type are former islands which were planed off by wave action and have since subsided, sometimes a thousand metres or so. Where coral growth has kept pace with subsidence the coral atolls, so readily associated with the Pacific, mark the location of sunken volcanic islands. But it is around the perimeter of that ocean, along the belts of intense volcanic and earthquake activity, that the Pacific Basin is most distinctive. Here, along this turbulent boundary between ocean and continents, is demonstration enough of the earth's continuing capacity to change and remould its own surface.

This outline of ocean floor scenery makes little reference to local variation. The oceans include within themselves many marginal seas, each possessing some of the characteristics described above, although they vary a great deal in detail, such as in the proportion of continental shelf to deep water and abyssal plain. An analogy between the appearance of ocean scenery and that of the land might be made as follows: the continental shelf is an extension beneath the waves of the coastal plain which occurs along some coasts and which is really an emerged part of the shelf and continuous with it; many parts of the continental slope resemble the canyoned badland topography of some desert and semi-desert areas of land; the continental rise is like the overlapping fans of alluvial debris prominent in desert regions where flood waters sporadically debouch along the base of an escarpment – though on the sea floor it is on a much larger and more gradual scale; the lower continental rise probably looks very like the abyssal plain, and both of these compare in flatness with the alluvial flood-plains of great rivers such as the Mississippi; the abyssal hills and sea-mounts are nearly always volcanoes, the tops of some of which can be observed as volcanic islands, and the mid-ocean ridge may be visualised as a compound chain of numerous mountains constructed from vast sub-sea volcanic effusion, with occasional sheer cliffs formed by faulting. Of course, the surface texture of the floor will look very different from the landscape familiar to us, particularly in the absence of vegetation.

Ocean floors of the past were probably not always like those at the present time. For instance, the amount of turbidity current activity may have varied widely, in which case those features largely controlled by this action – the canyons, continental rise, and abyssal plain – would also have been more or less prominent. Likewise the mid-ocean ridge and marginal trenches which are the surface manifestation of more deep-seated activity beneath the floor will also have varied in their position and prominence throughout the ages.

THE SOFT LAYERS

Much of the sands and gravels worn away from the land and carried to the sea are deposited near the coasts or on the continental shelf, while the clay-rich material including airborne dust is fine enough to be carried largely in suspension towards the continental slope and deep-sea floor. Here it settles out as soft muddy layers which will eventually become new sedimentary rock. This mud, or 'ooze' as it is usually called, also includes the residual hard parts of tiny marine animals and plants living in the near-surface waters and which on death have joined the slow but

255. Fossil coccoliths magnified 6000 times under the electron microscope (see also figures 126, 127). These calcareous platelets from tiny marine plants are the major constituent of many limestones, including chalk (after Black 1964)

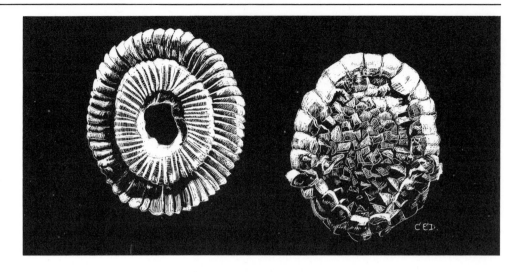

ceaseless rain of microscopic particles into the black depths. The remains of the relatively larger creatures (though still mainly less than one mm across) which have become incorporated in the mud have been used to name the sediment. Thus 'Globigerina ooze' contains a significant, though usually minor percentage of the calcareous shells of pelagic Foraminifera (figure 150), of which Globigerina species are the most abundant; 'radiolarian ooze' contains enough of the siliceous skeletons of the planktonic protozoans Radiolaria (figure 149) to make it distinctive; 'diatom ooze' contains the siliceous remnants of the tiny pelagic marine plants called diatoms (figure 122); and 'pteropod ooze' contains the calcareous shells of a small pelagic mollusc (figures 173, 193). The carbonate portion of the fine matter is mostly made up of the minute plates (coccoliths) of a tiny calcium carbonate-secreting planktonic plant called a coccolithophore. Although so insignificant in size, it is very important in deep-sea deposition because of the astronomical numbers of its coccoliths (figures 126, 255) shed into the general snowfall of organic and inorganic particles. They often comprise half the sediment, and sometimes even more, so that an almost pure deposit of coccoliths results, which, once compacted, would resemble the familiar chalk. The potential animal and plant content of the oceanic sediments depends on whether the water temperature and concentration of nutrient salts in the upper layers of the ocean are suitable for the growth and proliferation of Foraminifera or Radiolaria, or the sometimes prodigious blooming of diatoms and coccolithophorids.

Apart from these 'oozes' there are several other types of deep-sea deposit which generally exist in some kind of intergradation with them. Such are the glacial sediments shipped far from polar latitudes by melting icebergs, the fine volcanic ashes belched out from volcanoes and, in the deepest parts of the oceans, the 'red clay'. The latter accumulates below a depth of roughly 4-5000 m in the Pacific and somewhat deeper in the Atlantic and Indian oceans, where the combination of low temperature and high pressure tends to cause particles of calcium carbonate to be dissolved. Thus, the Foraminifera shells and the coccoliths are corroded away and finally disappear, together with even the toughest remains of the siliceous diatoms and Radiolaria, so that the various oozes give way to a reddish-brown inorganic residue in which small quantities of cosmic debris can be recognised (figure 258). This red clay accumulates exceedingly slowly at the rate of about one mm per thousand years, as compared with one cm or more per thousand years for Globigerina ooze. These rates, although typical of the deep-sea environment, will naturally vary greatly from place to place. For instance, off major rivers deposition can proceed a hundred times or so faster.

The pattern of the various deep-sea sediment types is reasonably clear-cut in space, and might perhaps be thought to be fairly constant in time. But as usual in geology, change is the rule, continuing stability the exception. Sediment cores, obtained by dropping a weighted tube into the sea floor (figure 256), usually indicate change of some kind (figure 257). One of the most common of these vertical variations (changes in time) is that in the calcium carbonate content. This may reflect variations in the degree of calcium carbonate solubility brought about by changes in water temperature, particularly during the frequent warmer and colder oscillations that occurred during the Glacial Period of the last one and a half million years or so. Or it may be due to changes in near-surface productivity brought about by shifts in the surface currents or changes in chemical content of the near-surface waters. Another series of changes, such as in

256. A corer being manoeuvred over the stern of RRS *Discovery*; the weights at the upper end drive the long tube into the sediments and the core is recovered within the tube

257. Three short sections of a long sediment core cut longitudinally; the banding effect of the deposition of different materials at different times is well marked

the relative abundance of certain species of planktonic Foraminifera, has been useful in giving a more precise chronology to the various stages within the Glacial Period. Some species prefer to live in warmer waters and some in cooler waters, and thus a count of their relative numbers at different levels in a core will give an idea of the relative water temperature at the different times. This, in conjunction with radiocarbon dating (Chapter 3), has allowed the decline of the last glaciation to be dated at about 15,000 years ago, and has also provided dates for the major, and many of the lesser, climatic oscillations of the entire Glacial Period (figure 259).

The most puzzling of the variations exhibited in deep-sea cores were the sand and silt layers found interbedded with the muds under the continental rise and abyssal plains. Until these were discovered it had been supposed that the ocean depths remained ever undisturbed, with at most a gentle creep of bottom waters. But here was evidence for strong, if occasional, currents, and with the frequent discovery of shallower water fauna in the sands it was realised that they are the deposits of turbidity currents. Discovery of these 'turbidites' made it necessary to reappraise many of the sedimentary rock sequences known on land, because it meant that many rocks which had been assigned to a shallow water origin were really deposited in far greater depths. However, the occasional and catastrophic turbidity currents are not the only appreciable water movement in the deep sea. Deep currents are certainly strong enough, at times, to cause differential deposition of the pelagic

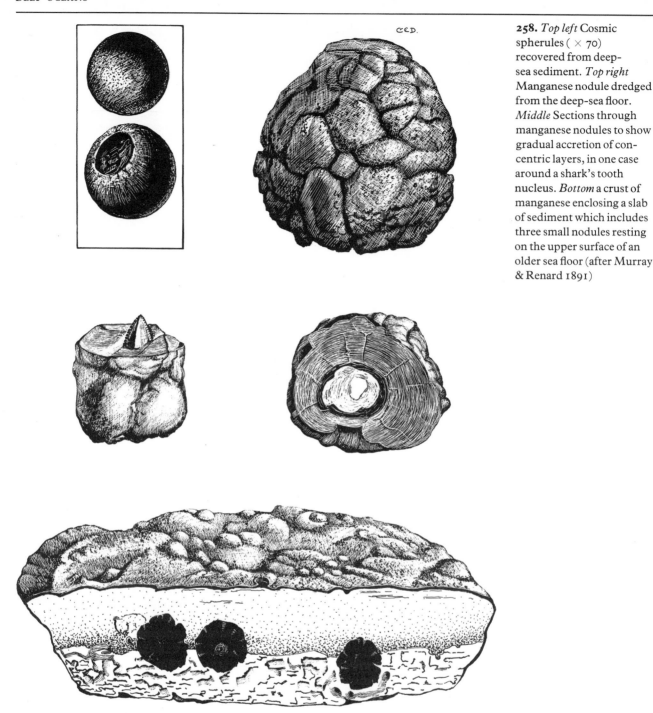

CED.

258. *Top left* Cosmic spherules (× 70) recovered from deep-sea sediment. *Top right* Manganese nodule dredged from the deep-sea floor. *Middle* Sections through manganese nodules to show gradual accretion of concentric layers, in one case around a shark's tooth nucleus. *Bottom* a crust of manganese enclosing a slab of sediment which includes three small nodules resting on the upper surface of an older sea floor (after Murray & Renard 1891)

muds in the hollows among abyssal hills, or in places to develop broad mound-like sedimentary features. Also, ripple and scour marks on sands are seen on many of the underwater photographs taken on the summits of sea-mounts at depths down to several thousand metres. Such evidence of relatively strong currents is less common on much of the continental slope, rise and abyssal plain, but ripple marks have been observed locally in abyssal depths, and reasonably strong and variable (though less than one knot, 50 cm/sec) currents

have been measured at great depths. Thus, not all silty layers in the deep sea are necessarily turbidites. It also seems probable that turbidity currents are not the only means of moving sand down the axes of submarine canyons. Observations by divers and from submersible craft have shown that a process of gravitational creep and sand flow sometimes takes place, and that this may also be responsible for causing deepening and undercutting of canyon walls.

Although the sea floor will be looked at through the

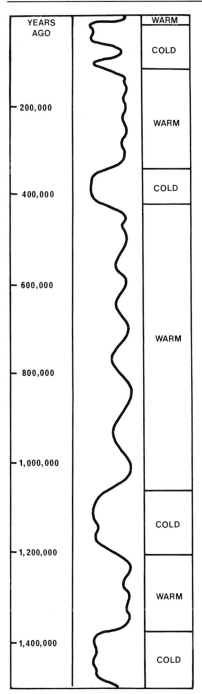

YEARS AGO		
		WARM
		COLD
— 200,000		WARM
— 400,000		COLD
— 600,000		WARM
— 800,000		
— 1,000,000		
— 1,200,000		COLD
		WARM
— 1,400,000		COLD

259. A general climatic curve of the cold and warm periods within the Ice Age, which has characterised the most recent era of earth history. The curve is based on a study of deep-sea sediment cores (after Ericson & others 1964)

mounds lie scattered like miniature volcanoes, and faecal pellets and worm tubes are frequently seen. Now and then a photograph shows one of the organisms responsible for the tracks – such as a sea cucumber eating its way over the mud surface. As the depth increases and *Globigerina* ooze gives way to red clay, so there is less life and the shapes of the tracks and burrows are influenced by the finer texture of the sediment. However, the signs of fauna do not disappear altogether, for living organisms exist in the bottom even in the deepest trenches.

A striking feature of many photographs of the ocean floor are the manganese nodules (figure 258). Enormous stretches of floor, particularly in the Pacific, are carpeted with these spheroidal concretions. They vary from a few cm or so in diameter, with the individuals in a particular region often closely resembling each other in size and composition. Manganese dioxide is generally the major constituent, but a large and varying amount of iron and much smaller proportions of a variety of more valuable metals are also included, as well as a considerable clay content. The spheroidal nodules are the most distinctive form, but manganese and iron oxides also occur as slab-shaped lumps, as encrustations and impregnations on and into various rocks, and as small grains within the sediment. There are several theories to account for the origin of these deposits. Of these, the favoured ones are direct precipitation from solution (sea-water is essentially saturated with manganese and iron), or possibly by means of a gradual agglomeration of hydrated colloidal particles around some object on the sea floor which acts as a centre for accumulation. The phenomenon of the continual creation of more manganese nodules at the present time represents a virtually inexhaustible potential bonanza waiting until it is needed, when no doubt a way will be found to recover the nodules at an economic price.

Powerful low frequency echo-sounders (Chapter 2), designed during the past ten years to achieve penetration into the sea floor have provided a sudden wealth of continuous profiles of the layered sediments. These show conclusively that the undersea foundations of the continents are covered in many regions by successive layers of younger sediments which have built the shelf upwards and extended the slope outwards, so that the continents have increased somewhat into the ocean (figures 267 and 268). Some slopes, however, are too steep for accumulation to take place (figure 260). The reflection profiles also show that a thick wedge of sediment underlies the continental rise, while flat-layered deposits underlie the abyssal plain in keeping with the turbidity current concept. The numerous sediment ponds amongst the abyssal hills and on mid-

portholes of deep submersible craft more commonly in the future, a general impression of the small-scale relief has already been obtained by the many thousands of photographs taken at numerous localities throughout the oceans. Ripple and scour marks have been mentioned above, but these are by no means typical. The great majority of the photographs show a muddy floor across which numerous convolute trails and tracks of animals wend their way, often in curious and systematic patterns (figures 182, 189). Various burrows and

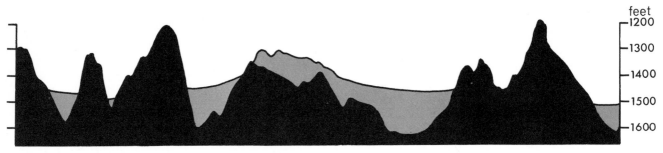

feet
1200
1300
1400
1500
1600

260. A continuous reflection profile on the flank of the Mid-Atlantic Ridge showing high volcanic peaks and valleys partly filled with sediment ponds which mask some of the volcanic topography (after Ewing & others 1964)

ocean ridges (figure 260) contain flat-layered turbidites, while it can be seen in general that the sediment cover is thinner towards the crests of the ridges. Beneath the younger sediments of the Atlantic there are obvious buried surfaces of great extent, but now somewhat distorted, which have been interpreted as older abyssal plains of much larger area than those of the present day. In addition to the submarine canyons, described from many parts of the world, it is now apparent from the new profiles that the continental slopes have also been scarred by numerous slumps (underwater landslips). The features range up to huge masses many kilometres in length and a hundred or so metres thick (figure 267). The amount of movement varies, from a hint of a slip, to blocks of oceanward dipping beds which have rotated backwards during movement so that the beds now slope towards the land, or in more extreme cases blocks have become detached and slid for many kilometres downslope, perhaps mixing pell-mell with the water and so initiating a turbidity current. They have been detected along escarpments caused by faulting and in the walls of submarine canyons, where they result from local steep gradients. Slumps are also induced on a regional scale where the gradient of an originally stable continental slope is increased during downwarping of the edge of a continent. Sediments with a high water content can be partially liquefied when agitated, as by an earthquake, because the grains become dispersed. They will then slump or even flow downslope. Soft muds, laid on a slope and buried rapidly, are also prone to yield when the load above them exceeds their strength. Other features produced by gravity, though in reverse as it were, have recently been found in the oceans. These are the so-called *diapiric* structures, such as salt domes, which have formed by the slow upward plastic migration of relatively lighter sediments through several kilometres of denser overburden and may project up to several hundred metres above the floor. These peculiar features, known for many years on land,

occur on some continental slopes and even on the deep-sea floor, such as that of the Gulf of Mexico (figure 261), where one has recently been drilled and oil found associated with it. Groups of large folds have also been described from the continental slope in several parts of the world. Some of these may be elongate diapiric features, but it seems likely that others are ruckles caused by the mass downslope creep of sediments.

THE VOLCANIC FOUNDATIONS

The mid-ocean ridges represent the largest exposures of volcanic rocks on the surface of the earth. Elsewhere, the oceans are littered with sea-mounts, which are also made of volcanic material, while in both areas active or recently active volcanoes reach above sea level as islands. The amount of present day activity below 2,000 m is not yet adequately known, as there may be no sign at the water surface which could be observed from passing ships. It has been estimated that there may be five times as many active submarine volcanoes as there are on land, many of which are concentrated along the crest of the mid-ocean ridges (figure 269). It seems likely that the volcanic layer extends under the abyssal plains that separate the mid-ocean ridge from the adjacent continents. There are many reasons for this belief, the most direct are that almost every sea-mount protruding through the abyssal plains is volcanic, as can be shown by sampling and by their magnetic signature, detectable by magnetometer (Chapter 2), and because similar magnetic patterns are also found even where no hills protrude. Seismic refraction shooting with explosive sound sources (Chapter 2 and figure 262) leads to the same conclusion, as rocks under the abyssal plains have a similar thickness and as the velocity of sound in them is about the same as in the exposed rocks. Indeed, such work with really big explosions as well as the behaviour of earthquake waves, indicates that the floors of the main oceans have a remarkably similar basic structure. At the top is the sedimentary layer which averages less than one kilometre in thickness, depending on whether there is ready access for sediment, particularly that moving as turbidity currents, which may be trapped near the land in trenches or be cut off by ridges. Below the undoubted sediments lies the so-called second layer which is one or two kilometres thick. In some regions this may partly consist of hard limestone but in general is more likely to be made of volcanic rocks of remarkably uniform basaltic composition, which have been sampled in many places where exposed, together with some interbedded sediments. Then comes about five and a half kilometres of 'oceanic crust'. This contrasts with 'continental crust' which is much thicker (about 35 km) and much more variable in composition but predominantly of granitic type. The base of both types of crust is defined by the Mohorovicic discontinuity [or 'Moho'], where there is a marked change in the speed of sound in the rocks. If a true scale section through the earth was drawn on this page the crust would be only a thin line, as the bulk of the earth consists of mantle rock around a relatively smaller central core. Just what the mantle is composed of, we are not yet sure. The favoured rock type is peridotite (a rock rather denser than basalt), and it has even been thought possible that the core is a much denser form of the same material, brought about by the great internal pressure of the earth. Similarly, the physical and chemical changes which rocks undergo, when subjected to different temperatures and pressures and change in water content, have been used to explain the existence of the Moho. This would mean that the material above the Moho (the 'oceanic crust') is the same as that below it (the mantle), so that the term 'oceanic crust' is not really suitable and could be replaced by the term 'rind'.

Three other geophysical approaches need to be mentioned because of the light they throw on movement of the upper layers of the earth. The first of these is concerned with earthquakes. Besides being able to pinpoint where these have occurred, on land or beneath the ocean, it is now possible in some cases even to determine the direction of motion on the two sides of the fault where the earthquake occurred. Measurements of gravity, which can now be made along a continuous profile from a surface ship, can reveal an excess or deficiency of crustal matter and hence indicate whether the floor is likely to rise or sink in order to restore equilibrium. A further indication of earth behaviour is provided by measurements of the rate at which heat is passing upwards through the ocean floor. From this it has been concluded that much larger heat flows are associated with the crest of a mid-ocean ridge than with its flanks and elsewhere in the oceans. This finding, together with the discovery

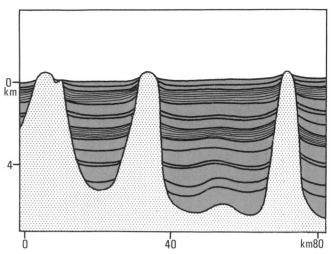

261. A heavier sheet of mud overlying a thick, less heavy layer of salt causes the latter to flow slowly upwards through the sediments as finger-like projections (after Talwani & Ewing 1966)

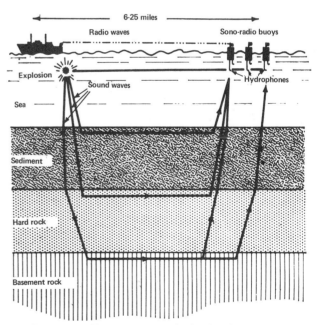

262. The sono-radio buoy system of seismic refraction shooting. Explosions are fired from the ship and the sound travels at different speeds through the different bottom layers. The signals received by the hydrophones beneath the buoy are transmitted back to the ship (from Hill 1963)

of anomalously low sound velocities in the mantle beneath the mid-ocean ridges suggests that the mantle material is hotter and more plastic thereabouts.

Two major conclusions follow from these studies. First, there are no large sunken continents beneath the oceans as was at one time thought, although there do appear to be some small, isolated fragments here and there. If 'phantom continents' did exist they must have been converted, somehow, from thick crust of continental type to thin crust of oceanic composition. Secondly, if the present oceans have existed as permanent features since the early days of the earth then the sediments on their floors are much thinner than would be expected. Both considerations support the idea that the continents have in some way moved laterally, 'drifted about', on the earth's surface.

THE MOVING FLOOR

The preceding part of this chapter showed how the sea floor varies, both in shape and composition, with its great underwater mountains, trenches, plains and canyons. A closer look at many such features shows them to be associated with large up-and-down and sideways movements of the earth's crust equalling, and possibly exceeding, similar movements detected on land. The geology of the land tells us that there have been repeated uplifts of the continents which must be balanced by down-bending of the ocean floor, for which there is some evidence. The many sunken volcanoes of the Pacific bear witness to subsidence over broad oceanic regions, while elsewhere there are islands with relatively young coral reefs that have been raised 1,000 m or more above present sea level. In the Mediterranean there are submarine canyons whose upper reaches (well below the depth due to sea level changes) are thought to be drowned river gorges, and there is also abundant additional evidence that rapid changes in the level of the adjoining land are going on now. Indeed, much of that sea, as we know it, is probably only ten million years or so old. Such changes are in keeping with the fact that the continents have periodically undergone large vertical displacements, that former sea floors have been deeply buried by sediments and these deposits have then been uplifted to a great height along mountain ranges. The overall vertical changes may be up to 15 km or so. In contrast, some of the horizontal displacements, up to some hundred times greater than the vertical ones, are so large that for the first 100 years of serious geological research they were not recognised. Even the concept of smaller lateral displacements of about 150 km only gained acceptance slowly. As knowledge of the southern continents increased it became obvious that the rocks of South America, Africa, India, Australia and Antarctica experienced remarkably similar geological evolution which could only be understood if they were at one time parts of a single land mass, known as Gondwanaland. The good fit of the two sides of the South Atlantic was further impressive evidence that this ocean was a huge gap resulting from relative opening of more than 3,000 miles (about 5,000 km).

The contribution of the marine geologists has been to produce evidence from the sea floor, every bit as compelling as any from the land, that both realms have been affected by vast lateral movements. Some of this movement has occurred along faults that are rather better seen than on land, for they have been followed as escarpments for many hundreds of kilometres, and offset the mid-ocean ridges by a few hundred kilometres. At the present time, numerous earthquakes around the Pacific Ocean and along the mid-ocean ridges show where substantial movements of some sort are taking place.

SOME MAJOR EVENTS

The portion of geological history that led to the present configuration of land and sea seems to have commenced about 200 million years ago. At that time, the lands that we know today were grouped in two much larger units, Laurasia in the north (Eurasia and North America) and Gondwanaland in the south. These were divided by the Sea of Tethys stretching from west to east between them (figure 263). Beyond the bounds of the two supercontinents must have extended an ocean far larger than the present Pacific.

Tethys seems to have been wide enough for the two sides to have their own distinctive assemblage of animals and plants. Gradually the western end of Tethys, including what is now the Gulf of Mexico and the Mediterranean, must have become more isolated from the main ocean, judging from the prevalence of salt deposits in and around it. The continued narrowing of Tethys was associated with the break-up of the super-continent of Gondwanaland into fragments as large as Australia, New Zealand and Antarctica.

The first indication of a new sea around Africa was provided by the salt deposits on both its eastern and western sides, which were formed about 150 million years ago. The early South Atlantic probably resembled the present East African rift valleys with their steep escarpments and lakes hundreds of metres deep, but when it widened and the sea broke in, perhaps temporarily at first, it began to resemble the Red Sea of the present time. Soon, however, the rift valley widened further and normal offshore, deep water, marine deposits were laid along its South American and African continental slopes. At about the same time, or perhaps as early as 180 million years or so ago, it seems, from

263. The site of the ancient sea of Tethys, between fragments of the two former super-continents

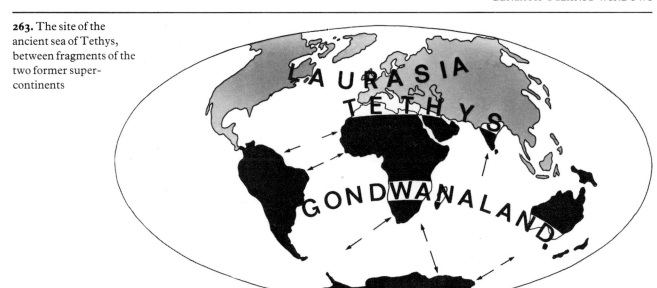

samples and from extensive continuous reflection profiles, that the North Atlantic had formed, leaving a scatter of continental fragments in shallow banks off the western side of Europe, rather in the way that the Seychelles and Madagascar are detached from eastern Africa. Around the perimeters of the two super-continents the soft continental rise and slope sediments were pushed up against, and welded to, the sides of the continents as these moved outwards over the proto-Pacific floor. In this way the young fold mountains such as the Andes and Rockies were formed.

Eventually, about 50 million years ago, the climax of events in the Sea of Tethys began to take place. Its soft deposits were crushed up along the whole line of its floor from the western Mediterranean between Africa and Europe to the Himalayas between India and Tibet and beyond, in the great collision between parts of Gondwanaland and Eurasia. Subsequently, Africa seems to have withdrawn somewhat from Europe, so forming the present Mediterranean, and leaving some of the Tethyan sediments welded to Europe. In contrast, India continued to plough into Asia, so giving rise to the exceptionally elevated regions of Tibet. Soon afterwards a gap between Africa and Arabia began widening along the line of the Red Sea. At the present time, the severest, most numerous and deep-seated earthquakes, most of the active volcanoes and the majority of the deep trenches, still occur around the edge of the Pacific; and there is also some activity along the Alpine-Himalayan belt. From this it can be concluded that the intense horizontal movements are still taking place.

THE HOT JAM-POT

Many geologists have looked upon the earth as a kind of 'superorganism' involved in a series of endless cycles. For example, old, hard rocks are broken up and the fragments conveyed elsewhere, and reconstituted into new sedimentary rocks which, in the course of time, themselves become exposed to decay and removal. This continual circulation of matter itself depends upon the conversion of seas and low-lying areas of land into uplands, and the bending down of former uplands at different times and places to form new sedimentary basins. A great deal is known about the effects of both local and regional earth movements – for instance, the squashed up sediments of the former Tethys are well exposed in the Alps, where they have been studied in great detail. But still the search goes on for a satisfactory fundamental explanation of the origin of these movements. The various hypotheses put forward include either an expanding or contracting earth, but the one which has gained most ground, and is currently most fashionable, is the theory of thermal convection within the earth's mantle. This supposes that the rocks of the mantle are more plastic than those of the crust are known to be, and can flow rather as does a pot of boiling jam, except of course that the process is very much in slow motion. Such convective movement within the earth implies uneven heat distribution – for which there have been a number of explanations.

The idea that convection cells might be able to move continents about originated with A.A.Holmes, although they had already been invoked to account for mountain building. The convection cell concept was refined by H.H.Hess, and the term 'sea floor spreading' was coined by R.S.Dietz. It is argued that a convection cell causes upwelling of rock beneath a mid-ocean ridge, and that much of the material spreads away from it to either side, thus causing a tensional crack, or rift valley, to appear along the medial line of the ocean floor (figure 265). The descending portions are located beneath adjacent continents, so that, to continue the jam-pot analogy, the continents represent patches of scum accumulated above the areas where cooler mantle rock is sinking. In this simple model, liquid basalt wells up and fills successive cracks extending along the rift valley, so that the most recently formed part of the ocean floor lies within the rift valley while the rest of the floor has been moved progressively further away. During the cooling of each strip of new floor the magnetic grains become aligned with the earth's magnetic field at the time of the intrusion. Magnetic profiles taken at right angles to the ridge show a series of alternating positive and negative anomalies (figure 264), from which the remarkable conclusion follows that the magnetic north pole has alternated from the Arctic to the Antarctic and back again many times during the last 200 million years or so. As this reversal pattern should be standard throughout the oceans, the relative age of large patches of mid-ocean ridge can be suggested by matching up their anomalies, from traverse to traverse, with those known from dated lavas on land and from deep sea sediments. Comparison of this sort seems to be rewarding for the medial parts of the mid-ocean ridges, probably formed during the past 10 million years, and suggests that the rate of sea floor spreading averages about 1 cm or so a year, but this must vary widely in space and time (figure 266). For instance, at the present time, spreading from the East Pacific Rise (at about 5 cm per year) appears to be several times faster than from the Mid-Atlantic Ridge. If we remember the boiling jam-pot, this is not a surprising finding. Here we see the rapidly varying intensity of individual convection cells, and also the lateral migration of the boundaries between the different cells, and that the more vigorous ones gain at the expense of their neighbours. The Pacific Ocean provides two possible examples of such variation on a much longer time scale. It seems likely that the East Pacific Rise and America have moved laterally with respect to one another, so that the rise which was at one time entirely in the open ocean is now partly under-neath California, from where it emerges northward before passing beneath Alaska. Such a position suggests that the Gulf of California will lengthen north-wards so that a strip of America will be torn away and thrust westward from the rest of that continent, which

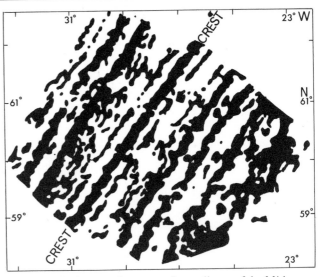

264. A detailed magnetic survey of a small area of the Mid-Atlantic Ridge, located south of Iceland. It shows essentially parallel anomalies of opposite polarity interpreted as successive vertical sheets of basalt injected as molten rock during sea-floor spreading (after Heirtzler & others 1966)

itself will perhaps move eastward to narrow the Atlantic of the future.

In view of what has been said already, it is not surprising that the middle of the Pacific is occupied by what appears to be an older mid-ocean ridge, known as the Darwin Rise, which has become inactive and has subsided from its originally more elevated position. It may be representative of ridges up to several hundred million years old which were responsible for closing up the earlier north Atlantic Ocean; this had something like the shape of the present ocean but existed up to 600 million years ago. Under such a mechanism the ocean basins have little chance of being filled up with sediment, since the floor is moving sideways about a thousand times as fast as the sediments are being deposited on it. This explains why sediment is thin near the crest of the Mid-Atlantic Ridge and progressively thicker away from it. The evolutionary stages of an ocean that seem to have occurred are shown in figure 272. A fissure develops and extends across the continent (e.g. Red Sea), the two buoyant parts are carried away from one another as on conveyor belts, the continued lateral movement of a continent is opposed by an adjacent oppositely moving convection cell, causing crumpling up of the continental rise sediments to give fold mountains to that side of it (e.g. South America); the stage then follows where material is being welded to both sides of the continent, with volcanoes resulting from the frictional heat, as

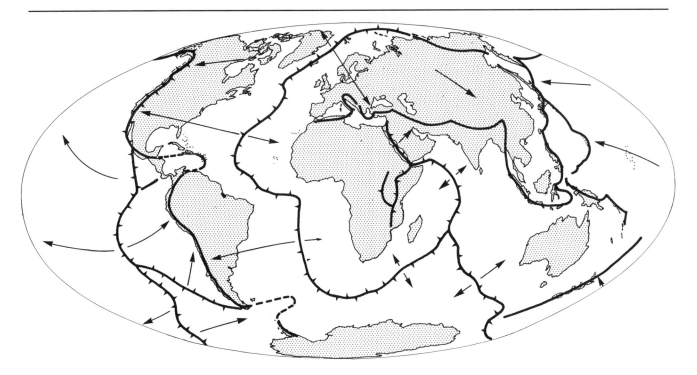

265. The mid-ocean ridges where the most recent ocean floor has been made (barbed line) and the latest main phase of mountain building on land. The arrows show one interpretation of the supposed direction of near surface flow of rock within the earth's mantle, which is thought to be responsible for causing continental drift (after Wilson 1963)

266. A recent interpretation of the position of the axes of the mid-ocean ridges (broad lines), the fractures which dissect them (thin lines) and the possible average rates of ocean floor spreading in cm per year for the past 10 million years. The dotted line shows the approximate distance moved away from the crest in the past 10 million years (Heirtzler & others 1968)

seems to be happening in south-east Asia at present and is supposed to have happened to North America in the very distant past; and the final stage where two continents are welded together. Thus, the expression 'sea floor spreading' is not synonymous with 'ocean widening' since the speed of spreading is determined by the speed of subcrustal convection flow, whereas the speed of ocean widening depends upon whether the continent is being transported with the subcrustal flow or not. And even if one ocean has widened it means another has been forced to contract (such as the Pacific) or has even largely disappeared (such as the Sea of Tethys).

Apart from this large-scale convection which is thought capable of moving entire continents around the face of the earth, there is also the possibility of smaller-scale convection cells exerting an influence on smaller areas. Thus, it has been suggested that some of the basins of sediment within the continents may have developed because part of the crust was worn away from beneath by local convection cells. In the more extreme cases, the intercontinental seas such as the Gulf of Mexico, Caribbean, Black and Caspian seas, each in part with an 'oceanic type' crust, could have developed in this way.

As is usual with a fashionable hypothesis, however, there is a danger of trying to use it to explain too much, and of glossing over the difficulties for which it cannot account. For example, there are one or two known patches of relatively older sediment near to the crest of the Mid-Atlantic Ridge, and what appear to be small pieces of continental material in the Atlantic and the Indian oceans which have somehow been partly left behind, instead of being carried off attached to their mother continents. Also, the youngest sedimentary fill of some of the deep trenches around the Pacific is practically undeformed. If the ocean floor is spreading out from the mid-ocean ridges and eventually diving underneath the continents, it would seem to be reasonable to find continental rise sediments crumpled up against the continent. Although this could very well have happened in the past in many places, there is little evidence of it taking place at the moment. A further

suspicious point is that the simple magnetic pattern recognised near the crestal zone of the mid-ocean ridges does not seem to be present on the flanks or beneath the abyssal plains.

NEW DISCOVERIES

Increasing numbers of geologists and geophysicists are working on many lines of sea floor research, and it seems likely that before long the present phase of major discovery will merge into one of consolidation, when the details will be filled in. Good contour maps of the entire ocean floor will be available eventually, and the new long range side-scan asdic (G.L.O.R.I.A. – see Chapter 2), first used recently in the Mediterranean, will provide wide plan views of the deep-sea floor giving valuable information which can be obtained in no other way.

A new dimension in the study of the history of the oceans and continents is now being added by the American Deep Sea Drilling Project. Numerous bore-holes are being drilled into the deep ocean floor from a specially equipped vessel. Those already completed have been of enormous benefit. For example, they have confirmed the relative youthfulness of the present ocean floor in comparison to the great age of the continents and proved that the earliest sediments, deposited on the basaltic basement, age progressively with distance from the crest of a mid-ocean ridge. The dating of these sediments is in close agreement with that suggested by the linear magnetic anomalies. This gives strong support for the hypothesis of sea floor spreading and thus, also, for continental drift. A further important result is the provision of an almost complete fossil marine sequence stretching back about 100 million years which will be of great use to geologists. It is to be hoped that this programme will continue, and that the planned much deeper hole will be bored to reach down through the whole sedimentary and basaltic sequence and even to penetrate beneath the Moho. Samples of the Earth's mantle rock should do much to set limits to speculation about processes operating deep within the earth.

267 and **268.** Simplified vertical sections between the continental shelf and upper continental rise off the western edge of Europe, based on continuous reflection profiles. Old rocks (brown) are overlain by the first deposits (orange) of the present Atlantic and these in turn by younger sedimentary rocks (yellow) representing the latest phase of outbuilding and upbuilding of the continental edge. The series from top to bottom shows increasing importance of slumping (S), faulting (F) and canyon dissection of the slope. (After Stride & others 1969)

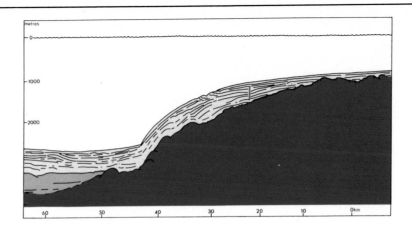

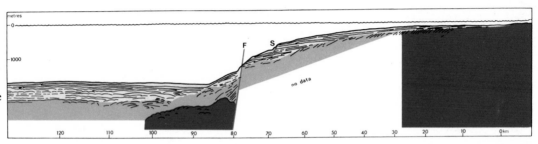

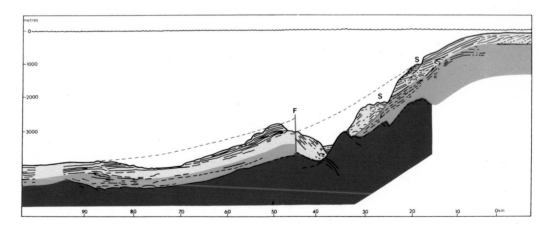

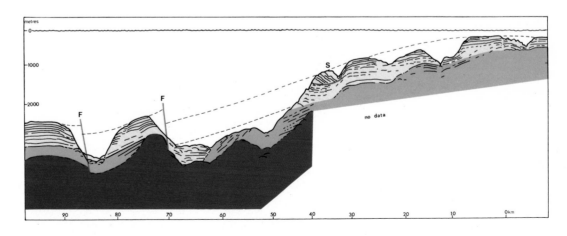

269. *overleaf* Submarine volcanoes are probably not uncommon in certain regions of the sea bed; the appearance of Surtsey Island in 1963 was a spectacular reminder of this activity

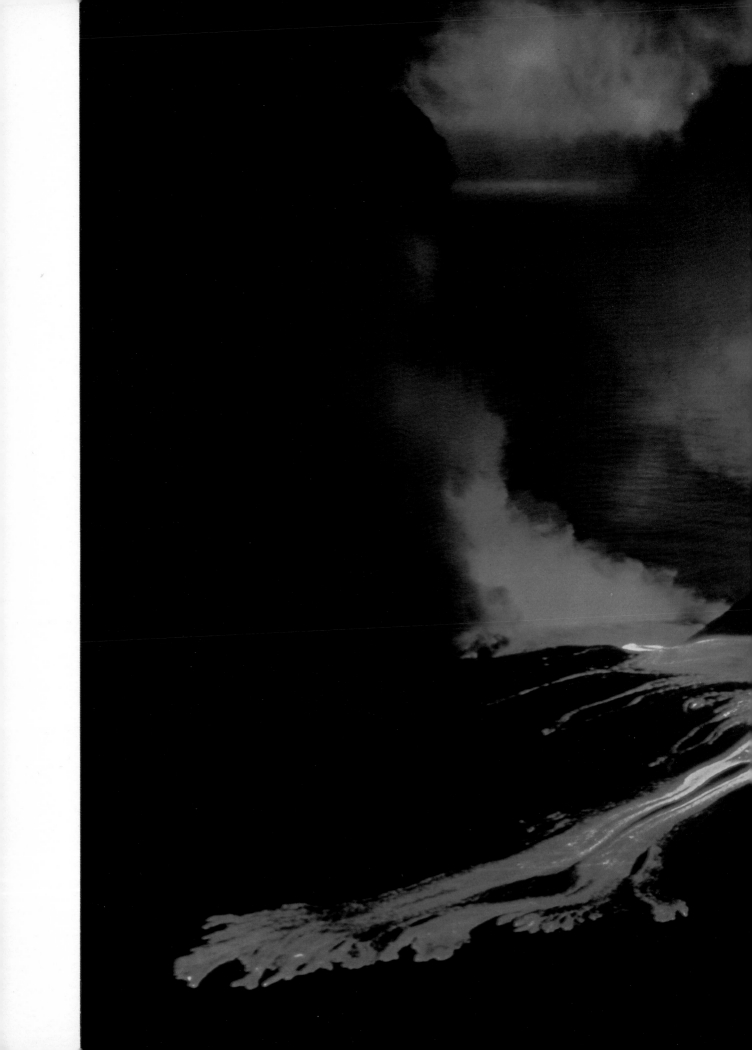

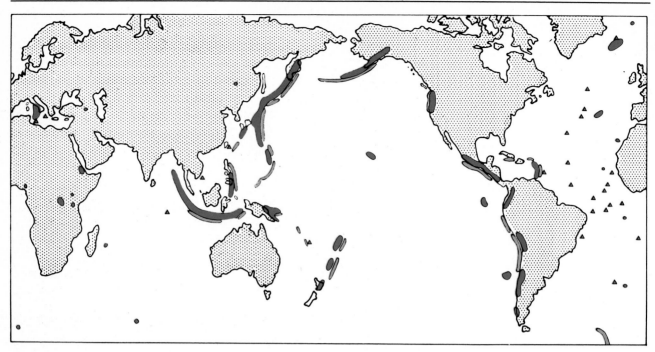

270. Regions of volcanic activity on land (red) and the few known sites of volcanic activity on the sea floor (red triangles), together with the location of long trenches (green) which are the deepest parts of the ocean. In the Pacific the volcanoes occur in curved island chains bordering these trenches

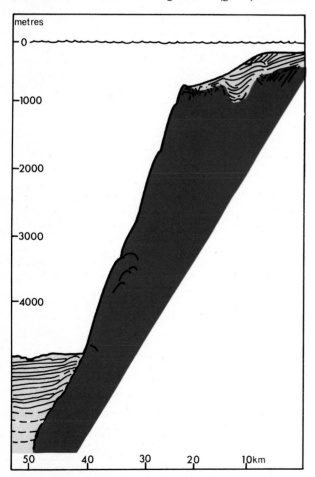

271. Continuous reflection profile of a continental slope too steep for present-day deposition. Youngest sediments (yellow) are deposited only on the continental shelf and at the base of the slope in a thick sequence which includes turbidites (after Stride & others 1969)

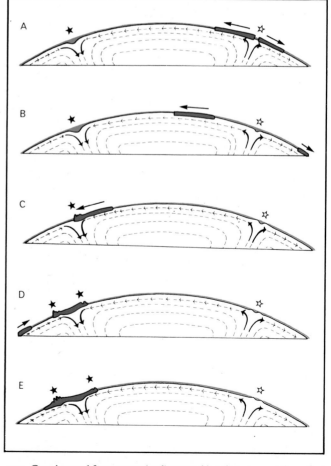

272. Continental fragments (red) moved by slow convection of the mantle rock (white). White stars show the location of tension induced by earthquakes (along mid-ocean ridges) and black stars indicate sites of compressional earthquakes (along island arcs and trenches)

11 Organization and Effort

Malcolm R. Clarke

To make any but the broadest generalisations concerning changes in the overall effort and relative national efforts in deep-sea research would be both difficult and imprudent, because most statistics are out of date before publication, are incomplete, and are open to grave errors of interpretation due to different methods of compilation. However, one thing is quite clear: mankind's interest in the sea has now reached a stage when large expenditure on research is amply justified by the sea's great potential as a source of food and minerals, its importance for both civil and military transport, and its influence on weather, climate, and the land. We stand on the threshold of a great expansion in the science of the sea; an expansion which is likely to become commensurate with that in space science. The overworked phrase 'Inner Space' to describe the ocean seems to have been coined with this expansion in view, and the claim that 'There is more of interest on the bottom of the sea than on the moon's backside' may prove more prophetic than controversial.

While avoiding further generalisations or conclusions, a few details on particular facets of this expanding and complicated field of endeavour may be of interest and may even indicate something of the present effort.

First, a word on research vessels. To gather data from deep water one can either use a small vessel with limited range and use it from a port near the place being investigated, or a large vessel with a much greater range and better sea-keeping qualities so that any weather can be endured without seeking port. A large vessel has many advantages for oceanography; its range allows it to work in the centre of the most inaccessible oceans, its size permits the use of large winches capable of paying out and hauling large quantities of wire measuring up to 15,000 m in length, extensive laboratories facilitate scientific work at sea and, as stability increases with size, work can be carried out in bad weather conditions. With such vessels, information can be gathered from wider afield and from deeper in the sea, so that scientific enquiry can be more broadly based and conclusions are more likely to be correct. Large ships have some disadvantages.

They are more expensive to run; a day may cost over a thousand pounds (2,400 dollars) and, as often no more than one instrument or net can be used at one time, the data is much more expensive than when obtained by a small ship. To be economical, large ships must be kept at sea for most of the time, and this means that they usually have to cater for a variety of functions, and their design is something of a compromise between the requirements of different sciences.

Before passing from the subject of ships let us consider how oceanographic expeditions are organised. The instigators of an expedition are either scientists who require certain information, and can justify and obtain the use of a research ship for their purposes, or they are international panels of scientists who consider the time is ripe for a concerted international effort in a particular field of work or region.

National expeditions often come about to satisfy the needs of a group within a department of a particular laboratory, and individuals from other departments who are studying related topics are often included. Thus, such expeditions are usually planned with particular targets in view and personnel are specially involved with the work and the results. Such personal involvement has proved essential for intelligent day-by-day planning of an expedition so that results may be assessed on the spot and programmes can be modified to take advantage of results obtained and conditions prevailing. It is, therefore, essential for the scientist to go to sea and to design the programme if high-quality results are to be obtained. Where possible, it is a great help if results can be constantly examined during the work, and this has led to the increasing use of computers or data loggers on board oceanographic vessels (Chapter 2).

International expeditions are usually planned by panels of scientists who may dictate the type and pattern of sampling in considerable detail, so that less initiative is left to the scientific complement on the vessels; such expeditions are consequently more suitable for broad studies which necessarily involve the use of a number of ships.

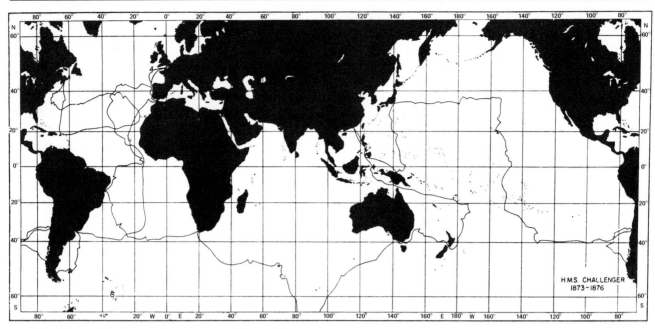

273. Track chart of the first major oceanographic expedition, made by HMS *Challenger* (from Wüst 1964)

Oceanographic expeditions have been divided into five eras (table 13). Modern or scientific oceanography commenced with the British *Challenger* expedition of 1873–6 (figure 273) and continued to the early 1900s. This era, which included great exploratory voyages of ships from Britain, Germany, Monaco, Norway, and the USA, laid the foundations for all the various sciences considered in this book. Startling discoveries were made in every field and the large expedition reports resulting are just as valuable today as they were when they first appeared. The decades between the two world wars saw an era of national systematic surveys (table 12 and figure 274). The Second World War led to great technological advance which was incorporated into the greatly accelerated oceanographic research. In 1957 international cooperation began on a large scale with the International Geophysical Year (IGY), involving Argentinian, British, Russian and United States vessels. Further international programmes followed, including the Norpac programme in the Pacific, the North Atlantic Polarfront Programme (figure 275), and the Indian Ocean Expedition. While these co-operative programmes have been carried out, it must be remembered that nearly all oceanographic ships are necessarily employed the whole time on innumerable minor expeditions (figure 276). In addition, one must not overlook the great contribution of all the more locally orientated laboratories, which do much to provide detailed knowledge of the deep ocean in their immediate vicinity and its effects upon their fisheries and inshore waters.

Actual numbers of oceanographic ships may be a poor guide to the scale of national contributions because 'oceanographic' is often ill-defined. However, the major expeditions up to 1955 can be grouped according to nationality (table 12). Details of a selection of the larger oceanographic ships are given in table 17. This only includes ships over 100 feet in length with laboratory space, power and winches to permit general or specialized oceanographic work. Many fisheries' vessels, purely hydrographic survey vessels and naval vessels are not included.

LABORATORIES

Having scanned briefly the major expeditions and the majority of ships contributing to *deep-sea* studies we have a clue to the main centres of oceanography because such centres nearly always possess ships. To these ship-controlling establishments we may add a few more which are known to be mainly involved in deep-sea studies (included in table 14), but to appreciate the total effort in the deep ocean one must take into account the numerous university and museum personnel who work on ships belonging to other laboratories, and the oceanographic work carried out in fisheries' laboratories and by hydrographic ships. Such effort is difficult to assess.

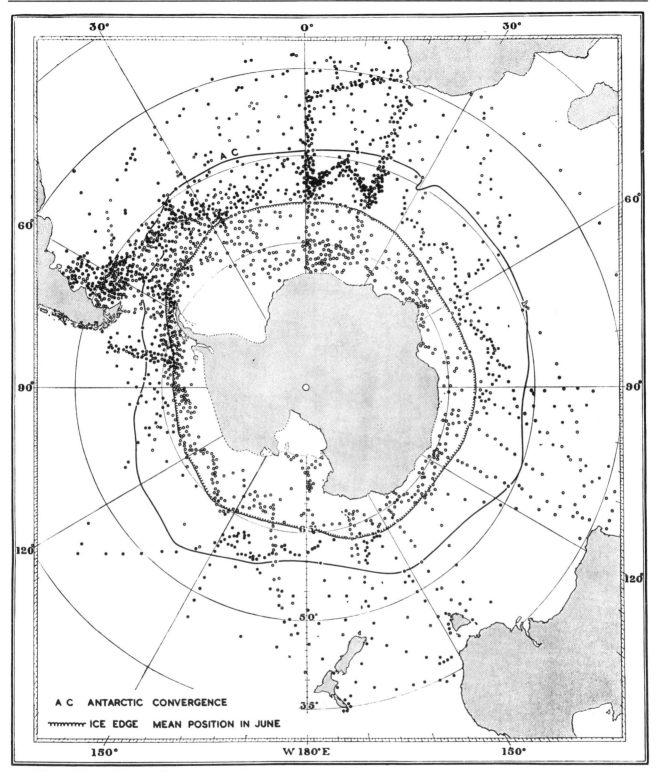

274. The positions of about half the stations worked by
Discovery II in the Antarctic (from Marr 1962)

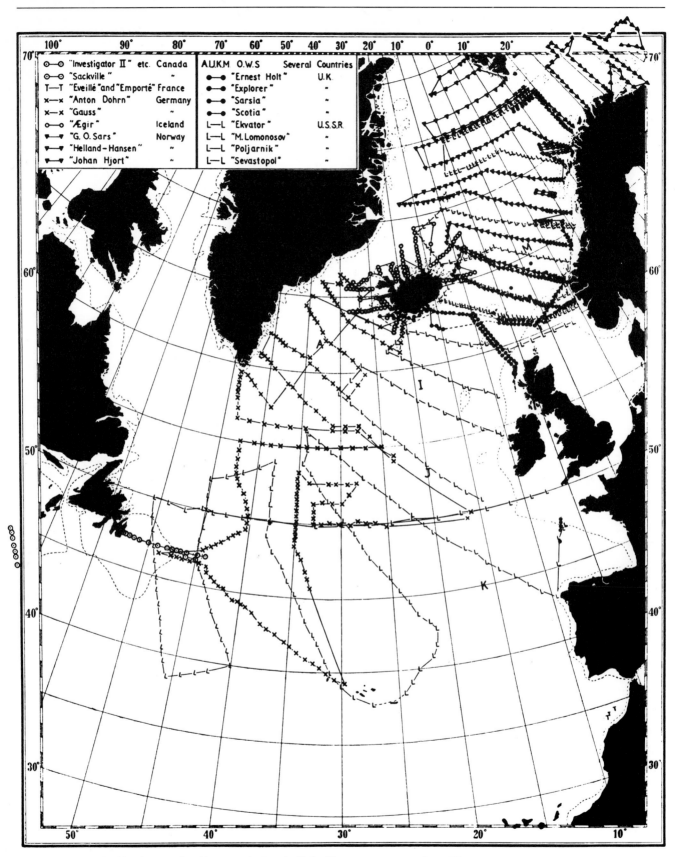

275. Stations worked by the vessels taking part during the Polar Front programme (from Wüst 1964)

276. Interlacing of ships' tracks is indicative of the hundreds of thousands of miles logged by vessels of the Scripps Institution of Oceanography, University of California, San Diego, from 1950 to 1967 (from S.I.O. Annual Report 1967)

FINANCE

Where does the money come from and how is it dispensed? In the past, certain public benefactors were able to finance deep-sea exploration. Even now, such names as Rockefeller and Carlsberg occur in the funds' lists of some laboratories; but, in the main, only governments have been able to afford such apparent extravagance. To the layman, the way in which government funds are allotted to one or another deserving cause is surrounded by a fog of historical accident and political mystique. The ways such funds are dispensed are as various as the magnanimity of the funding bodies. Some examples will indicate a few of the methods of giving money to oceanography as well as something of the financial complexities involved.

In Britain, the Natural Environment Research

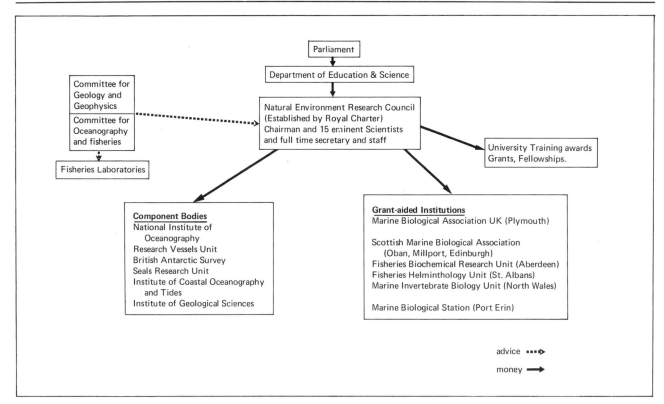

277. Simplified scheme to show the main influence and functions of the United Kingdom Natural Environment Research Council

Council (NERC) was established by Royal Charter to encourage, plan and conduct research relating to man's environment, including 'the seas and oceans, their behaviour and their living and mineral resources (oceanography, marine biology and ecology, and fisheries)'. Committees appointed by this Council recommend the scale of financial backing to laboratories involved in (among other things) deep-sea studies. A simplified picture of NERC's influence and functions is given in figure 277. More government money is dispensed to deep-sea research via the Ministry of Defence for such things as hydrography, diving physiology and underwater acoustics; via the Ministry of Agriculture, Fisheries and Food for physical oceanography and extension of fisheries into deep water; via the Ministry of Technology for development of techniques and instruments for use in deep waters.

In the United States, financing of oceanography is even more complex. The 1969 programme of expenditure on marine resources, requested by the National Council on Marine Resources and Engineering Development, included 516.2 million dollars of which 228.7 million dollars was for non-defence activities. The non-military contribution into *deep-sea* work can only

be estimated, but the departments concerned and their slice of the dollar cake can be seen in table 16 with some details of their intended expenditure on deep-sea research.

In France, eight Ministries give financial support to marine science in over 100 private laboratories, universities and other establishments. Recently, the National Council for the Exploitation of the Oceans (CNEXO) has been set up to outline policy and co-ordinate the national programme in marine sciences.

In the Federal Republic of Germany, responsibility for and funding of marine science is dispersed, and the National Committee for the Intergovernmental Oceanographic Commission provides some interagency co-ordination and is advised by the German Research Associations' (DFG) Commission of Oceanography. The Commission is advisory and encouraging in function. The eleven States of the Federal Republic finance a large part of the research by maintaining the independent science universities. Maximum co-operation is encouraged by the Ministry of Scientific Research of the Federal Government, which plans a unified marine science programme with the aid of the DFG's Commission on oceanography.

In Japan, marine science is co-ordinated through the Marine Science and Technology Council, which has representatives from universities, private industries, and government agencies involved in marine science, of which there are eight with major marine research programmes. Budgeting of the large number of establishments is too complex to analyse. Japan leads the world in the application of oceanographic information to the development of ocean commercial fisheries (see Chapter 9), and further development in this field has high priority.

National effort is now supplemented by international effort, and several important international organisations provide finance, information and co-ordination facilities for research into the deep ocean (table 16). To use the data and samples resulting from the recent accelerating work in the deep ocean, data centres have been created and use of computers has become commonplace. There are now two world data centres and more than twenty national data centres. Data and samples entering data centres will provide answers to many problems concerning the deep ocean, but the collection of such information is necessarily planned with particular aims in view. Although such aims may be broad, all requirements cannot be foreseen and there will always be a place for the individual personal approach of a scientist who goes to sea to wrestle with a particular problem in his own particular way.

RESULTS

The great majority of results accruing from the civil investigation of the deep ocean are written up by the scientists concerned and published in scientific journals. Such publications are read by interested scientists throughout the world, who incorporate other people's results into their own work so that a great corpus of knowledge develops and indirectly affects the life of the ordinary citizen, who is often quite unsuspecting of the energetic scientific effort on his behalf.

There are about 50 scientific journals dealing almost exclusively with marine science. Examples are *Deep Sea Research, Marine Biology, Marine Geology, Kieler Meeresforschungen.* In addition, large numbers of scientific papers concerned with the deep sea appear in journals relating to particular subjects such as *Crustaceana, Sedimentology, Journal of Geophysical Research, Journal of Fluid Mechanics* or pure research in general such as *Philosophical Transactions of the Royal Society, Proceedings of the Royal Society, Nature* and *Science.*

Reports and papers relating to the sea appear in over 2,400 journals, and one year now involves the publication of over 10,000 scientific papers on the sea. It is essential for the scientist to be aware of developments in his own field, and he can only do this effectively by using indexes and abstracts published for the purpose. *Oceanic Index* lists scientific papers on the sea; *Chemical Abstracts* includes chemistry of sea water and animals in the sea; *Zoological Record* lists most zoological papers; *Arctic Bibliography* is an example of a bibliography concerned with a region; *Current Contents* gives titles of all papers appearing in a large list of journals as they appear.

Besides all these research publications a number of journals of a semi-popular nature are available for the interested public. These include *Sea Frontiers, Under Sea Technology, World Fishing, Scientific American, Science Journal, New Scientist,* and the *National Geographic Magazine.*

MANPOWER

As explained elsewhere, oceanographers are specialists in a diversity of subjects who apply the techniques of their subject to the ocean. In the process of their work they learn a little of their colleagues' subjects and methods and often rely heavily upon them for information and help. They are supported royally by scientific assistants and the officers and crew of the research ships, whose efforts can make possible the impossible.

The life of an oceanographer is divided between a practical life at sea occupying up to half his time and a scientific, near-academic life on shore.

TABLE 12

Major national oceanographic expeditions up to 1955 (after Bruns 1958)

	No.	%
USSR	76	24·9
UK	60	20·0
USA	45	14·7
Germany	37	11·1
France	20	6·5
Norway	17	5·6
Denmark	9	2·9
Sweden	8	2·6
Austria	6	1·9
Japan	5	1·6
Canada	5	1·6
Holland	3	1·0
Monaco	3	1·0
Belgium	3	1·0
Italy	3	1·0
Finland	2	0·7
Egypt	1	0·3
Argentina	1	0·3
Australia	1	0·3
Total:	305	100

TABLE 13
Representative vessels in oceanographic deep-sea research (1873–1960) From Wüst 1964

Name of vessel	Nationality	Commissioned	Tonnage	Ocean	Main studies
Ia. Era of exploration (1873–1914)					
Challenger	British	1873–6	2306	Atlantic, Indian, Pacific	Biology, Physics, Sediments
Gazelle	German	1874–6	1900	Atlantic, Indian, Pacific	Physics
Blake	USA	1877–86	400	Atlantic (Gulf Stream system)	Physics, Currents
Albatross	USA	1887–8	1000	North Atlantic	Biology, Physics
National	German	1889	854 (gross)	North Atlantic	Plankton
Fram	Norwegian	1893–6	530	North Polar Sea	Physics
Valdivia	German	1898–9	2176 (gross)	Atlantic, Indian	Biology of great depths
Princess Alice I and II Hirondelle I and II	Monaco	1888–1922	—	North Atlantic	Biology
Gauss	German	1901–3	1332	Atlantic, Indian, Antarctic	Physics, Geography
Planet	German	1906–7	650	Atlantic, Indian, West Pacific	Physics
Deutschland	German	1911–12	598 (gross)	Atlantic, Antarctic	Physics
Ib. Transition to systematic research (1904–24)					
Michael Sars	Norwegian	1904–13	226	Norwegian Sea, North Atlantic	Physics
Armauer Hansen	Norwegian	Since 1913	53	Norwegian Sea, North Atlantic	Physics
II. Era of National Systematic Surveys (1925–40)					
Meteor	German	1925–7	1178	German Atlantic Exp. 65°S–20°N	Physics, Chemistry, Meteorology, Geology, and Biology
Meteor	German	1929–38	1178	Iceland–Greenland waters, some N. Atl. cruises	Physics, Chemistry, Biology
Dana I and II	Denmark	1921–35	Dana II 360 (gross)	Atlantic, Indian, Pacific	Biology
Carnegie	USA	1928–9	568	Pacific	Physics, Biology, Sediments
Willebrord Snellus	Dutch	1929–30	1055	Indonesian Seas	Physics, Sediments
Discovery and Discovery II	British	1930	(Discovery II) 2100	Antarctic	Physics, Biology
Atlantis	USA	1931	460 (gross)	North Atlantic	Physics, Chemistry
Ryofu Maru	Japanese	1937	1206	Pacific	Physics
E. W. Scripps	USA	1938	140	Pacific	Physics, Chemistry, Sediments
Altair	German	1938	4000 (gross)	Gulf Stream N. Atlantic	Physics, Currents
Sedov	Russian	1938–9	1538	North Polar Sea	Physics
IIIa. Period of New Geological, Geophysical, Biological and Oceanographic Methods (1947–1956)					
Albatross	Sweden	1947–8	1450	Atlantic, Indian	Sediments, Physics
Galathea	Denmark	1950–2	1600	Pacific	Biology of abyssal depths
Atlantis	USA	1947	460	Atlantic	Physics, Geophysics
Vema	USA	1953	743	Atlantic, Indian, Pacific	Geophysics, Geology
Anton Dohrn	German	1955–6	999	North Atlantic	Biology, Physics
Spencer F. Baird	USA	1956	143 ft	Pacific (operation Chinook)	Geophysics, Physics
IIIb. Transition to synoptic research in smaller areas (1950)					
Atlantis and other vessels	USA	1950, (1956)	—	Gulf Stream	Physics, Dynamics (Operation Cabot and second survey)

TABLE 13
continued

Name of vessel	Nationality	Commissioned	Tonnage	Ocean	Main studies
IV. Era of International Research Cooperation since 1957					
Crawford	USA	1957–9	260	Atlantic IGY Programme	Physics, Chemistry
Atlantis	USA	1957–9	460	Atlantic IGY Programme	Physics, Chemistry
Discovery II	British	1957–9	2100	Atlantic IGY Programme	Physics, Chemistry
Capitan Canepa	Argentina	1957–60	1000	Atlantic IGY Programme	Physics
Gauss	W. Germany	1957–60	845	North Atlantic Polar Front and Overflow Programmes	Physics
Anton Dohrn	W. Germany	1957–60	1000	N. Atlantic Polar Front and Overflow Programmes	Biology, Physics
Explorer and other vessels	Scottish	1957–60	(*Explorer*) 204 ft	N. Atlantic Polar Front and Overflow Programmes	Biology, Physics
Vitiaz	USSR	1957–60	5710	All Seas (IGY Programme)	Physics, Chemistry, Biology and Geology Trenches
Numerous research vessels	Japan, USA, Canada	1955–60	—	Pacific (Norpac Programme)	Physics, Biology
Mikhail Lomonosov	USSR	1957–60	5960	All Seas (IGY Programme)	Physics, Chemistry, Biology
Ob	USSR	1957–60	6400 (gross)	Southern Seas (IGY Programme)	Physics, Chemistry, Biology
Horizon	USA	1957–60	900	Pacific	Geophysics, Physics
Chain	USA	1958–60	1324	Atlantic, Indian	Geophysics, Physics
Argo	USA	1959–60	1400	Pacific, Indian	Geophysics, Physics
Shokalski	USSR	1960	3600	All Seas	Physics, Meteorology
Woeikof	USSR	1959	3600	All Seas	Physics, Meteorology
Pole	USSR	1959	5000	—	Physics, etc
Lena	USSR	1958	12,600	Polar Seas	Physics, Meteorology
Severyanka	USSR	1958	1050	All Seas	Submarine for oceanographic research
Numerous research vessels	International	Since 1958	—	Indian Ocean Expedition Programme	Physics, Geophysics, Biology, Chemistry, Geology, Meteorology

TABLE 14
Selected centres of marine research

Country	Location	Organization	Activities	Number professional staff
Argentina	Puerto Deseado	Buenos Aires University	B	6
Australia	Cronulla	Commonwealth Scientific & Industrial Research Organization	FP	30
Brazil	Recife	University	BPGF	12
Bulgaria	Vaena	Institute of Fishery Research (in Black Sea)	BP	23
Canada	Various	Fisheries Research Board (8 laboratories)	FP	142
	Dartmouth	Bedford Institute of Oceanography	BFGMF	146
Ceylon	Colombo	Fisheries Research Station	F	9
Chile	Santiago	University Marine Biological Station	B	8
	Conception	University Central Institute of Biology	B	6

TABLE 14
continued

Country	Location	Organization	Activities	Number professional staff
Cuba	Havana	Fisheries Research Centre	BPF	10
Dahomey	Cotonon	Centre for Scientific Study & Applied Fishery Technology	BPF	7
Denmark	Charlottenlund	Institute for Fishery & Marine Research	BPF	21
Federal Rep. Germany	Hamburg	Institute for Sea Fisheries	F	9
	Kiel	University, Institute for Marine Science	PMB	7
France	Villefranche	Zoological Station	B	12
	Endoume	Marine Station	BP	30
	Roscoff	Biological Station	B	6
	Banyuls-sur-Mer	Arago Laboratory	B	5
	(9 laboratories)	Scientific & Technical Institute of Marine Fishers	BP	25
	Paris	Oceanographical Institute (4 laboratories)	BP	14
	Paris	Oceanographic Research Centre	P	10
	Biarritz	Centre of Scientific Studies & Research	B	7
Hong Kong	—	Fisheries Research Station	F	11
Iceland	Reykjavik	Fisheries Research Institute	FP	8
India	Manadapam	Central Marine Fisheries Research Station	FBP	50
	Ernakulam	Kerala State Sub-Station	F	10
	Andhra	University Marine Laboratory	BP	14
Indonesia	Jakarta	Institute of Marine Research	BP	7
Israel	Haifa	Sea Fisheries Research Station	F	5
Italy	Naples	Zoological Station	B	10
Japan	Kagoshima	University	B	41
	Kobe	Marine Observatory	PB	2
	Hakodate	Marine Observatory	PB	5
	Nagasaki	Marine Observatory	MPB	4
	Tokyo	University of Fisheries	FP	82
Malaysia	Penang	Fisheries Research Laboratory	PFB	7
New Zealand	Wellington	Oceanographic Institute (DSIR)	BGP	13
Norway	Bergen	Directorate of Fisheries, Institute of Marine Research	FBP	25
Peru	Callao	Marine Resources Research Institute	FBP	15
South Africa	Cape Town	Department of Commerce & Industries. Sea Fisheries	FP	23
	Cape Town	University Department of Oceanography	BP	12
Spain	Madrid	Institute of Oceanography (6 branches)	PF	40
UK	Lowestoft	Ministry of Agriculture, Fisheries & Food Laboratory	FBP	84
	Surrey	National Institute of Oceanography	BPG	88
	Plymouth	Marine Biological Association	B	30
	Aberdeen	Marine Laboratory	FBP	99
	Oban & Millport	Scottish Marine Biological Association	B	25
	Edinburgh	Oceanographic Laboratory	B	22
USA	Los Angeles	Allan Hancock Foundation	BGP	7
	New Haven	Bingham Oceanographic Laboratory	PB	9
	San Francisco	California Academy of Sciences	B	30
	California	California State Department of Fish & Game (Marine)	BF	38
	La Jolla	Inter-American Tropical Tuna Commission	F	20
	North Carolina	Duke University Marine Laboratory	B	10
	Hawaii	Hawaii Marine Laboratory	B	7
	New York	Lamont Geological Laboratory	GB	50
	La Jolla	Scripps Institute of Oceanography	PBG	147
	(19 laboratories)	Fish & Wildlife Service, Bureau of Commercial Fisheries	F	290
	Washington DC	US Naval Oceanographic Office, Marine Sciences	BP	15
	Miami	University of Miami, Marine Laboratory	BGPM	120
	Seattle	University of Washington, Department of Oceanography	BP	24
	Virginia	Virginia Institute of Marine Science	PBGF	35
	Woods Hole	Oceanographic Institution	PBG	134
USSR	Moscow	Institute of Oceanology	PGB	120
Venezuela	Cumana	University of Oriente, Oceanographic Institute	BP	15

B = Biology   F = Fisheries   P = Physical Oceanography   G = Geology   M = Meteorology

TABLE 15

Expenditure (incurred or proposed) by the principal financing bodies in the United States in deep-sea research. These figures may include backing for some shallower sea work but they indicate the main sources. Expressed in millions of dollars

	Total budget 1968	1969	Industrial co-operation	National security	Health	Budget for 1969 Non-living resources	Oceanographic research	Ocean exploration mapping, charting & geodesy	General purpose ocean emergency development	environmental observation Prediction & Services
State Department	5·0	5·2	5·2*							
Department of the Interior	73·5	75·5				4·0*				
National Science Foundation	38·3	41·0					36·0			
Department of Commerce	38·4	38·1					4·2	19·5		6·3
Department of Transportation	10·7	33·1					15·7		5·3	6·8
Atomic Energy Commission	12·7	11·8					4·4		6·6	0·8
Department of Health Education & Welfare	6·4	7·5			3·0*					
Smithsonian Institution	1·6	1·7					1·4			
National Aeronautics & Space Administration	1·6	1·6						0·3		1·3
Department of Defence				82·8			38·0	72·3	14·9	11·3

*mainly not *deep-sea*.

TABLE 16

International organizations

Name	Aims	HQ
Food and Agriculture Organization (FAO) Committees—Advisory Committee on Marine Resources Research (ACMRR) General Fisheries Council for the Mediterranean (GFCM) Indo-Pacific Fisheries Council (IPFC) Regional Fisheries Advisory Commission for SW/Atlantic (CARPAS)	Sponsors world meetings on fishing, research vessels etc. Assists developing nations	Rome
UNESCO Intergovernmental Oceanographic Commission (IOC)	Promote scientific investigation of oceans. Several publications	Paris
UN Intergovernmental Maritime Consultative Organisation (IMCO)	Consultative and advisory body on shipping etc. including pollution	London
International Association for the Physical Sciences of the Ocean (IAPSO)	Promote scientific study of oceanographic problems by publications, initiation and co-ordination	Woods Hole
International Council for the Exploration of the Sea (ICES)	Promote and encourage research for the study of the sea particularly that relating to living resources	Denmark
Scientific Committee on Oceanic Research (SCOR)	To further international scientific activity in all branches of oceanic research	California
International Commission for the Scientific Exploration of the Mediterranean Sea (ICSEMS) Several relating to fisheries, whaling, sealing ICNAF, NEAFC, INPFC, IATTC, IWC	Promote co-operation in international projects	Monaco

TABLE 17

Details of a range of oceanographic vessels of the world involved in deep-sea research. Only vessels over 100 ft in length are included. Different sources often quote slightly different figures; these are based largely upon *Undersea Technology Handbook Directory 1969* and indicate something of the capacity of the ships for deep water research

Country	Owner	Name	Length ft.	Tonnage	Complement Crew	Complement	Built	Range miles	Propulsion	H.P.	Winches No.	Max. wire m
Australia	CSIRO	Diamantina	300	1490d	140	8	1945	3,800	st	5,500	4?	10,000
Brazil	Navy Diretoria de Hidrografia e Navegacao	Almirante Saldanha	300	2079g	150	39	1934	10,800	D	2,100	4	6,000
Canada	Bedford Institute of Oceanography	Hudson	296	3721g	65	28	1962	11,000	D–E	8,400	12	
		Kapuskasing	222	1250d	56	9	1943	3,300	st	2,400	2+	
		Maxwell	115	230d	13	7	1961	2,770	D	700	2	
	Hydrographic Service	Parizeau	212		45		1967	12,000	D	3,400	6	7,000
Fed. Rep. Germany	Hydrographic Institute	Gauss	186	845g	40	13	1941	10,000	D	840	4	6,000
		Meteor	250	2615d	54	25	1964	12,000	D–E	2,000	5	12,000
France	Min. Nat. Educ.	Calypso	141	360d	12	10	1942	5,000	D	600		
	COMEXO	Jean Charcot	230	2200d	34	29	1965	12,000	D	1,120	6	
Italy	Consiglio Nazionale delle Richerche	Bannock	205	1750d	28	26	1943	8,000	D	3,000B	4	3,000
Japan	Nagasaki Marine Observatory	Chofu Maru	136	266g	23	15	1960	6,000	D	500	3	8,000
	Kagoshima University	Kagoshima Maru	200	635d	42	53	1960	12,900	D	1,700	6	9,800
	Tokyo University of Fisheries	Shinyo Maru	130	236g	23	2	1937	7,000	D	450	2	5,000
		Umitaka Maru	230	2100d	51	13	1955	14,000	D	2,100	6	8,500
	Tokyo University Oceanographic Res. Inst.	Tansei Maru	130	258g	23	10	1963	7,500	D	550B	5	10,000
New Zealand	Department of Scientific & Industrial Res.	Taranui	200	840d	18	6	1935	7,000	D	950B	3	8,000
		Lachlan	300	1490d	140	8	1944	3,800	st	5,500i	4	10,000
Norway	Bergen University Geophysics Inst.	Helland Hansen	113	186g	8	8	1957	4,800	D	400	3	10,000
	Inst. Marine Res. Bergen	G.O. Sars	170	595g	38	10	1950		D	1,200	4	6,000
		Johan Hjort	171	697g	38	10	1958		D	1,300	2	6,000
Peru	Inst. del Mar del Peru	Unanue	143	858	10	29	1944	3,500	D–E	1,500	3	
USSR	Academy of Sciences	Acad. Kurchatov	407	6828d	84	82	1966	15,000	D	8,000	3+	
		Mikhail Lomonosov	336	5960d	70	60	1957	11,000	S	2,450	10	15,000
		Petr Lebedev	300	4600d			1960		D	2,400	5	11,000
		Sergey Vavilov	300	4600d			1960		D	2,400	5	11,000
		Vitjaz	359	5710d	84	82	1949	21,000	D	3,000	11	
	Murman Arctic Marine Association	Ob	420	6400g	72	140	1955	16,000	D–E	7,000	5	5,000
	Polar Sci. Res. Inst. of Sci. Fish. & Ocean.	Sevastopol	230	937g	66	26	1951		D	1,080	5	1,000

Country	Operator	Vessel										
USA	Coast & Geodetic Survey	*Oceanographer & others*	303	3701g	116		1966	16,000	D-E	5,000	7	15,000
	Hopkins Marine Station Stanford Univ.	*Proteus*	96	192g	7	9	1946	4,000	D	400	3	6,000
	Scripps Inst. of Oceanography	*Alexander Agassiz*	180	825g	18	13	1944	6,000	D	150+500	3	
		Alpha Helix	133	294g	12	10	1966	6,500	D	820B	1	
		Argo	213	1400g	32	22	1944	7,000	D	3,800	4	
		Flip	355	1500g	3	3	1962	towed				
		Thomas Washington	209	1151g	25	17	1965	7,000	D	1,000	5	15,000
	California Fish and Game	*N.B. Scofield*	100	168g	9	3	1938	6,000	D	350	2	2,500
	Texas A & M University	*Alaminos*	180	573g	17	14	1945	4,500	D	1,000	4	400
	Navy	*Bowditch*	454	7600g	59	47	1945	24,000	st	8,500	4	10,000
		Eltanin	266	2486g	48	38	1962	10,000	D	1,600	5	
	Miami University	*Catamaran*	141	300g	17	17	1963	9,000	D	950	5	
	Washington University	*J. E. Pillsbury*	176	560g	22	14	1944	5,200	D	1,000	4	
		Brown Bear	116	300g	11	19	1934	5,400	D	400	3	
	Woods Hole Oceanographic Inst.	*Thomas G. Thompson*	208	1151g	23	18	1965	10,000	D-E	1,125	5	
		Atlantis II	210	1529g	31	25	1963	8,000	st		6	
		Chain	214	1324g	31	26	1944	10,500	D-E	3,000	4	10,000
	Duke University	*Crawford*	125	260g	15	8	1927	5,000	D	400	1	10,000
		Eastward	117	276g	15	15	1964	5,000	D	640	3	10,000
	Rhode Island University	*Trident*	180	856g	18	13	1944	7,500	D	1,000	3	8,000
	Lamont Geological Observatory	*Vema*	202	743g	19	17	1944	6,000	D-sail		2	11,000
United Kingdom	National Institute of Oceanography	*Discovery*	260	2800d	40	21	1962	15,000	D-E	2,000	7	10,000
	Natural Environment Research Council	*John Murray*	133	441g	19	8	1963	4,000	D	2×394	4	9,000
	Marine Biological Assoc. of the UK	*Sarsia*	128	319d	16	5	1953	4,000	D	290B	3	5,000
	Scottish Marine Biological Assoc.	To be built (specification)	163		23	9	1970	3,000	st	1,000B	7	13,700
	Ministry of Agriculture, Fisheries & Food	*Corella*	136	580d	17	5	1967		D	1,060B	3	2,600
		Clione	140	496g	24	5	1961		D	960B	6	2,200
		Ernest Holt replacement	235				1970		D-E			
	Min. Agric. & Fish. for Scotland, Aberdeen	*Explorer*	204		30	9	1956	3,000	st	1,200	4	
		Scotia replacement	225			12						
Union of South Africa Fisheries Board		*Africana II*	206	882g	28	6	1949	6,000	st	1,120	4	6,000

D = Diesel   D-E = Diesel-Electric   S = Sail   st = Steam   d = displacement   g = gross   B = Brake horsepower

Bibliography

(for further reading;
additional book titles may be found in References)

Aspects of Deep Sea Biology, N.B.Marshall. Hutchinson, London, 1954

Aspects of Marine Zoology, Ed. N.B.Marshall, Symp. Zool. Soc. London No. 19, Academic Press, London, 1967

Bioluminescence, E.N.Harvey. Academic Press, New York, 1952

Bioluminescence in Progress, Eds. F.H.Johnson and Y.Haneda. Princeton Univ. Press, Princeton, N.J., 1966

Buoyancy Mechanisms of Sea Creatures, E.J.Denton. *Endeavour 22,* 3–8, 1963

Chemical Oceanography, Vols 1 and 2, Eds J.P.Riley and G. Skirrow. Academic Press, 1965

Chemistry and Fertility of Sea-waters, Ed. H.G.Harvey. Cambridge Univ. Press, 1955

Deep Challenge, H.B.Stewart Jr. D. Van Nostrand Co. Inc., Princeton, N.J., 1966

The Deep Submersible, R.D.Terry, (North Hollywood Calif.) Western Periodicals Co., North Hollywood, Calif., 1966

The Depths of the Ocean, J.Murray and J.Hjort. MacMillan & Co. Ltd., London, 1912

A Discussion on Progress and Needs of Marine Science, G.E.R.Deacon and F.S.Russell (Chairmen), *Proc. Roy. Soc. A, 265,* 285–406, 1962

Earth, Sky and Sea, A.Piccard, Oxford Univ. Press, New York, 1956

Effective Use of the Sea—Report of the panel on Oceanography, President's Science Advisory Committee, The White House, 1966

The Encyclopaedia of Marine Resources, Ed. F.E.Firth. Van Nostrand Reinhold Co., New York, 1969

Exploring the Ocean Depths, E.H.Shenton. W.W. Norton & Co. Inc., New York, 1968

The Eye, Vol. 2, The Visual Process, Ed. H.Davson. Academic Press, New York and London, 1962

Half Mile Down, W.Beebe. Harcourt, Brace & Co., New York, 1934

Harvest of the Sea, John Bardach. George Allen & Unwin, 1969

The Last Resource, T.Loftas. Hamish Hamilton, London, 1969

The Life of Fishes, N.B.Marshall. Weidenfeld & Nicolson, London, 1965

Light and Animal Life, G.L.Clarke and E.J.Denton, In *The Sea,* Ed. M.N.Hill, Vol. 1, pp 456–68, Interscience, New York, London, 1962

The Living Sea, J.Y.Cousteau and J.Dugan. Harper Bros., New York, 1963

Marine Geology of the Pacific, H.W.Menard. McGraw-Hill Book Co., New York, 1964

Marine Microbial Ecology, E.J.Ferguson Wood. Chapman & Hall Ltd., London, 1965

Marine Microbiology, C.E.ZoBell. Chronica Botanica Press, Waltham, Mass., 1946

Marine Pharmacology, Morris H.Baslow. William & Wilkins, Baltimore, Maryland, 1969

Marine toxins and venomous and poisonous marine animals, F.E.Russell, Adv. Mar. Biol. 3, 255–384, 1965

The Mineral Resources of the Sea, Ed. J.L.Mero. Elsevier Pub. Co., New York, 1965

The Ocean (reprint of articles published in Sept. 1969 issue of Scientific American), W.H.Freeman & Co., San Francisco, 1969

Oceanography – an introduction to the marine environment, P.K.Weyl. John Wiley & Sons Inc., New York, 1970

Oceans, Ed. G.E.R.Deacon. Paul Hamlyn, London, 1962

The Open Sea : its Natural History—Fish and Fisheries, A.C.Hardy. Collins, London, 1959

The Open Sea: its Natural History—The World of Plankton, A.C.Hardy. Collins, London (also Houghton, Mifflin Co., Boston, 1959), 1956

Plankton and Productivity in the Oceans, J.E.G. Raymont. Pergamon Press, London and New York, 1963

Plankton bioluminescence, B.P.Boden and E.M.Kampa, Oceanogr. Mar. Biol. Ann. Rev. 2, 341–71, 1964

The Plankton of the Sea, R.S.Wimpenny. Faber & Faber Ltd., London, 1966

Questions about the Oceans, US Naval Oceanographic Office, Washington, D.C., 1968

Report on Marine Science and Technology, HMSO, 1969

Resources and Man, Committee on Resources and Man, National Academy of Sciences–National Research Council. W.H.Freeman & Co., San Francisco, 1969

Responses of marine animals to changes in hydrostatic pressure, E.W.Knight-Jones and E.Morgan, Oceanogr. Mar. Biol. Ann. Rev. *4*, 267–99, 1966

The Sea, R.C.Miller. Thomas Nelson & Sons Ltd., London, 1966

The Stocks of Whales, N.A.Mackintosh. Fishing News (Books) Ltd., London, 1965

Submarine Canyons, F.P.Shepard and R.F.Dill. Rand McNally & Co., New York, 1966

The surface fauna of the ocean, P.M.David, Endeavour, *24*, 95–100, 1965

2,000 Fathoms Down, G.S.Houot and P.H.Willm. E.P.Dutton & Co., New York, 1955

Undersea Frontiers, G.S.Soule. Rand McNally & Co., New York, 1969

Under the Deep Oceans, T.F.Gaskell. Eyre & Spottiswoode, London, 1960

Underwater sound: biological aspects, J.M.Moulton, Oceanogr. Mar. Biol. Ann. Rev. *2*, 425–54, 1964

Uses of the Seas, Ed. E.A.Gullion. Prentice-Hall Inc., New Jersey, 1968

The Voyage of the "Challenger", H.S.Bailey Jr. Scientific American, May 1953, 87–94

Whales, E.J.Slijper. Hutchinson, London, Basic Books, New York, 1962

Whales, Dolphins, and Porpoises, K.S.Norris. Univ. California Press, Los Angeles, 1966

Figure references

The name of each of the species illustrated in Chapter 5, and most of those in Chapter 6, is that given in the source of each illustration despite the fact that there may have been subsequent alterations at generic or specific level

Agassiz, A. 1881 *Challenger* Zoology *3* Part 9

Apstein, C. 1900 Ergebn. Atlant. Ozean Plankton-Exped. Bd 2 1–62

Barnes, H. 1959 *Oceanography and Marine Biology* – a book of techniques. George Allen & Unwin Ltd, London

Bé, A.H.W. 1962 Deep-Sea Res. *9* 144–151

Berrill, N.J. 1961 Scient. Am. *204* 150–160

Beverton, R.J.H.&Tungate, D. S. 1969 J. Cons. perm. int. Explor. Mer *31* (2) 145–157

Bigelow, H.B. & Schroeder, W.C. 1953 in *Fishes of the Western North Atlantic* Mem Sears Fn Mar. Res. *1* (2) 1–514

Bigelow, H.B. & Schroeder, W.C. 1948 in *Fishes of the Western North Atlantic* Mem Sears Fn Mar. Res. *1* (1) 59–546

Boden, B.P. & Kampa, E.M. 1967 Symp. zool. Soc. London No. 19 15–26 ed. N.B.Marshall, Academic Press, London

Boden, B.P., Kampa, E.M. & Snodgrass, J.M. 1960 J. mar. biol. Ass. UK *39* 227–238

Bouvier, E.L. 1917 Résult. camp.scient. *Prince Albert I* Fasc 51

Brady, H.B. 1884 *Challenger* Repts Zoology *9* Part 22

Brauer, A. 1906 Wiss. Ergebn. dt. Tiefsee-Exped. *Valdivia* Bd 15

Bruns, E. 1958 *Ozeanologie* Bd 1 Veb Deutscher Verlag der Wissenschaften, Berlin

Carey, F.G.& Teal, J.M. 1969 Comp. Biochem. Physiol *28* 205–213

Carlisle, D.B.& Denton, E.J. 1959 J. mar. biol. Ass. UK *38* 97–102

Carpenter, P.H. 1884 *Challenger* Repts Zoology *11* Part 32

Chun, C. 1910 Wiss. Ergebn. dt. Tiefsee-Exped. *Valdivia* Bd 18

Clarke, W.D. 1964 J. mar. Res. *22* (3) 284–287

Collett, R. 1896 Résult. camp. scient. *Prince Albert I*, Fasc 10

Conrad, W.& Kufferath, H. 1954 Mém. Inst. r. Sci. nat. Belg. No. 127 1–346

Copeland, D.E. 1968 Biol. Bull. *135* 486–500

Dakin, W.J. & Colefax, A.N. 1940 *Plankton of the Australian coastal waters off New South Wales,* Part 1

Deflandre, G. 1952 in *Traité de Zoologie* 1 439–470 ed. P. Grassé, Masson et Cie, Paris

Denton, E.J. & Marshall, N.B. 1958 J. mar. biol. Ass. UK *37* 753–767

Denton, E.J.& Nicol, J.A.C. 1964 J. mar. biol. Ass. UK *44* 219–258

Denton, E.J.& Warren, F.J. 1957 J. mar. biol. Ass. UK *36* 651–662

Desikachary, T.V. 1959 *Cyanophyta.* Academic Press, New York

Elliot, F.E. 1960 General Electric Advanced Electronics Center at Cornell University Rept No. R60ELC45

Ericson, D.B., Ewing, M.& Wollin, G. 1964 Science *146* (3645) 723–732

Ewing, M., Ewing, J.I. & Talwani, M. 1964 Bull. geol. Soc. Am. *75* 17–36

Eyriès, M. 1968 Cah. oceanogr. *20* (5) 355–368

Foxton, P.F. 1966 *Discovery* Repts *34* 1–116

Fye, P.M., Maxwell, A.E., Emery, K.O. & Ketchum, B.H. 1968 'Ocean Sciences and Marine Resources' in *Uses of the Seas* ed. E.A.Gullion, The American Assembly, Columbia University. Prentice-Hall Inc. New Jersey

Garman, S. 1899 Mem. Mus. comp. Zool. Harv. *24* 1–431

Garrick, J.A.F. 1960 Trans. R. Soc. N.Z. *88* (3) 489–517

Giesbrecht, W. 1892 *Fauna und Flora des Golfes von Neapel 19* Pelagische Copepoden

Goode, G.B. & Beane, T.H. 1895 *Oceanic Ichthyology.* Smithsonian Institution, City of Washington

Gran, H.H. 1912 in *Depths of the Ocean* pp 307–386 ed. Murray, J.& Hjort, J. MacMillan & Co., London

Groen, P. 1967 *The Waters of the Sea.* D. Van Nostrand & Co. Ltd, London

Gulland, J.A. 1968 Science Journal *4* (5) 81–86

Haeckel, E. 1888 *Challenger* Repts Zoology *28* Part 77

Haeckel, E. 1887 *Challenger* Repts Zoology *18* Part 40

Haeckel, E. 1882 *Challenger* Repts Zoology *4* Part 12

Hardy, A.C. 1936 *Discovery* Repts *11* 457–510

Hardy, A.C. 1956 *The Open Sea: the World of Plankton.* Collins, London

Heezen, B.C., Tharp, M.& Ewing, M. 1959 Geol. Soc. Am. Spec. Paper No. 65

Heirtzler, J.R, Le Pichon, X.& Baron, G.J. 1966 Deep-Sea Res. *13* 427–443

Heirtzler, J.R., Dickson, G.O., Herron, E.M., Pitman, W.C.& Le Pichon, X. 1968 J. geophys. Res. *73* (6) 2119–2136

Hendey, N.I. 1964 Fishery Invest., Lond. Series 4 1–317

Herdman, W.A. 1882 *Challenger* Repts Zoology *6* Part 17

Herring, P.J. 1967 Symp. zool. Soc. London ed. N.B.Marshall No. 19, 215–235 Academic Press, London

Hertwig, R. 1882 *Challenger* Repts Zoology *6* Part 15

Hill, M.N. 1963 in *The Sea* ed. M.N.Hill *3* 39–46. John Wiley & Sons New York & London

Houghton, H.G. 1954 J. Met. *11* (1) 1–9

Ivanov, A.V. 1963 *Pogonophora.* Academic Press, London

Jerlov, N.G. 1951 Rep. Swed. deep Sea Exped. *3* (1) 3–57

Jones, B.W.& van Eck, T.H. 1967 The South African News and Fishing Industry Review, November 1967

Kampa, E.M.& Boden, B.P. 1957 Deep-Sea Res. *4* 73–92

Kamenaga, T. 1967 Studies trop. Oceanogr. No. 5 412–422

Kemp, S.W. 1906 Scient. Invest. Fish. Brch Ire. 1905 *1* 28pp

Kolliker, A. von 1880 *Challenger* Repts. Zoology *1* part 2

Knauss, J.A. 1960 Deep-Sea Res. *6* 265–286

Lafond, E.C. 1962 in *The Sea* ed. M.N.Hill *1* 731–751. John Wiley & Sons New York & London

Lebour, M.V. 1925 *The dinoflagellates of northern seas.* Plymouth Marine Biological Association UK

Lemmerman, E. 1908 Nordisches Plankton, Botanischer Teil (ed. K.

Brandt & C.Apstein) *21* 1–40

Lohmann, H. 1896 Ergebn. Atlant. Ozean Plankton-Expd. Bd 2 1–148

Lythgoe, J.N. 1966 in *Light as an Ecological Factor*, ed. R. Bainbridge, G.C. Evans & O. Rackham, pp 375–391. Blackwell Scientific Publications, Oxford

M'Intosh, W.C. 1885 *Challenger* Repts Zoology *12* Part 34

McLellan, H.J. 1965 *Elements of Physical Oceanography*. Pergamon Press, London

Marr, J.W.S. 1962 *Discovery* Repts *32* 33–464

Marshall, S. 1934 Scient. Rep. Gt Barrier Reef Exped. *4* (15) 623–644

Mayer, A.G. 1910 *Medusae of the World* Vols. 1–3, Carnegie Institution, Washington

Meals, W.D. 1969 in *Handbook of Ocean and Underwater Engineering* pp 4–32 – 4–42, ed. J.J.Myers, C.H. Holm & R.F.McAllister, McGraw-Hill Book Co., New York

Meisenheimer, J. 1905 Wiss. Ergebn. dt. Tiefsee-Exped. *Valdivia* Bd 9 314pp

Montgomery, R.B. 1958 Deep-Sea Res. *5* (2) 134–148

Moseley H.N. 1881 *Challenger* Repts Zoology *2* Part 7

Motoda, S. 1967 Bull. Fac. Fish. Hokkaido Univ. *18* (1) 3–8

Müller, G.W. 1894 Wiss Ergebn. dt. Tiefsee-Exped. *Valdivia* Bd 8

Munk, W.H. 1950 J.Met 7 (2) 79–93

Munk, W.H.& Carrier, G.F. 1950 Tellus *2* (3) 158–167

Murray, J.& Renard, A.F. 1891 *Challenger* Repts, Deep Sea Deposits

Nafpaktatis, B.G. 1968 *Dana* Report No. 78 1–131

Neumann, G. 1913 Wiss. Ergebn. dt. Tiefsee-Exped. *Valdivia* Bd 12 93–243

Pelseneer, P. 1887 *Challenger* Repts Zoology *19* Part 58

Pierson, W.J., Neumann, G.& James, R.W. 1955 US Navy Hydrographic Office Publication 603

Polejaeff, N. 1884 *Challenger* Repts Zoology *11* Part 31 and *8* Part 24

Pickard, G.L. 1964 *Descriptive Physical Oceanography*, Pergamon Press Ltd., London

Rhumbler, L. 1909 Ergebn. Atlant. Ozean Planton-Exped. Bd 3 1–331

Roule, L. 1919 Résult. Camp. scient. *Prince Albert I* Fasc 52 1–171

Ryther, J.H. 1959 Science *130* (3376) 602–608

Sars, G.O. 1899 *Crustacea of Norway 2*

Sars, G.O. 1893 *Crustacea of Norway 1*

Sars, G.O. 1885 *Challenger* Repts Zoology *13* Part 37

Schiller, J. 1937 Dinoflagellatae *10* (2) 1–589 in Dr L.Rabenhorsts Kryptogamen-Flora, Leipzig

Schiller, J. 1930 Coccolithineae *10* 89–273 in Dr L.Rabenhorsts Kryptogamen-Flora, Leipzig

Schütt, F. 1895 Ergebn. Atlant. Ozean Plankton-Exped. Bd *4* 1–170

Schütt, F. 1892 Ergebn. Atlant. Ozean Plankton-Exped. Bd *1*A 243–314

Schulze, F.E. 1887 *Challenger* Repts Zoology *21* Part 53

Sladen, W.P. 1889 *Challenger* Repts Zoology *30* Part 51

Smayda, T.J. 1965 Bull. inter-Amer. trop. Tuna Commn 9 (7) 465–531

Smith, W.& McIntyre, A.D. 1954 J. mar. biol. Ass. UK *33* 257–264

Snodgrass, F.E. 1968 Science *162* (3849) 78–87

Steeman-Nielsen, E.& Hansen, V.K. 1959 Physiologia Pl. *12* 353–370

Stride, A.H., Curray, J.R., Moore, D.G. & Belderson, R.H. 1969 Phil. Trans. R. Soc. A *264* (1148) 31–75

Studer, T. 1889 *Challenger* Repts Zoology *32* Part 81

Sutcliffe, R.C. 1966 *Weather and Climate*. Weidenfeld & Nicolson, London

Sverdrup, H.U. 1954 in *The Earth as a Planet* ed. C.P.Kuiper pp 215–257 University of Chicago Press, Chicago

Sverdrup, H.U., Johnson, M.W.& Fleming, R.H. 1942 *The Oceans – Their Physics, Chemistry and Biology*. Prentice-Hall Inc., New York

Swallow, M. 1961 New Scient. *9* (227) 740–743

Talwani, M.& Ewing, M. 1966 J. geophys. Res. *71* (18) 4434–4438

Théel, H. 1882 *Challenger* Repts Zoology *4* Part 13

Trewartha, G.T. 1954 *An Introduction to Climate*. McGraw-Hill Publishing Co. Ltd, London

Vayssière, A. 1915 Résult. Camp. scient. *Prince Albert I*. Fasc 47

Vayssière, A. 1904 Résult. Camp. scient. *Prince Albert I*. Fasc 26

Wheeler, J.F.G. 1934 *Discovery* Repts 9 215–294

Wilson, J.T. 1963 Scient. Am. *208* 86–100

Winge, O. 1923 Rep. Dan. oceanogr. Exped. Mediterr. *3* (2) 1–34

Wüst, G. 1964 in *Progress in Oceanography* ed. M.Sears *2* 1–52 Pergamon Press Ltd, London

Wüst, G. 1957 Paper at Toronto meeting of I.A.P.O. 1957

Wüst, G. 1928 Jubilaums-Sonderband Z. Ges. Erdk. Berl. pp 507–534

Yentsch, C.S. 1966 Proc. I.B.P. Symp. held in Amsterdam and Nieuwersluis pp 255–270

Yentsch, C.S. 1962 in *Physiology and Biochemistry of Algae* ed. R.A. Lewin, 52, 771–797. Academic Press, New York & London

Zugmayer, E. 1911 Résult. Camp. scient. *Prince Albert I*. Fasc 52

Photo copyrights

National Institute of Oceanography: Figures 4, 9, 10, 12, 15, 20, 21, 22, 23, 24, 25, 29, 31, 39, 46, 48, 49, 58, 59, 110, 112, 255, 256

P.M.David: 18, 19, 42, 117, 118, 119, 128, 129, 131, 145, 146, 147, 149, 158, 163, 164, 165, 167, 168, 170, 171, 172, 174, 176, 179, 180, 181, 187, 190, 191, 193, 194, 195, 196, 197, 198, 199, 200, 209, 210, 211, 213, 226, 227, 243

John D.Isaacs: 244, 245, 246, 247, 248

National Aeronautical and Space Administration: 64, 81, 241

Deutsches Hydrographisches Institut: 6

G.L.Clarke, Harvard University, Cambridge, Mass.: 26

D.R.Houghton, Central Dockyard Laboratory, Portsmouth: 11

P.J.Herring: 17, 36, 100, 101, 130, 228

H.H.Angel: frontispiece, 62, 116

Woods Hole Oceanographic Institution: 5, 47, 71, 72, 201 (R.Pratt), 188 (D.M.Owen)

J.D.Woods (contributed by permission of the Director-General of the Meteorological Office): 92

Sidney C.Reynolds: 98

Icelandic Press and Photo Service: 78, 261

Paul Hargraves, Narragansett Marine Laboratory, University of Rhode Island: 124

A.McIntyre, Lamont-Doherty Geological Observatory, Columbia University, New York: 127

G.Berge (reproduced from *Sarsia* by kind permission of Dr H. Brattström): 79

D.P.Wilson: 134, 135

Lamont-Doherty Geological Observatory, Columbia University, Palisades, New York: 182, 189

H.Edgerton, Massachusetts Institute of Technology: 192

W. Greve, Biologische Anstalt, Heligoland: 212 (part)

J.A.C. Nicol: 212 (part)

US Bureau of Commercial Fisheries, Biological Laboratory, Honolulu, Hawaii: 234

S.G.Brown: 238

N.R.Merrett: 242

G.A.MacDonald, University of Hawaii: 97

US Naval Oceanographic Office: 43, 44, 67, 68, 69, 70

US Navy Electronic Laboratory: 65, 66, 202

Acknowledgements

The editors are particularly indebted to Dr G.E.R. Deacon, Director of the National Institute of Oceanography, Surrey, England, for advice, encouragement and understanding which greatly facilitated the planning and preparation of this book. Mr P.M.David, head of the biological department, provided not only the opportunity to undertake the project but also much invaluable advice and assistance, in addition to his unique photographs detailed below. A great many of our colleagues have provided help, information, ideas, suggestions and criticism without which the book would have been much the poorer, and Dr J.N. Carruthers kindly provided a translation of the foreword as well as helping with the literature. Dr J.S.M. Rusby and the engineering and design staff assisted materially with sections of Chapter 2, and the assembly of the many figures is a tribute to the skill of the photographic staff.

Mr N.Satchell produced figures 7, 8, 13, 16, 27, 30, 32, 33, 34, 35, 37, 38, 40, 41, 51, 52, 53, 54, 57, 63, 120, 148, 231 and 235.

Mrs C. Darter prepared figures 117, 149, 150, 152, 153, 154, 155, 157, 159, 160, 161, 162, 166, 169, 173, 175, 177, 178, 183, 184, 185, 186, 222, 250, 251, 252, 253, 254, 257, 258, 259, 260, 262, 263, 264, 265, 266, 267, 268, 270, 271.

Miss Blanche Coyne at the Narragansett Marine Laboratory, University of Rhode Island, kindly prepared figures 121, 122, 123, 125, 126, 132, 133.

We are most grateful to the original authors and publishers of many of the figures, as enumerated in the list of Figure References, for permission to reproduce them. Among these are the McGraw-Hill Book Co., Pergamon Press, D. van Nostrand and Co., Prentice-Hall Inc., Academic Press, John Wiley and Sons, MacMillan and Co., University of Chicago Press, Cambridge University Press, Her Majesty's Stationery Office, Blackwell Scientific Publications, the University of Miami, the Biological Bulletin, New Scientist, the American Meteorological Society, the American Association for the Advancement of Science, the US Naval Oceanographic Office, the Council of the Marine Biological Association, the Zoological Society of London and the Challenger Society. Goode Base Map Series, Department of Geography, University of Chicago kindly gave their permission to reproduce their map (which is copyright by the University of Chicago) in figure 93, and we are grateful to Prentice-Hall Inc. for permission to quote from *Uses of the Seas*, ed. E.A. Gullion, 1968.

Our grateful thanks are also due to all those who made available photographic material for the illustrations (see Photographic Copyrights).

Lastly we are deeply grateful to our contributing authors for their ready acquiescence to, and tolerance of, the many editorial foibles to which they have been subjected.

Index

Numbers in italics, e.g. *229,* denote
page numbers on which a relevant
illustration or figure occurs